Age-Related Macular Degeneration and Diabetic Retinopathy

Age-Related Macular Degeneration and Diabetic Retinopathy

Editors

Peter D. Westenskow
Andreas Ebneter

MDPI • Basel • Beijing • Wuhan • Barcelona • Belgrade • Manchester • Tokyo • Cluj • Tianjin

Editors

Peter D. Westenskow
Opthalmology
F. Hoffmann-La Roche Ltd.
Basel
Switzerland

Andreas Ebneter
Ophthalmology
F. Hoffman-La Roche Ltd
Basel
Switzerland

Editorial Office
MDPI
St. Alban-Anlage 66
4052 Basel, Switzerland

This is a reprint of articles from the Special Issue published online in the open access journal *Journal of Personalized Medicine* (ISSN 2075-4426) (available at: www.mdpi.com/journal/jpm/special_issues/Macular_Retinopathy).

For citation purposes, cite each article independently as indicated on the article page online and as indicated below:

LastName, A.A.; LastName, B.B.; LastName, C.C. Article Title. *Journal Name* **Year**, *Volume Number*, Page Range.

ISBN 978-3-0365-4210-2 (Hbk)
ISBN 978-3-0365-4209-6 (PDF)

Contents

About the Editors

Peter D. Westenskow

Peter Westenskow is a neurovascular biologist by training with a strong portfolio in vision research. Professional skills include stem cells, ophthalmology, neurodegeneration, angiogenesis, neuroinflammation, and hypoxia.

Andreas Ebneter

Andreas Ebneter is an experienced vitreo-retinal surgeon aware of and closely following the latest innovations in ophthalmic surgery and biomedical research, and committed to providing and improving retinal care for patients

Journal of
Personalized
Medicine

MDPI

Editorial

Age-Related Macular Degeneration and Diabetic Retinopathy

Andreas Ebneter and Peter D. Westenskow *

Roche Pharma Research and Early Development, F. Hoffmann-La Roche Ltd., 4070 Basel, Switzerland;
andreas.ebneter@roche.com
* Correspondence: peter.westenskow@roche.com

Citation: Ebneter, A.; Westenskow, P.D. Age-Related Macular Degeneration and Diabetic Retinopathy. *J. Pers. Med.* **2022**, *12*, 581. https://doi.org/10.3390/jpm12040581

Received: 28 March 2022
Accepted: 29 March 2022
Published: 5 April 2022

Publisher's Note: MDPI stays neutral with regard to jurisdictional claims in published maps and institutional affiliations.

More than 15 years ago, the results of the pivotal trials supporting the intravitreal use of ranibizumab were published [1,2]. The intravitreal application of anti-vascular endothelial growth factor (anti-VEGF) drugs has subsequently revolutionized the treatment of several common retinal diseases [3]. Suddenly, widespread conditions such as neovascular age-related macular degeneration have become treatable, whereas previously, physicians could only observe and document the rapid deterioration of vision due to irreversible structural retinal damage.

This important paradigm shift in treatment has been further driven by the simultaneous emergence of novel, powerful, non-invasive imaging capabilities. Indeed, the advent of anti-VEGF and optical coherence tomography (OCT) at roughly the same time has advanced retinal treatment in unprecedented ways. OCT has evolved rapidly due to incredible technological advances that have led to a dramatic increase in resolution and acquisition speed. Combined with ingenious software, this has led to the advent of OCT-Angiography, which has the potential to replace the more invasive fluorescein angiography in the not-too-distant future.

In terms of intravitreal treatments, we are now at an inflection point with new treatment modalities emerging. While treatment decisions for intravitreal medications are currently limited to choosing between anti-VEGF and corticosteroids, the need for informed decision making by physicians and choosing the best drug for an individual patient will likely complicate the lives of retina specialists but, hopefully, add more value for patients. We now need to start building the tools to tailor treatment decisions for every individual patient.

For this issue, we approached world-class experts in the fields of medical retina, ocular imaging, and drug development. We have assembled a collection of valuable articles that contain innovative ideas and suggestions for future individualized treatment decisions. We hope you enjoy browsing and reading the articles and that you will be inspired to contribute further and ultimately feel empowered to advance the way we treat patients.

1. Imaging Biomarkers

The first section of papers focuses on imaging biomarkers. With dramatic progress in technology for OCT and fundus imaging, we anticipate that more sophisticated data analysis will lead to more tailored treatments for patients and better outcomes.

Sen et al. provide a comprehensive overview of established imaging biomarkers for predicting visual outcomes in diabetic macular edema (DME), which is an important cause of vision loss secondary to diabetes mellitus [4].

Babluch et al. report the 2-year results from the RECOVERY Study in proliferative diabetic retinopathy using an innovative automatic segmentation algorithm to quantify leakage and microaneurysm count, which have been shown to decrease under intravitreal aflibercept treatment. The results suggest that these biomarkers reflect changes in diabetic retinopathy severity score (DRSS) and central subfield thickness well [5].

The 2-year outcomes from the PRIME trial confirm that leakage on ultra-widefield fluorescein angiography can indeed guide intravitreal aflibercept treatment for non-proliferative diabetic retinopathy and help provide individualized treatments to patients. Of note, change in peripheral leakage preceded DRSS worsening [6].

Gadde et al. conducted a very comprehensive volumetric analysis on spectral-domain OCT in patients with diabetic retinopathy. The edema volume correlated with diabetic retinopathy severity and predicted the response to intravitreal treatment. Edema volume increased with disease severity and was the best predictor for response to treatment [7].

To reflect the current state of development, the review by Kalra et al. provides a comprehensive review of currently used quantitative imaging biomarkers in age-related macular degeneration (AMD) and diabetic eye disease. Various approaches to extracting the pertinent information are also presented [8].

The findings by Lai et al. illustrate that morphological aspects can be directly related to function. This real-world data study confirms previous reports from clinical trials that greater fluctuations in retinal thickness are associated with worse visual outcomes [9].

2. Systemic Biomarkers

Existing evidence suggests that glucose variability might be an independent risk factor for the development of diabetic retinopathy. However, quantification of glucose variability is costly. Hsing et al. investigated the possibility of using the glycemic gap as a surrogate for glucose variability. Findings suggest that a negative glycemic gap is associated with progression and show the importance of minimizing glucose fluctuations [10].

Prasuhan et al. explored the predictive value of specific autoantibodies in neovascular AMD. Unfortunately, the five autoantibodies investigated were not useful in predicting the clinical course. However, the selection was limited, and the outcome highlights the importance of a more holistic approach [11].

In addition, Vader et al. reported results of a study designed to determine if circulating microRNAs (miRs) levels could predict patient responsiveness to anti-VEGF therapies in patients with DME. Their findings suggest that miR-181a may be negatively associated with the central area thickness of the retina at baseline. Work focused on miRs is emerging, and it is exciting to think similar trends can be determined as we identify more systemic biomarker candidates [12].

Finally, Acar et al. report that multiplex profiling of complement proteins, demographic, factors, genetic variants, and systemic metabolites are correlated in patients with AMD. This holistic approach is exciting as it may ultimately provide clinicians with more options when treating patients [13].

3. Public Health Aspects

In the paper by Nugawela et al., a different angle is taken to highlight the importance of ethnic and socio-economic aspects in diabetic eye disease care. The group investigated the influence of ethnicity on the risk of DR and sight-threatening complications in the UK. The outcomes show the importance of improving prevention, screening and treatment for specific ethnic minorities [14].

4. Innovative Technologies

This section of this Special Issue focuses more on technical aspects and innovative technology.

Maunz et al. show that their machine learning/artificial intelligence approach based on spectral-domain OCT images from the HARBOR trial was accurate in classifying choroidal neovascular lesions in exudative AMD. Results highlight the power of machine learning and suggest that the different OCT modalities have the potential to substitute fluorescein angiography in the very near future [15].

Cunha-Vaz et al., in their pilot study, experimented with an advanced OCT-Angiography-based grading system to substitute DRSS grading on color fundus images [16].

Frizziero et al. report interesting results from a subthreshold micropulse laser treatment for diabetic macular edema. Different groups have used this innovative method to treat macular edema, but its role is still not entirely clear. These results suggest that the subthreshold micropulse laser is safe and repeatable for mild diabetic macular edema. Such treatment might be considered for stabilizing macular edema in settings where other alternatives are lacking [17].

In addition, Maurissen et al. reviewed the current and future opportunities for modeling neurovascular interactions using organ-on-a-chip and other physiological microsystems. They argue that optimized models of the inner brain–retinal–blood barrier (iBRB) can be generated using the knowledge obtained from studies of the neurovascular unit in the retina and brain. Developing robust iBRB models could allow vision researchers to test key hypotheses about iBRB integrity and identify novel targets to prevent devastating features of diabetic retinopathy and other neurovascular diseases [18].

Author Contributions: Conceptualization, A.E. and P.D.W.; writing—original draft preparation, A.E and P.D.W.; writing—review and editing, A.E and P.D.W. All authors have read and agreed to the published version of the manuscript.

Funding: This research received no external funding.

Institutional Review Board Statement: Not applicable.

Informed Consent Statement: Not applicable.

Data Availability Statement: Not applicable.

Conflicts of Interest: There is no conflict of interest to declare in this Special Issue.

References

1. Brown, D.M.; Kaiser, P.K.; Michels, M.; Soubrane, G.; Heier, J.S.; Kim, R.Y.; Sy, J.P.; Schneider, S.; ANCHOR Study Group. Ranibizumab versus verteporfin for neovascular age-related macular degeneration. *N. Engl. J. Med.* **2006**, *355*, 1432–1444. [CrossRef] [PubMed]
2. Rosenfeld, P.J.; Brown, D.M.; Heier, J.S.; Boyer, D.S.; Kaiser, P.K.; Chung, C.Y.; Kim, R.Y.; MARINA Study Group. Ranibizumab for neovascular age-related macular degeneration. *N. Engl. J. Med.* **2006**, *355*, 1419–1431. [CrossRef] [PubMed]
3. Pham, B.; Thomas, S.M.; Lillie, E.; Lee, T.; Hamid, J.; Richter, T.; Janoudi, G.; Agarwal, A.; Sharpe, J.P.; Scott, A.; et al. Anti-vascular endothelial growth factor treatment for retinal conditions: A systematic review and meta-analysis. *BMJ Open.* **2019**, *9*, e022031. [CrossRef] [PubMed]
4. Sen, S.; Ramasamy, K.; Sivaprasad, S. Indicators of Visual Prognosis in Diabetic Macular Oedema. *J. Pers. Med.* **2021**, *11*, 449. [CrossRef] [PubMed]
5. Babiuch, A.S.; Wykoff, C.C.; Yordi, S.; Yu, H.; Srivastava, S.K.; Hu, M.; Le, T.K.; Lunasco, L.; Reese, J.; Nittala, M.G.; et al. The 2-Year Leakage Index and Quantitative Microaneurysm Results of the RECOVERY Study: Quantitative Ultra-Widefield Findings in Proliferative Diabetic Retinopathy Treated with Intravitreal Aflibercept. *J. Pers. Med.* **2021**, *11*, 1126. [CrossRef] [PubMed]
6. Yu, H.J.; Ehlers, J.P.; Sevgi, D.D.; O'Connell, M.; Reese, J.L.; Srivastava, S.K.; Wykoff, C.C. Real-Time Diabetic Retinopathy Severity Score Level versus Ultra-Widefield Leakage Index-Guided Management of Diabetic Retinopathy: Two-Year Outcomes from the Randomized PRIME Trial. *J. Pers. Med.* **2021**, *11*, 885. [CrossRef] [PubMed]
7. Gadde, S.G.K.; Kshirsagar, A.; Anegondi, N.; Mochi, T.B.; Heymans, S.; Ghosh, A.; Roy, A.S. Correlation of Volume of Macular Edema with Retinal Tomography Features in Diabetic Retinopathy Eyes. *J. Pers. Med.* **2021**, *11*, 1337. [CrossRef] [PubMed]
8. Kalra, G.; Kar, S.S.; Sevgi, D.D.; Madabhushi, A.; Srivastava, S.K.; Ehlers, J.P. Quantitative Imaging Biomarkers in Age-Related Macular Degeneration and Diabetic Eye Disease: A Step Closer to Precision Medicine. *J. Pers. Med.* **2021**, *11*, 1161. [CrossRef] [PubMed]
9. Lai, T.Y.Y.; Lai, R.Y.K. Association between Retinal Thickness Variability and Visual Acuity Outcome during Maintenance Therapy Using Intravitreal Anti-Vascular Endothelial Growth Factor Agents for Neovascular Age-Related Macular Degeneration. *J. Pers. Med.* **2021**, *11*, 1024. [CrossRef] [PubMed]
10. Hsing, S.C.; Lin, C.; Chen, J.T.; Chen, Y.H.; Fang, W.H. Glycemic Gap as a Useful Surrogate Marker for Glucose Variability and Progression of Diabetic Retinopathy. *J. Pers. Med.* **2021**, *11*, 799. [CrossRef] [PubMed]
11. Prasuhn, M.; Hillers, C.; Rommel, F.; Riemekasten, G.; Heidecke, H.; Nassar, K.; Ranjbar, M.; Grisanti, S.; Tura, A. Specific Autoantibodies in Neovascular Age-Related Macular Degeneration: Evaluation of Morphological and Functional Progression over Five Years. *J. Pers. Med.* **2021**, *11*, 1207. [CrossRef] [PubMed]
12. Vader, M.J.C.; Habani, Y.I.; Schlingemann, R.O.; Klaassen, I. miRNA Levels as a Biomarker for Anti-VEGF Response in Patients with Diabetic Macular Edema. *J. Pers. Med.* **2021**, *11*, 1297. [CrossRef] [PubMed]
13. Acar, I.E.; Willems, E.; Kersten, E.; Keizer-Garritsen, J.; Kragt, E.; Bakker, B.; Galesloot, T.E.; Hoyng, C.B.; Fauser, S.; van Gool, A.J.; et al. Semi-Quantitative Multiplex Profiling of the Complement System Identifies Associations of Complement Proteins with Genetic Variants and Metabolites in Age-Related Macular Degeneration. *J. Pers. Med.* **2021**, *11*, 1256. [CrossRef] [PubMed]

14. Nugawela, M.D.; Gurudas, S.; Prevost, A.T.; Mathur, R.; Robson, J.; Hanif, W.; Majeed, A.; Sivaprasad, S. Ethnic Disparities in the Development of Sight-Threatening Diabetic Retinopathy in a UK Multi-Ethnic Population with Diabetes: An Observational Cohort Study. *J. Pers. Med.* **2021**, *11*, 740. [CrossRef] [PubMed]

15. Maunz, A.; Benmansour, F.; Li, Y.; Albrecht, T.; Zhang, Y.P.; Arcadu, F.; Zheng, Y.; Madhusudhan, S.; Sahni, J. Accuracy of a Machine-Learning Algorithm for Detecting and Classifying Choroidal Neovascularization on Spectral-Domain Optical Coherence Tomography. *J. Pers. Med.* **2021**, *11*, 524. [CrossRef] [PubMed]

16. Cunha-Vaz, J.; Mendes, L. Characterization of Risk Profiles for Diabetic Retinopathy Progression. *J. Pers. Med.* **2021**, *11*, 826. [CrossRef] [PubMed]

17. Frizziero, L.; Calciati, A.; Torresin, T.; Midena, G.; Parrozzani, R.; Pilotto, E.; Midena, E. Diabetic Macular Edema Treated with 577-nm Subthreshold Micropulse Laser: A Real-Life, Long-Term Study. *J. Pers. Med.* **2021**, *11*, 405. [CrossRef] [PubMed]

18. Maurissen, T.L.; Pavlou, G.; Bichsel, C.; Villaseñor, R.; Kamm, R.D.; Ragelle, H. Microphysiological Neurovascular Barriers to Model the Inner Retinal Microvasculature. *J. Pers. Med.* **2022**, *11*, 148. [CrossRef] [PubMed]

Journal of
*Personalized
Medicine*

Review

Microphysiological Neurovascular Barriers to Model the Inner Retinal Microvasculature

Thomas L. Maurissen [1], Georgios Pavlou [2], Colette Bichsel [3,4], Roberto Villaseñor [5], Roger D. Kamm [2,6,*] and Héloïse Ragelle [1,*]

1   Roche Pharma Research and Early Development, Immunology, Infectious Diseases and Ophthalmology, Roche Innovation Center Basel, F. Hoffmann-La Roche Ltd., Grenzacherstrasse 124, 4070 Basel, Switzerland; thomas.maurissen@roche.com
2   Department of Biological Engineering, Massachusetts Institute of Technology, 77 Massachusetts Ave., MIT Building, Room NE47-321, Cambridge, MA 02139, USA; gpavlou@mit.edu
3   Roche Pharma Research and Early Development, Pharmaceutical Sciences, Roche Innovation Center Basel, F. Hoffmann-La Roche Ltd., Grenzacherstrasse 124, 4070 Basel, Switzerland; colette.bichsel.cb1@roche.com
4   Roche Pharma Research and Early Development, Institute for Translational Bioengineering, Roche Innovation Center Basel, F. Hoffmann-La Roche Ltd., Grenzacherstrasse 124, 4070 Basel, Switzerland
5   Roche Pharma Research and Early Development, Neuroscience and Rare Diseases, Roche Innovation Center Basel, F. Hoffmann-La Roche Ltd., Grenzacherstrasse 124, 4070 Basel, Switzerland; roberto.villasenor_solorio@roche.com
6   Department of Mechanical Engineering, Massachusetts Institute of Technology, 77 Massachusetts Ave., MIT Building, Room NE47-321, Cambridge, MA 02139, USA
*   Correspondence: rdkamm@mit.edu (R.D.K.); heloise.ragelle@roche.com (H.R.)

Citation: Maurissen, T.L.; Pavlou, G.; Bichsel, C.; Villaseñor, R.; Kamm, R.D.; Ragelle, H. Microphysiological Neurovascular Barriers to Model the Inner Retinal Microvasculature. *J. Pers. Med.* **2022**, 12, 148. https://doi.org/10.3390/jpm12020148

Academic Editors: Peter D. Westenskow and Andreas Ebneter

Received: 15 December 2021
Accepted: 18 January 2022
Published: 24 January 2022

Publisher's Note: MDPI stays neutral with regard to jurisdictional claims in published maps and institutional affiliations.

Abstract: Blood-neural barriers regulate nutrient supply to neuronal tissues and prevent neurotoxicity. In particular, the inner blood-retinal barrier (iBRB) and blood–brain barrier (BBB) share common origins in development, and similar morphology and function in adult tissue, while barrier breakdown and leakage of neurotoxic molecules can be accompanied by neurodegeneration. Therefore, pre-clinical research requires human in vitro models that elucidate pathophysiological mechanisms and support drug discovery, to add to animal in vivo modeling that poorly predict patient responses. Advanced cellular models such as microphysiological systems (MPS) recapitulate tissue organization and function in many organ-specific contexts, providing physiological relevance, potential for customization to different population groups, and scalability for drug screening purposes. While human-based MPS have been developed for tissues such as lung, gut, brain and tumors, few comprehensive models exist for ocular tissues and iBRB modeling. Recent BBB in vitro models using human cells of the neurovascular unit (NVU) showed physiological morphology and permeability values, and reproduced brain neurological disorder phenotypes that could be applicable to modeling the iBRB. Here, we describe similarities between iBRB and BBB properties, compare existing neurovascular barrier models, propose leverage of MPS-based strategies to develop new iBRB models, and explore potentials to personalize cellular inputs and improve pre-clinical testing.

Keywords: microphysiological systems; blood-neural barriers; neurovascular unit; disease modeling; 3D models; organ-on-a-chip; inner blood-retinal barrier

1. Introduction

The inner blood-retinal barrier (iBRB) is a selective endothelial barrier that restricts molecular transport from the retinal microvasculature, located in between neuronal nuclear layers, to neighboring neural tissue [1]. Likewise, the blood–brain barrier (BBB) controls molecular transport to brain neural tissue. These blood–neural barriers are composed of perivascular mural and glial cells that share a common origin in the central nervous system (CNS) vascular development [2,3], form the retinal and brain neurovascular units (NVUs) and play multiple roles in barrier stabilization and maintenance to protect against neurotoxicity [4–6]. Additionally, both neurovascular barriers have similar cellular properties [7], including the formation of specialized intercellular junctions and downregulation of transcytosis [8,9], which limit paracellular and transcellular permeability, respectively. Microvascular diseases linked to chronic or genetic conditions often lead to barrier breakdown and neurodegeneration in the retina or brain. iBRB dysfunction and vascular leakage are associated with some of the most predominant ocular diseases such as diabetic retinopathy (DR) and age-related macular degeneration (AMD) [10,11]. BBB breakdown was suggested to be involved in the pathophysiology of several neurological disorders such as Alzheimer's disease, Parkinson's disease, Huntington's disease, amyotrophic lateral sclerosis and multiple sclerosis [12]. However, the mechanisms driving barrier breakdown and their link to pathophysiology of neurodegenerative diseases are not yet fully characterized [13]. Pre-clinical studies in animal models poorly translate to humans and represent a major hurdle in the drug discovery pipeline, since 80% of potential treatments fail in clinical trials [14]. Therefore, there is a need for human-based models that faithfully recapitulate the pathophysiology of human diseases, accelerate drug discovery processes, and better predict patient response to treatments.

Human advanced in vitro models are becoming widespread tools to study cellular interactions and function in many organ-specific contexts [15]. This is made possible through technological advances in bioengineered microphysiological systems (MPS) that support three-dimensional (3D) cellular self-organization [16], such as microfluidics or 3D patterning [17,18] in combination with biocompatible hydrogels [19,20]. While these individual technological advances have progressed incrementally over decades, their combination was only achieved recently. Resulting MPS now facilitate the formation, maturation and maintenance of complex organotypic structures in well-defined and customizable microenvironments [18,21,22]. Human-based MPS have already been developed for tissues such as lung, gut, brain and tumors, yet few integrative models exist for ocular tissues and iBRB modeling [23,24]. Considering the implications of barrier breakdown and vascular leakiness in ocular and neurodegenerative diseases, there is an urgent need to elucidate the pathophysiological mechanisms of vascular instability and accelerate therapeutic interventions.

MPS are especially important to model the NVU, where close interactions between endothelial cells (ECs), pericytes and astrocytes are crucial for proper function [24–26]. Various in vitro models and assays are being developed to investigate BBB function and pathogenesis. ECs in mono-culture or in co-culture together with pericytes and astrocytes have been grown in different configurations as 2D monolayers and more recently in 3D MPS to interrogate barrier properties in standard culture or in response to treatments disrupting or maintaining vascular permeability [27,28]. Only recently, the assembly of endothelial and perivascular cells was accomplished in a manner that resembles NVUs found in CNS micro-vasculatures, by self-organization of cells into microvascular networks (MVNs) in MPS [29]. Deeper characterization of vascular models using primary cell sources alone, or in combination with induced pluripotent stem cell (iPSC)-derived ECs, demonstrated acquisition of tissue-specific cellular identities representative of physiological human blood-neural barriers. Furthermore, human-based advanced cellular models have the potential to be customized to specific patient groups/populations and to be scalable for drug discovery purposes. Taken together, these BBB in vitro models showed physiologically relevant morphology and permeability values, and reproduced pathophysiological phenotypes such as vascular leakage and neurodegeneration [30–32].

Here, we review similarities in blood–neural barrier and NVU properties in the retina and brain, and focus on recent developments in MPS-based neurovascular barrier models. While bioengineering approaches to model BBB features have already been extensively reviewed [22,25,33–38], we propose to harness the strengths of existing MPS-based strategies to generate better iBRB models. Finally, we explore potentials to further personalize cellular inputs and improve pre-clinical testing using MPS.

2. Neurovascular Units in Health and Disease

CNS micro-vasculatures contain specialized endothelial and perivascular cells that, together with neurons and glial cells, are organized into NVUs. Well-orchestrated interactions between these different cell groups secure the healthy function of the CNS while dysfunction of any of them can lead to inflammation, neurotoxicity and are associated with neurodegeneration [6,12]. Here, we provide a brief summary of the key components and properties of the iBRB and BBB, before highlighting similarities in their development. We then describe the impact of neurovascular barrier breakdown on pathophysiological outcomes.

2.1. Barrier Properties

NVUs consist of specialized endothelial and supporting pericytes and astrocytes that closely interact to establish and maintain barrier properties. Increasing evidence suggests that ECs in NVUs possess tissue-specific characteristics that distinguish them from peripheral ECs and contribute to barrier properties. First, retinal and brain ECs express specific tight junction proteins, such as Claudin-5 [39] and Occludin [40], that are indispensable to restrict paracellular permeability [41]. Second, transcellular transport across the retinal and brain endothelium is lower compared to peripheral endothelia [42,43]. This is partly due to high expression of Mfsd2a, a lipid transporter that inhibits transcytosis in retinal and brain ECs [8]. In addition, intracellular regulators of transcytosis described in epithelial cells (e.g., Rab25 and Rab17) are absent in CNS-ECs [5]. Third, a specific set of transporters regulates the influx and efflux of substances. For example, ECs at the iBRB and BBB have high expression of the glucose transporter GLUT-1 [44] and the efflux pump P-gp [45]. Finally, the retinal and brain endothelia express low levels of leukocyte adhesion molecules (LAMs) such as ICAM-1 and VCAM-1 [46], which restricts the migration of peripheral immune cells across the healthy iBRB and BBB.

Surrounding and closely interacting with ECs are pericytes and astrocytes [47,48]. Pericytes line the micro-vessels, share and contribute to the basement membrane of those vessels and regulate endothelial proliferation [49]. The ratio of ECs to pericytes in the retina and brain is approximately 1:1–3:1, whereas this ratio is much lower in the lung (10:1) and skeletal muscle (100:1) [50,51]. Interactions between pericytes and ECs induce TGF-β activation, which increases basement membrane synthesis and reduces its degradation via proteinase inhibitors such as TIMP-1 [51,52]. During angiogenesis, PDGF-B secreted by sprouting ECs binds to heparin sulfate proteoglycans of the extracellular matrix (ECM) and forms a concentration gradient that leads to the recruitment and attachment of PDGFRβ-expressing pericytes, which promotes vascular stabilization and maturation, and plays a role in controlling vascular permeability [3].

Astrocytes support iBRB and BBB integrity through end-feet projections, which are highly specialized and polarized structures that cover almost the entire abluminal surface of the blood vessels [48]. Astrocytes secrete Sonic hedgehog (Shh) that increases tight junction protein expression in ECs and reduces expression of leukocyte adhesion molecules [53,54]. Co-cultures of ECs with astrocytes also increased tight junction expression in ECs and reduced paracellular permeability to sucrose [55].

Important and often neglected components of the NVU are the surrounding extracellular matrix molecules of the glycocalyx and basement membrane, located at the luminal and abluminal side, respectively. The basement membrane is composed of collagen IV, laminin, fibronectin, nidogen and perlecan [56], with which the cells interact primarily via integrins. Proteoglycans and glycoproteins of the glycocalyx are also involved in mediating cellular signaling. Astrocyte-derived laminin, for example, promotes BBB integrity by increasing EC tight junction proteins and regulating pericyte maturation, and conditional knock-out of laminin gamma-1 in mice leads to BBB breakdown [57].

2.2. Neurovascular Development

Blood vessels in the brain originate from the perineural vascular plexus (PNVP). In humans, around 5 weeks after conception, endothelial sprouts from the PNVP invade the CNS and start to form a vascular network [58] (Figure 1a). In parallel, the retina is vascularized from the optic nerve outward [4] (Figure 1b). During this angiogenic process, signaling pathways that drive angiogenesis in peripheral tissues are activated. These include VEGF and Notch signaling, while PDGF-B/PDGFRβ [52] and ANG/TIE2 [51] are involved in perivascular cell recruitment. In contrast, the Wnt signaling pathway is activated specifically during brain angiogenesis but not in other vascular beds, and controls expression of CLDN5 and PLVAP [59,60]. Early signs of blood–CNS barrier formation are observed, with new vessels exhibiting barrier properties such as the expression of proteins involved in the formation of specialized intercellular junctions (e.g., CLDN5), expression of efflux transporters (e.g., MDR1/ABCB1), and reduced transcellular transport [42,61,62].

Important signal transduction pathways involving ECs and pericytes at the NVU are PDGF-B/PDGFRβ, ANG/TIE2 and TGF-β/TGFβR2. The PDGF-B/PDGFRβ axis is indispensable in pericyte recruitment and BBB stabilization as well as astrocyte end-feet polarization, as demonstrated in pericyte-deficient mouse models [63,64]. Ang1 in both iBRB and BBB is thought to originate from pericytes and astrocytes [65,66], although some studies argue that Ang1 is not produced by pericytes in retinal vessels, and that Tie2 may be activated by blood flow-mediated shear stress instead [2,67–69]. Finally, TGF-β signaling is multifaceted, affecting endothelial and pericyte proliferation and barrier integrity [41,70–72].

Astrocyte networks populate the developing retina and form a network to guide the retinal vascularization process by radial outgrowth from the optic nerve to the periphery [73] (Figure 1b). In the brain, astrocyte development and maturation take place late in embryonic development and continue after birth [74]. Since developing vessels in the retina are leaky despite the presence of astrocytes [62], and astrocytes are not present in the early stages of brain development [75], they are not likely play a key role in initial barrier formation, but rather in barrier maintenance through secreted factors [57,76,77].

2.3. Pathophysiology

Disruption of iBRB/BBB promoting signaling pathways, progressive loss of vascular cells and dissolution of endothelial junctions result in barrier breakdown and neurotoxicity (Figure 1c).

Diabetic retinopathy. In diabetic conditions, different mechanisms were identified to cause pericyte loss and other diabetic retinopathy-(DR)-associated changes. Hyperglycemia induces pericyte-specific activation of PKC-δ and SH-1, leading to PDGFR-β dephosphorylation and resistance to PDGF-B secreted by ECs, and resulting in pericyte apoptosis [78]. In addition, retinal capillary pericytes were found to accumulate vasculotoxic molecules in diabetic conditions, leading to their detachment from ECs or selective pericyte degeneration [79]. For instance, soluble epoxide hydrolase (sEH) expression and accumulation of 19,20-dihydroxydocosapentaenoic acid prevented the formation of N-cadherin and VE-cadherin adherens junctions, thus disrupting pericyte-EC and inter-EC junctions [80]. How these different mechanisms of pericyte loss are connected remains unclear, but sEH inhibition was suggested to increase PDGF-B expression thereby restoring PDGF-mediated survival actions, besides rescuing the formation of adherens junctions and preventing the dissociation of vascular cells. Similarly, Notch3 expression in pericytes promotes survival and EC-pericyte interactions through expression of N-cadherin, so interruption of Notch signaling can result in vascular instability [81,82]. Activation of the Notch pathway was also shown to modulate PDGFRβ signaling independently of PDGF-B. Moreover, pericytes play a direct role in regulating the iBRB by influencing gene expression patterns in ECs and polarization of astrocyte end-feet [64]. Pericyte loss thus leads to losing astrocyte-derived components secreted from the end-feet. In sum, mechanisms causing pericyte loss in DR interfere with paracrine signaling axes between pericytes and ECs such as PDGF-B/PDGFR-β, or with junctions binding pericytes and ECs such as N-Cadherin and Notch3.

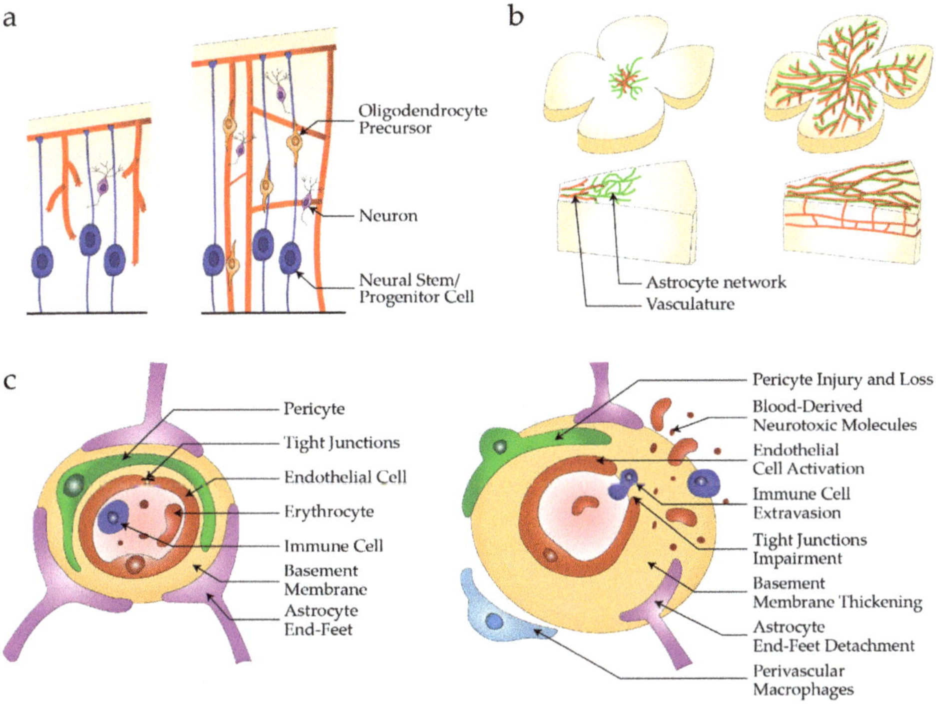

Figure 1. Healthy and diseased NVU. (**a**) Vascularization in brain development originates from the PNVP with sprouting vessels extending along neural stem and progenitor cells and oligodendrocyte precursors that give rise to differentiated neuronal cell types; (**b**) In retinal development, an astrocyte network forms radially from the optic nerve to the periphery followed by sprouting vessels that vascularize the neuroretina following the astrocyte network, as shown in the retinal flat mounts and cross section of the retinal layers; (**c**) Neurovascular unit composed of endothelial cells, pericytes and astrocytes with structural features in healthy conditions (**left**), and during barrier breakdown (**right**). Adapted from [73,83].

Neurodegenerative diseases. Neuronal pathologies such as Alzheimer's disease (AD), Parkinson's disease, Huntington's disease and amyotrophic lateral sclerosis are highly associated with compromised BBB integrity. Specifically, in AD, BBB breakdown is correlated with amyloid-β (Aβ) protein accumulation on the vascular walls and the reduction in pericyte coverage [12,84,85]. Interestingly, increased BBB permeability, which can be measured in vivo using several techniques (PET, MRI, near-infrared time-domain optical imaging) [86] correlates well with cognitive decline in humans [87–89], and may actually precede it. Furthermore, the expression of transporters, tight junction proteins, and membrane transport regulators is also altered in AD [12,90], suggesting that multiple aspects of BBB function are affected during the course of the disease. Whether these changes have a causative role on the pathophysiology of disease or are instead a secondary result of neurotoxicity and inflammation in the brain is still being investigated.

Monogenic diseases. Given the importance of signaling pathways on blood vessel formation and maintenance, as well as on BBB integrity, the improper function of a single protein can be expected to result in devastating outcomes. Indeed, a large number of human inherited monogenic diseases have been identified as disrupting BBB function [65,85].

Typical examples include mutations in *PDGFB* or *PDGFRB* growth factor/receptor causing primary familial brain calcification (idiopathic basal ganglia calcification, Fah's disease), mutations in receptor *NOTCH3* associated with cerebral autosomal-dominant arteriopathy with subcortical infarcts and leukoencephalopathy (CADASIL) and tight junction *OCLN* in band-like calcification with simplified gyration and polymicrogyria (BLC-PMG), and mutations in basement membrane *COL4A1* and *COL4A2* underlying cerebral small vessel disease or *LAMA2* in congenital muscular dystrophy 1A (MDC1A). The resulting dysfunctional protein often disrupts normal function in specific cell types, including pericytes, vascular smooth muscle cells and ECs. Deeper study of these genes vulnerable to mutations might provide new insights into disease-related vascular dysfunction and potential new therapeutic strategies.

3. Strategies to Mimic NVU Architecture and Function

The retinal and brain NVU share functional characteristics and develop in parallel, and the retinal in vivo model has been extensively used to study vessel formation and maturation [62]. In contrast, in vitro models with human cells have largely focused on recapitulating features of the brain NVU. Here, we present both retinal and brain barrier models with applications in drug development, and suggest that approaches developed in connection with BBB models might also be applied to iBRB modeling, despite some region-specific signaling differences [91].

3.1. Preliminary Considerations

Cellular sources. In the development of BBB models, one of the major challenges is to maintain consistency in critical functional parameters such as permeability. In this regard, the cell source has been identified as a major cause of variability since cells obtained from different donors can show substantially different behaviors in culture, such as self-organization capacity, response to treatment and/or barrier properties. Models of the iBRB or BBB have been developed with ECs, and also with pericytes, astrocytes and neurons obtained from primary tissue or differentiated from iPSC lines [24,25]. In order to recapitulate in vitro the key properties of CNS micro-vasculatures, tissue-specific features, such as higher expression of tight junction proteins and transporter receptors, are required for input cells.

Human primary vascular and perivascular cells obtained from donors have been widely used in different endothelial barrier culture models, and while primary ECs express high levels of barrier markers, the nature of the model system (2D or 3D) and the other cell types used (e.g., pericytes and astrocytes) can dramatically influence expression levels, and as a result, the models only partially recapitulate physiological barrier function, as shown by permeability assays [25,92]. Moreover, their barrier properties vary and are limited by cell passage number and donor-associated heterogeneity [93]. In particular, human umbilical vein endothelial cells (HUVECs) have been reliably used as a source for vascular models, and in the presence of supporting human lung fibroblasts showed effective self-organization into microvascular networks with low permeability values [94–97].

Human iPSCs are becoming a preferred cell source as differentiation protocols are being improved and cellular identity and maturity are becoming better characterized [98]. Donor-specific iPSCs are used to generate multiple cell types in the same genetic background, and reconstitute complex in vitro models such as the NVU, corresponding to individual patients. Various protocols are available to induce ECs and other components of the NVU [99]. Direct conversion of induced brain microvascular endothelial-like cells (iBECs) were previously described [100], yet recent work showed that the identity of these cells do not correspond to ECs but are instead neuroepithelial cells [101]. These findings raise questions on the applicability of these cells for modeling BBB-specific processes such as immune cell migration or transcytosis. On the other hand, in a recent publication, data were presented showing that primary brain ECs and iPSC-derived ECs share similar gene expression levels of key genes involved in the regulation of BBB function when they are in tri-culture 3D systems [28]. Primary cells and iPSCs are both promising tools to generate in vitro models suitable for physiological and pharmaceutical studies of the iBRB and BBB (Figure 2). For this, cell lines need to be validated for the expression of relevant markers, response to angiogenic stimuli, low permeability values, and capacity to form microvascular networks. In this regard, authors should strive to analyze and report the same genes measured in previous studies and use comparable metrics for function in order to facilitate direct comparisons.

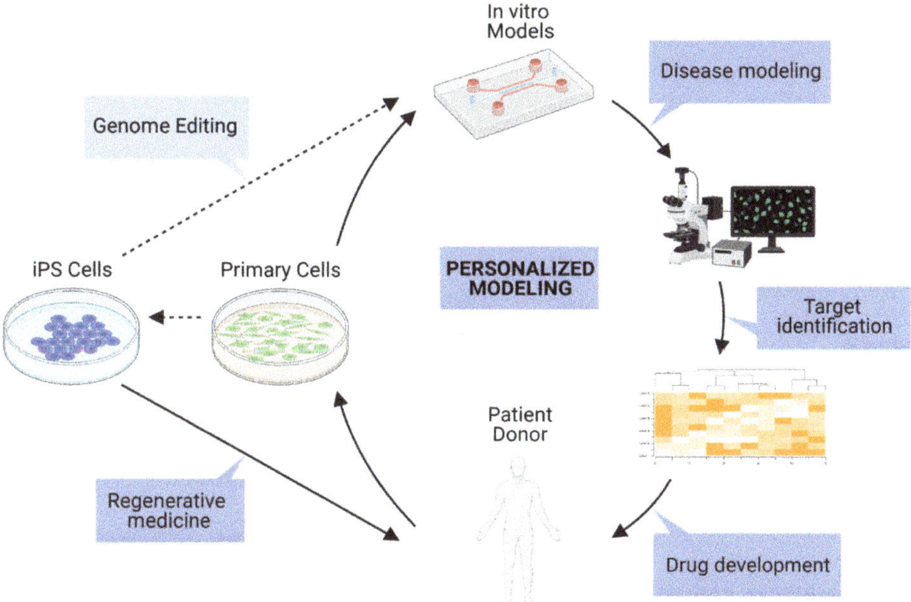

Figure 2. Personalized approaches for neurovascular barrier models. Patient or donor tissue is a source for cellular and genetic manipulation with the aim of treating patients through regenerative medicine or disease modeling approaches. Self-organization of relevant cell types in MPS models provide valuable tools to test targets in a personalized manner and develop drugs that are pre-validated in vitro on autologous cells. Created with BioRender.com.

Gene Editing. Alongside the development of organotypic in vitro systems, the relevance of disease models can be further enhanced by genetic manipulation, to investigate target mechanisms or pathogenic mutations associated with diseases such as retinal dystrophies. Taking into consideration cell culture requirements when making genetic modifications, including the selection, characterization and expansion of stable cell lines, immortalized cell lines are better suited than freshly isolated primary cells or iPSC-derived cells that have a very limited proliferation capacity. Ideally, genetic modification and stable cell line generation are carried out in a suitable parent iPSC line before differentiation. Nevertheless, genetic modification and associated experimental steps of clonal selection could impact the differentiation efficiency of iPSC lines and downstream phenotyping [102]. Therefore, it is important to test in parallel several iPSC lines or, if unavailable, iPSC clones to distinguish signal from noise during assay measurements.

In vitro modeling. Significant advances have been achieved in the development of in vitro BBB models. A variety of models has indeed been referenced and can serve different applications in particular with the engineering of drugs/compounds that could (i) overcome the BBB restricted permeability and target the brain parenchymal cells such as neurons, glial cells or even brain stem cells, but also metastatic brain tumors or (ii) protect from and repair BBB lesions or dysfunctions. In addition to their potential for predictive drug studies and therapeutic applications, such models represent an important asset for dissecting the complex interplay between the different NVU components in steady-state homeostatic conditions as well as in disease, which is often characterized by a primary defect in BBB permeability and protective properties. Indeed, by combining co- and tri-cultures of relevant cell types in a gel matrix, well-calibrated platforms can be designed to better recapitulate the physiological function of the human microvasculature at the capillary level. Since numerous neuronal and vascular diseases have been linked to disrupted neurovascular interactions and BBB failure, multi-component in vitro systems, although challenging to validate, represent promising approaches in tackling these ever increasing health issues.

Current efforts to evaluate transport of biologics to the CNS rely primarily on in vivo studies with rodents. However, recent evidence shows clear differences in the molecular composition of the human BBB compared to mice [103]. In addition, animal studies are not amenable to extensive screening. Therefore, human in vitro models of the BBB can support drug development by improving confidence in targets, studying drug mechanism of action and/or enabling lead optimization campaigns.

Progress in BBB modeling is also likely to benefit related research into the development of iBRB models in health and disease. Further, these developments could provide a rationale for engineering a more complex multicellular system that would include both the inner and the outer components of the BRB for better functional integration.

3.2. 2D Models

2D layers. Historically, 2D models exemplified with Transwell systems were the first approaches aiming to recreate endothelial barrier properties in vitro. In these settings, ECs placed on top of a porous flat membrane, in mono- or co-culture with pericytes, and with supporting cells such as pericytes and/or astrocytes grown on the other side of the membrane or at the bottom of the well, form monolayers. These multi-component systems enable direct measurement of permeability, but also the spatiotemporal behavior of (i) cell migration through the cell layers and (ii) dynamic interactions between the different cell types of the reconstituted platforms [104–112]. A major advantage of these systems is the direct electrical assessment of the trans-endothelial electrical resistance (TEER) and solute values for permeability [113]. Additionally, the use of each side of the membrane support, which can be controlled in its composition, porosity and thickness in conjunction with advanced transcriptomic analysis, provide simplified but informative 2D experimental frameworks [114,115]. Monolayers thus provided a first approach to investigation of the barrier properties of the iBRB and BBB using various readouts.

2D organs-on-a-chip. To further improve the relevance of these widely used systems and recreate the tissue and organ complexity within a dynamic microenvironment with physiological flows, 2D organ-on-a-chip models were introduced by seeding different cell types inside a microfluidic device under more controlled conditions that would be fully compatible with cell growth and critical multi-lineage network development. The addition of media flow and control of shear stresses provide a more realistic biomechanical framework which can still be combined with TEER measurement while allowing high-resolution live imaging [116,117]. Various 2D systems have managed to partially recapitulate iBRB or BBB functions such as permeability and high expression of tight junction proteins [30–32,117]. Despite the 2D organ-on-a-chip "biomimetic" models offering several advantages over the 2D monolayer models, they are still limited in their morphological relevance, in particular because of the large size of their vasculature-like structures, which largely exceeds the human retinal or brain microvasculature. 2D organs-on-a-chip nonetheless improve barrier properties and increase the co-culture possibilities by introducing additional cellular compartments, which could be applied to modeling the iBRB with additional cell types, such as neurons, in different configurations.

3.3. 3D Models

Blood-brain barrier organoids. Instead of co-culturing NVU cell types spatially separated on a trans-well, Urich et al. noticed that when primary human brain ECs, pericytes and astrocytes are mixed in hanging drop cultures, they spontaneously self-assemble into organoids [118]. The organoid core mostly contains astrocytes, while pericytes wrap around the core and the ECs form a continuous layer on the organoid surface. Interestingly, ECs and pericytes also form spheroids when cultured individually, while astrocytes alone or together with ECs are not able to form clusters. Co-culturing ECs, pericytes and astrocytes in organoids also upregulated surface VE-Cadherin, CD31, and P-gp [119] while ICAM-1 and -2 were downregulated. This change in endothelial adherens junction molecules, efflux pump and leucocyte adhesion molecule (LAM) expression was not found in a trans-well setup [118,119], which is likely due to the lack of, or very limited, direct cellular contact in trans-wells.

Various methods have since been used for BBB organoid formation, from hanging drops [118,120,121] to agarose [119], ultra-low attachment plates [122], micro-patterned hydrogels [123] and custom-made microfluidic spheroid arrays [124]. Cho et al. assessed the BBB tightness and functionality by demonstrating P-gp activity and LRP-1 mediated transcytosis of Angiopep-2 into the core of a tri-culture organoid made with primary human brain microvascular ECs (BMECs), pericytes and astrocytes. To demonstrate the aptness of the BBB organoid model for drug screening purposes, they evaluated barrier penetration of fluorescently labeled peptides, and verified their ability to cross the BBB into the brain parenchyma in a mouse model. Simonneau et al. scaled up the BBB organoid assay and used CRISPR/Cas9 edited ECs to identify critical regulators of endocytosis across the BBB model. Using this model, they demonstrated that clathrin, but not caveolin, is required for transferrin receptor-dependent transcytosis of antibody shuttles.

Others have focused on increasing the cellular complexity of BBB organoid models. Nzou et al. added oligodendrocyte progenitors, microglia and neuronal progenitors to the organoid core, all of which were iPSC-derived. They demonstrated selective neurotoxicity by exposing the organoid to 1-methyl-4-phenyl-1,2,3,6-tetrahydropyridine (MPTP) and 1-methyl-4-phenylpyridinium (MPP+). The neurotoxin MPP+ is hydrophilic and thus not able to penetrate an intact BBB, while MPTP can cross the barrier and become metabolized to MPP+ in the presence of glial cells. Nzou et al. demonstrated these different cytotoxic effects in BBB organoids [120]. Instead of adding iPSC-derived cells, Kumarasamy & Sosnik combined primary human BMECs, pericytes and astrocytes with rat cortical neurons and microglia, and used this model to screen for nanoparticle penetration [125].

The overall advantages of BBB organoid models are their ease of assembly and culture, cost-effectiveness and scalability. They can be used for screening small molecule BBB penetration [119] and large molecule transcytosis [123]. By combining this model with genetic engineering on selected cell types [123], BBB organoids can also help to investigate mechanisms of transcytosis. While no similar model exists using retinal microvascular cells, this approach could be applied to model and compare transport mechanisms across the iBRB with medium to high throughput.

3D tubular channels. The constant need for better recapitulating organ formation and their vasculature led to a shift of current microfabrication technologies to 3D systems where cells are embedded in hydrogels. The development of 3D-like vascular systems, together with co-culturing different cell types adjacent to each other in hydrogel scaffolds, takes a step closer to mimicking the in vivo complexity of the vasculature. More specifically, microfluidic channels are coated with a layer of ECs to form a tube-like structure, which is adjacent to a gel mixed with different cells of interest. For modeling the BBB, astrocytes and neuronal cells are embedded in adjacent gels to establish a BBB-like organization that can have various applications such as disease modeling [126–128]. For the iBRB, an endothelial barrier was used to probe permeability in response to disease triggers [129]. Similar models were achieved using the viscous fingering method or by casting the gel around needles or wires that can subsequently be removed leaving hollow tubular channels, all within a petri dish or microfluidic device, which can subsequently be seeded with cells of interest [130–132]. While these models improved BBB properties such as permeability values, morphology and tight junction proteins expression levels, they fail to fully recapitulate the morphology (diameter, branching pattern, etc.) and permeability of the BBB observed in vivo. In particular, the size of the lumens are much larger than in vivo microvasculature and permeability values generally tend to be one to two orders of magnitude higher than in vivo measurements [25].

3D microvascular networks. 3D self-assembled MVNs are a recent approach used to recapitulate features of the human brain vasculature more precisely. In this approach, cells of interest are mixed with a hydrogel and acquire BBB-like structure and function following self-assembly. Several published models showed improved relevance for mimicking the human system in terms of vasculature morphology and barrier properties in comparison with the previously described 2D and 3D in vitro models [29,133,134] and better recapitulated pathophysiological features of neurodegenerative diseases than in other in vitro models. For example, in Alzheimer's disease, pericytes showed a detrimental role in APOE4-mediated amyloid accumulation in the cerebral vasculature [135], and in cancer astrocytes secreting the C-C motif chemokine ligand 2 promoted cancer cell transmigration in the BBB [136]. Advanced transcriptomics and proteomics analysis can be readily obtained after cell extraction from the devices [28]. Barrier permeability can be measured by perfusing different sizes of fluorescent tracers or compounds of interest tagged with a fluorescent marker and measuring accumulated signal in the extra-vascular space [25,27,97]. Overall, direct comparison between a 2D co-culture trans-well system and a 3D co-culture vascular network, all using the same cells, show uniformly lower permeabilities with the 3D network, and these values are comparable to in vivo permeability measurements [27].

Introduction of media flow in a continuous circulating direction was described in similar models [137], but most 3D self-assembled BBB MVNs have not been tested yet under flow conditions. It would be beneficial for these systems to be continuously perfused with media flow using circulating pumps since the shear stresses acting on the vascular walls are known to decrease permeability values, in comparison to static models, and to enhance barrier function [137–140]. Furthermore, applying 3D self-assembly to recapitulate the iBRB would allow for better models with a microvasculature-like morphology, multicellular organization of co-cultured cell types with direct cellular interactions, and enable a more accurate representation of disease phenotypes.

Vascularized brain organoids. While there are several cortical organoids for development and disease modeling that have formed features of brain tissue [141,142], there is a need to vascularize these systems to ensure proper growth and better functionality. BBB organoids have been developed and grafted onto mice for perfusion. They mimic in vivo BBB properties such as high expression levels of tight junctions and adherens junctions, low permeability of different molecules and expression of efflux transporters [143–145]. Vascularization of brain organoids has recently benefitted by the incorporation of human embryonic stem cells in the organoid that ectopically express the human ETS variant 2 (ETV2), but it is unclear to what extent these ECs acquire a functional BBB identity [146]. However, there are challenges that the field needs to overcome by improving the organoid formation and culture methods and by increasing their reproducibility. Similarly, retinal organoids have been differentiated but lack vascularization, and thus tissue-specific iBRB properties, for further interrogation [147,148]. This remains a promising direction for increasing the cellular diversity and complexity of the in vitro models to an anatomical level.

Overall, each approach has strengths but also limitations (Table 1). Research labs can combine these approaches depending on the readouts they are interested in. Current tools can provide morphological and functional characteristics of the iBRB/BBB but further improvement is still necessary. Additional cellular and tissue components can be added to the systems such as lymphatic vasculature and microglia that have been shown to be important components of the NVU and the retina or brain in general. In sum, the majority of novel human-based in vitro model developments focus on the BBB while fewer iBRB models are being established. Given the physiological similarities between the iBRB and BBB, we propose that many of the novel approaches developed for the BBB can be applied to accelerate model development for the iBRB. Indeed, retinal and brain ECs share similar CNS vasculature-specific markers, such as Claudin-5, that make assay parallelization possible for this type of readout [41]. Likewise, pericytes and astrocytes are organized similarly in the retinal and brain NVU, so that in vitro iBRB and BBB platform designs could be used interchangeably depending on the degree of complexity and the variety of cellular components [81]. Generally, it has been shown that iPSC-derived ECs tend to acquire a CNS-like identity when cultured in the corresponding BBB-specific niche [28,29], which could also be potentially applicable to the iBRB. However, it is unclear if tissue-specific primary cells could be used interchangeably between iBRB and BBB models since their cellular identities despite being similar are not identical, and the acquisition of tissue-specific gene expression patterns has not been demonstrated. Finally, the iBRB and BBB have additional tissue-specific cell types, such as Müller glia and neurons that would only be introduced in their respective model.

Table 1. Summary of in vitro neurovascular barrier models.

Model	Application	Limitations	iBRB Refs.	BBB Refs.
2D layers (e.g., Trans-well)	Tight junction immunostaining TEER Permeability Imaging Transcriptomic analysis	Stiffness of tissue culture plastic 2D monolayer morphology Tendency for high permeability	[104–108]	[109–112]
2D organ-on-a-chip (e.g., Emulate)	Tight junction immunostaining TEER Permeability Imaging Transcriptomic analysis	Poor morphology Tendency for high permeability Low throughput	[117]	[30–32]
3D organoids	Immunostaining Permeability Transcytosis Imaging medium-high throughput screening	TEER not available Limited cellular self-organization	N/A	[118–123]
3D tubular channels (e.g., Mimetas)	Tight junction immunostaining Permeability Imaging	Morphologically large networks Limited cellular self-organization Moderate throughput	[129]	[126–128,130,131]
3D self-assembled micro-vasculatures (e.g., AIM Biotech)	Immunostaining Permeability Imaging Transcriptomic analysis	TEER not available Morphological variability Moderate throughput	N/A	[28,29,133–140]

4. Outlook

A significant number of drugs fail during Phase II, during which the efficacy of a given molecule is evaluated in a disease population [149]. This is due to inadequate prediction of human efficacy during preclinical development. Therefore, developing methods that better capture drug performance in patients may help to bring more drugs to clinical approval. In vitro human models that recapitulate critical aspects of human biology and disease are one class of tool that can screen therapeutics early in the development process and inform decisions as to whether to move forward with a molecule, or stop its development if the selection criteria are not met. Information gathered from human models can span the investigation of drug mechanism of action, prediction of on- and off-target effects, and evaluation of transport properties or safety. Altogether these data will improve compound optimization, clinical candidate selection and human dose prediction accuracy before clinical trials in humans.

More specifically, microphysiological neurovascular barrier models that contain human NVU components and capture specific NVU function (e.g., permeability, transport) can provide a better understanding of human barrier biology and interrogate barrier dysfunction in disease. When the pathophysiology of the disease is well-documented, as for DR, it is easier to validate the model by the observation of critical pathological events (e.g., pericyte loss, basement membrane thickening, vascular leakage). While in vitro models of the NVU comprise a powerful tool to understand disease processes and screen potential therapeutics, major challenges for broad implementation remain. These include reproducibility, scalability and utility of the readouts, matching the complexity of the model for the specific question, ease of use and throughput.

4.1. Readouts

In vitro iBRB/BBB models are well-suited to measure barrier permeability and transport of molecules. These readouts can be used both for compound ranking and selection, as well as for investigating pathways and molecular mechanisms. Barrier permeability can be assessed by TEER or fluorescently labeled molecules such as dextrans in combination with time-lapse imaging, depending on the model and platform. Permeability readouts can be used to assess whether barrier integrity is affected by disease triggers (e.g., VEGFA-induced permeability [129]) and test compound efficacy to prevent and/or restore barrier dysfunction. As previously discussed, crossing of molecules at the BBB can be quantified by adding a fluorescent tag to the compound of interest and measuring the accumulation of signal in the perivascular compartment (or the organoid core in case of iBRB/BBB spheroids [119,123]). This measurement can help optimizing compound properties, or studying mechanisms of receptor-mediated transcytosis [123]. In addition, more complex models that contain cell types of the parenchyma can be used to assess which cell types a molecule of interest targets after crossing the iBRB/BBB, and whether the molecule of interest is effective on its target (i.e., pharmacodynamics) or also affects other cell types (i.e., neurotoxicity [120]). Advanced readouts on vascular functionality are also possible with in vitro models containing blood vessels, for example, to measure vascular contractility [150].

4.2. Reproducibility

Within in vitro models, the least controllable parameter remains cellular behavior, which originates and differs from various unpredictable factors. Batch-to-batch variability is common when using biological materials. In the case of primary cell lines and iPSCs, differences in donor profile (age, sex, genetic background) and isolation procedure (cell sorting, media, quality control) can lead to variable cellular behavior. In addition, for a given cell batch in culture, external factors (e.g., extracellular matrix, serum) can lead to cellular heterogeneity, such as astrocyte reactivity in response to serum in the cell culture medium [151]. Systematic cell phenotyping, validation via expression of specific relevant markers and functional assays can help to control heterogeneity and define standards for the cells to be incorporated in 3D models. In self-assembled micro-vessels or self-organized organoids, variations in cell–cell communication during the assembly process can lead to substantial morphological and functional differences from one batch to another. This variability needs to be accounted for in advanced in vitro iBRB and BBB models and defining relevant functional readouts for each model as quality control parameters will facilitate the implementation and regular use of complex models. Further, the parameters of interest, for example permeability, need to be consistent across experiments. As observed with patient populations, some cells and/or models might respond differently to the same trigger. Classification and stratification of models might help understand disease heterogeneity, response to treatment and develop treatments better suited for a given patient population.

4.3. Scalability

The more complex the model is in terms of cell types, matrix and structure, the closer we get to accurately mimic organotypic functions, but the more difficult it is to guarantee reproducibility, ease of use, and throughput. Typically in preclinical development, high-throughput systems such as microtissues/spheroids or organoids in suspension would be used for primary screens while lower throughput systems such as organs-on-chips or assembloids that require more direct handling would be employed for target validation or toxicity assessment. Complex models that include perfusion, additional cell types (e.g., microglia) or self-renewal potential (organoids and assembloids) could also be useful in disease modeling over longer periods of time. While they may provide crucial insights into human biological mechanisms in the health and disease of the iBRB/BBB, they require a high investment in resources and time, and require careful handling.

In contrast, models of intermediate complexity, such as tubular NVUs and self-assembled micro-vessels, have the potential to be improved upon in terms of readout reproducibility. Once this is achieved, they are tools that add value in the drug discovery and target validation process.

4.4. Personalization

The promise of personalized medicine revolves around using a patient's autologous cells or a donor's allogeneic cells with matching human leukocyte antigens (HLA) for drug discovery or regenerative medicine, to test specific drug response and dosage, or to repair or replace damaged tissue [98]. For this reason, cellular sources, genetic manipulation and the accuracy of downstream phenotyping have the potential to personalize pre-clinical research (Figure 2). Primary cells can be isolated, and iPSCs reprogrammed from healthy and diseased individuals, taking into consideration specific characteristics of population groups such as age, sex, genetic background and predisposition to disease.

In addition, genetic factors are known to influence the onset and progression of ocular diseases such as diabetic retinopathy [152] and age-related macular degeneration [153]. Advances in CRISPR/Cas9 template-mediated gene editing are facilitating the generation of single-nucleotide polymorphisms (SNPs) and single-nucleotide variants (SNVs) in iPSCs [154,155], in order to recreate or rescue associated phenotypes in vitro (Figure 2). Similar strategies have been applied for retinal disease modeling or gene therapy, by knock-in of patient mutations, or correction or knock-out of mutant alleles, respectively [156].

5. Conclusions

The retinal and brain micro-vasculatures share specific properties and ensure neuroprotective barrier function in health, and their dysfunction in disease can influence neurodegenerative outcomes. Different retinal and brain micro-vasculature in vitro models have been generated that reassemble native tissue morphology and function, mimicking endothelial barrier properties and multicellular NVU architecture, and recapitulating physiological neurovascular barrier permeability values. MPS are a powerful tool to recapitulate the complex cellular microenvironments and experimental boundary conditions required to obtain relevant multicellular organization and interactions. iBRB MPS models are still lacking and can benefit from development in parallel to new BBB models that leverage this technology. In vitro neurovascular barrier models will allow addressing of the mechanisms of disease initiation and progression causing barrier breakdown, vascular leakage and neurodegeneration in ocular and neurological diseases. MPS-based models have the potential to accelerate pathophysiological insights relevant to human biology and accelerate the development of new therapies, by increasing the controllability over cellular inputs and by matching the level of complexity of a model with its capacity for high-throughput and reproducible testing.

Author Contributions: Conceptualization, T.L.M., H.R.; Writing—Original Draft Preparation, T.L.M., G.P., C.B.; Writing—Review & Editing, T.L.M., G.P., C.B., R.V., R.D.K., H.R.; Supervision, R.V., R.D.K., H.R. All authors have read and agreed to the published version of the manuscript.

Funding: T.L.M. was supported by a Roche Postdoctoral Fellowship (2020–2022, RPF-ID:565), C.B. was supported by a grant from the Roche Institute for Translational Bioengineering (2020–2022). Funding to R.K. and G.P. from the Cure Alzheimer's Fund and the US National Institutes of Health, NINDS (R21 NS105027, R01 NS121078) are gratefully acknowledged.

Acknowledgments: We thank Sakurako Iribe for digitalizing drawings in Figure 1.

Conflicts of Interest: T.L.M., C.B., R.V. and H.R. are employees of F. Hoffmann-La Roche Ltd. R.K. is co-founder of AIM Biotech, and receives research support from F. Hoffmann-La Roche Ltd., AbbVie, Amgen, Glaxo-Smith-Kline, Novartis, and Boehringer Ingelheim.

References

1. Chen, J.; Liu, C.-H.; Sapieha, P. Retinal Vascular Development. In *Anti-Angiogenic Therapy in Ophthalmology*; Stahl, A., Ed.; Springer International Publishing: Cham, Switzerland, 2016; pp. 1–19. ISBN 978-3-319-24097-8.
2. Winkler, E.A.; Bell, R.D.; Zlokovic, B.V. Central nervous system pericytes in health and disease. *Nat. Neurosci.* **2011**, *14*, 1398–1405. [CrossRef] [PubMed]
3. Trost, A.; Lange, S.; Schroedl, F.; Bruckner, D.; Motloch, K.A.; Bogner, B.; Kaser-Eichberger, A.; Strohmaier, C.; Runge, C.; Aigner, L.; et al. Brain and retinal pericytes: Origin, function and role. *Front. Cell. Neurosci.* **2016**, *10*, 20. [CrossRef] [PubMed]
4. Sapieha, P. Eyeing central neurons in vascular growth and reparative angiogenesis. *Blood* **2012**, *120*, 2182–2194. [CrossRef] [PubMed]
5. Villaseñor, R.; Kuennecke, B.; Ozmen, L.; Ammann, M.; Kugler, C.; Grüninger, F.; Loetscher, H.; Freskgård, P.O.; Collin, L. Region-specific permeability of the blood–brain barrier upon pericyte loss. *J. Cereb. Blood Flow Metab.* **2017**, *37*, 3683–3694. [CrossRef] [PubMed]
6. Hussain, B.; Fang, C.; Chang, J. Blood–Brain Barrier Breakdown: An Emerging Biomarker of Cognitive Impairment in Normal Aging and Dementia. *Front. Neurosci.* **2021**, *15*, 688090. [CrossRef]
7. Spadoni, I.; Fornasa, G.; Rescigno, M. Organ-specific protection mediated by cooperation between vascular and epithelial barriers. *Nat. Rev. Immunol.* **2017**, *17*, 761–773. [CrossRef]
8. Andreone, B.J.; Chow, B.W.; Tata, A.; Lacoste, B.; Ben-Zvi, A.; Bullock, K.; Deik, A.A.; Ginty, D.D.; Clish, C.B.; Gu, C. Blood-Brain Barrier Permeability Is Regulated by Lipid Transport-Dependent Suppression of Caveolae-Mediated Transcytosis. *Neuron* **2017**, *94*, 581–594.e5. [CrossRef]
9. Wang, Z.; Liu, C.H.; Huang, S.; Fu, Z.; Tomita, Y.; Britton, W.R.; Cho, S.S.; Chen, C.T.; Sun, Y.; Ma, J.X.; et al. Wnt signaling activates MFSD2A to suppress vascular endothelial transcytosis and maintain blood-retinal barrier. *Sci. Adv.* **2020**, *6*, eaba7457. [CrossRef]
10. Klaassen, I.; Van Noorden, C.J.F.; Schlingemann, R.O. Molecular basis of the inner blood-retinal barrier and its breakdown in diabetic macular edema and other pathological conditions. *Prog. Retin. Eye Res.* **2013**, *34*, 19–48. [CrossRef]
11. Kusuhara, S.; Fukushima, Y.; Ogura, S.; Inoue, N.; Uemura, A. Pathophysiology of diabetic retinopathy: The old and the new. *Diabetes Metab. J.* **2018**, *42*, 364–376. [CrossRef]
12. Sweeney, M.D.; Sagare, A.P.; Zlokovic, B.V. Blood-brain barrier breakdown in Alzheimer disease and other neurodegenerative disorders. *Nat. Rev. Neurol.* **2018**, *14*, 133–150. [CrossRef] [PubMed]
13. Little, K.; Llorián-Salvador, M.; Scullion, S.; Hernández, C.; Simó-Servat, O.; del Marco, A.; Bosma, E.; Vargas-Soria, M.; Carranza-Naval, M.J.; Van Bergen, T.; et al. Common pathways in dementia and diabetic retinopathy: Understanding the mechanisms of diabetes-related cognitive decline. *Trends Endocrinol. Metab.* **2021**, *33*, 50–71. [CrossRef] [PubMed]
14. Perrin, S. Preclinical research: Make mouse studies work. *Nature* **2014**, *507*, 423–425. [CrossRef] [PubMed]
15. Shamir, E.R.; Ewald, A.J. Three-dimensional organotypic culture: Experimental models of mammalian biology and disease. *Nat. Rev. Mol. Cell Biol.* **2014**, *15*, 647–664. [CrossRef] [PubMed]
16. Garreta, E.; Kamm, R.D.; Chuva de Sousa Lopes, S.M.; Lancaster, M.A.; Weiss, R.; Trepat, X.; Hyun, I.; Montserrat, N. Rethinking organoid technology through bioengineering. *Nat. Mater.* **2021**, *20*, 145–155. [CrossRef]
17. Low, L.A.; Mummery, C.; Berridge, B.R.; Austin, C.P.; Tagle, D.A. Organs-on-chips: Into the next decade. *Nat. Rev. Drug Discov.* **2021**, *20*, 345–361. [CrossRef]
18. Vunjak-novakovic, G.; Ronaldson-bouchard, K.; Radisic, M. Organs-on-a-chip models for biological research. *Cell* **2021**, *184*, 4597–4611. [CrossRef]
19. Osaki, T.; Sivathanu, V.; Kamm, R.D. Vascularized microfluidic organ-chips for drug screening, disease models and tissue engineering. *Curr. Opin. Biotechnol.* **2018**, *52*, 116–123. [CrossRef]
20. Roth, J.G.; Huang, M.S.; Li, T.L.; Feig, V.R.; Jiang, Y.; Cui, B.; Greely, H.T.; Bao, Z.; Paşca, S.P.; Heilshorn, S.C. Advancing models of neural development with biomaterials. *Nat. Rev. Neurosci.* **2021**, *22*, 593–615. [CrossRef]

21. Ronaldson-Bouchard, K.; Vunjak-Novakovic, G. Organs-on-a-Chip: A Fast Track for Engineered Human Tissues in Drug Development. *Cell Stem Cell* **2018**, *22*, 310–324. [CrossRef]
22. Offeddu, G.S.; Shin, Y.; Kamm, R.D. Microphysiological models of neurological disorders for drug development. *Curr. Opin. Biomed. Eng.* **2020**, *13*, 119–126. [CrossRef]
23. Haderspeck, J.C.; Chuchuy, J.; Kustermann, S.; Liebau, S.; Loskill, P. Organ-on-a-chip technologies that can transform ophthalmic drug discovery and disease modeling. *Expert Opin. Drug Discov.* **2019**, *14*, 47–57. [CrossRef] [PubMed]
24. Ragelle, H.; Goncalves, A.; Kustermann, S.; Antonetti, D.A.; Jayagopal, A. Organ-On-A-Chip Technologies for Advanced Blood-Retinal Barrier Models. *J. Ocul. Pharmacol. Ther.* **2020**, *36*, 30–41. [CrossRef] [PubMed]
25. Hajal, C.; Le Roi, B.; Kamm, R.D.; Maoz, B.M. Biology and Models of the Blood-Brain Barrier. *Annu. Rev. Biomed. Eng.* **2021**, *23*, 359–384. [CrossRef]
26. Caffrey, T.M.; Button, E.; Robert, J. Toward three-dimensional in vitro models to study neurovascular unit functions in health and disease. *Neural Regen. Res.* **2021**, *16*, 2132–2140. [CrossRef]
27. Offeddu, G.S.; Haase, K.; Gillrie, M.R.; Li, R.; Morozova, O.; Hickman, D.; Knutson, C.G.; Kamm, R.D. An on-chip model of protein paracellular and transcellular permeability in the microcirculation. *Biomaterials* **2019**, *212*, 115–125. [CrossRef]
28. Hajal, C.; Offeddu, G.S.; Shin, Y.; Zhang, S.; Morozova, O.; Hickman, D.; Knutson, C.G.; Kamm, R.D. Engineered human blood-brain barrier microfluidic model for vascular permeability analyses. *Nat. Protoc.* **2022**, *17*, 95–128. [CrossRef]
29. Campisi, M.; Shin, Y.; Osaki, T.; Hajal, C.; Chiono, V.; Kamm, R.D. 3D self-organized microvascular model of the human blood-brain barrier with endothelial cells, pericytes and astrocytes. *Biomaterials* **2018**, *180*, 117–129. [CrossRef]
30. Maoz, B.M.; Herland, A.; Fitzgerald, E.A.; Grevesse, T.; Vidoudez, C.; Pacheco, A.R.; Sheehy, S.P.; Park, T.E.; Dauth, S.; Mannix, R.; et al. A linked organ-on-chip model of the human neurovascular unit reveals the metabolic coupling of endothelial and neuronal cells. *Nat. Biotechnol.* **2018**, *36*, 865–877. [CrossRef]
31. Park, T.E.; Mustafaoglu, N.; Herland, A.; Hasselkus, R.; Mannix, R.; FitzGerald, E.A.; Prantil-Baun, R.; Watters, A.; Henry, O.; Benz, M.; et al. Hypoxia-enhanced Blood-Brain Barrier Chip recapitulates human barrier function and shuttling of drugs and antibodies. *Nat. Commun.* **2019**, *10*, 2621. [CrossRef]
32. Vatine, G.D.; Barrile, R.; Workman, M.J.; Sances, S.; Barriga, B.K.; Rahnama, M.; Barthakur, S.; Kasendra, M.; Lucchesi, C.; Kerns, J.; et al. Human iPSC-Derived Blood-Brain Barrier Chips Enable Disease Modeling and Personalized Medicine Applications. *Cell Stem Cell* **2019**, *24*, 995–1005.e6. [CrossRef] [PubMed]
33. Gastfriend, B.D.; Palecek, S.P.; Shusta, E.V. Modeling the blood–brain barrier: Beyond the endothelial cells. *Curr. Opin. Biomed. Eng.* **2018**, *5*, 6–12. [CrossRef] [PubMed]
34. Oddo, A.; Peng, B.; Tong, Z.; Wei, Y.; Tong, W.Y.; Thissen, H.; Voelcker, N.H. Advances in Microfluidic Blood–Brain Barrier (BBB) Models. *Trends Biotechnol.* **2019**, *37*, 1295–1314. [CrossRef] [PubMed]
35. Tan, H.Y.; Cho, H.; Lee, L.P. Human mini-brain models. *Nat. Biomed. Eng.* **2021**, *5*, 11–25. [CrossRef] [PubMed]
36. Lee, C.S.; Leong, K.W. Advances in microphysiological blood-brain barrier (BBB) models towards drug delivery. *Curr. Opin. Biotechnol.* **2020**, *66*, 78–87. [CrossRef] [PubMed]
37. Browne, S.; Gill, E.L.; Schultheiss, P.; Goswami, I.; Healy, K.E. Stem cell-based vascularization of microphysiological systems. *Stem Cell Rep.* **2021**, *16*, 2058–2075. [CrossRef]
38. Ding, Y.; Shusta, E.V.; Palecek, S.P. Integrating in vitro disease models of the neurovascular unit into discovery and development of neurotherapeutics. *Curr. Opin. Biomed. Eng.* **2021**, *20*, 100341. [CrossRef]
39. Nitta, T.; Hata, M.; Gotoh, S.; Seo, Y.; Sasaki, H.; Hashimoto, N.; Furuse, M.; Tsukita, S. Size-selective loosening of the blood-brain barrier in claudin-5-deficient mice. *J. Cell Biol.* **2003**, *161*, 653–660. [CrossRef]
40. Saitou, M.; Furuse, M.; Sasaki, H.; Schulzke, J.D.; Fromm, M.; Takano, H.; Noda, T.; Tsukita, S. Complex phenotype of mice lacking occludin, a component of tight junction strands. *Mol. Biol. Cell* **2000**, *11*, 4131–4142. [CrossRef]
41. Roudnicky, F.; Zhang, J.D.; Kim, B.K.; Pandya, N.J.; Lan, Y.; Sach-Peltason, L.; Ragelle, H.; Strassburger, P.; Gruener, S.; Lazendic, M.; et al. Inducers of the endothelial cell barrier identified through chemogenomic screening in genome-edited hPSC-endothelial cells. *Proc. Natl. Acad. Sci. USA* **2020**, *117*, 19854–19865. [CrossRef]
42. Langen, U.H.; Ayloo, S.; Gu, C. Development and cell biology of the blood-brain barrier. *Annu. Rev. Cell Dev. Biol.* **2019**, *35*, 591–613. [CrossRef] [PubMed]
43. Villaseñor, R.; Lampe, J.; Schwaninger, M.; Collin, L. Intracellular transport and regulation of transcytosis across the blood–brain barrier. *Cell. Mol. Life Sci.* **2019**, *76*, 1081–1092. [CrossRef]
44. Pardridge, W.M.; Triguero, D.; Farrell, C.R. Downregulation of blood-brain barrier glucose transporter in experimental diabetes. *Diabetes* **1990**, *39*, 1040–1044. [CrossRef] [PubMed]
45. Ambudkar, S.V.; Kimchi-Sarfaty, C.; Sauna, Z.E.; Gottesman, M.M. P-glycoprotein: From genomics to mechanism. *Oncogene* **2003**, *22*, 7468–7485. [CrossRef] [PubMed]
46. Rössler, K.; Neuchrist, C.; Kitz, K.; Scheiner, O.; Kraft, D.; Lassmann, H. Expression of leucocyte adhesion molecules at the human blood-brain barrier (BBB). *J. Neurosci. Res.* **1992**, *31*, 365–374. [CrossRef] [PubMed]
47. Frank, R.N.; Turczyn, T.J.; Das, A. Pericyte coverage of retinal and cerebral capillaries. *Investig. Ophthalmol. Vis. Sci.* **1990**, *31*, 999–1007.
48. Mathiisen, T.M.; Lehre, K.P.; Danbolt, N.C.; Ottersen, O.P. The perivascular astroglial sheath provides a complete covering of the brain microvessels: An electron microscopic 3D reconstruction. *Glia* **2010**, *58*, 1094–1103. [CrossRef]

49. Gaengel, K.; Genové, G.; Armulik, A.; Betsholtz, C. Endothelial-mural cell signaling in vascular development and angiogenesis. *Arterioscler. Thromb. Vasc. Biol.* **2009**, *29*, 630–638. [CrossRef]

50. Díaz-Flores, L.; Gutiérrez, R.; Madrid, J.F.; Varela, H.; Valladares, F.; Acosta, E.; Martín-Vasallo, P.; Díaz-Flores, J. Pericytes. Morphofunction, interactions and pathology in a quiescent and activated mesenchymal cell niche. *Histol. Histopathol.* **2009**, *24*, 909–969. [CrossRef]

51. Armulik, A.; Genové, G.; Betsholtz, C. Pericytes: Developmental, Physiological, and Pathological Perspectives, Problems, and Promises. *Dev. Cell* **2011**, *21*, 193–215. [CrossRef]

52. Hellström, M.; Kalén, M.; Lindahl, P.; Abramsson, A.; Betsholtz, C. Role of PDGF-B and PDGFR-β in recruitment of vascular smooth muscle cells and pericytes during embryonic blood vessel formation in the mouse. *Development* **1999**, *126*, 3047–3055. [CrossRef] [PubMed]

53. Janzer, R.C.; Raff, M.C. Astrocytes induce blood-brain barrier properties in endothelial cells. *Nature* **1987**, *325*, 253–257. [CrossRef] [PubMed]

54. Alvarez, J.I.; Cayrol, R.; Prat, A. Disruption of central nervous system barriers in multiple sclerosis. *Biochim. Biophys. Acta Mol. Basis Dis.* **2011**, *1812*, 252–264. [CrossRef] [PubMed]

55. Lee, S.W.; Kim, W.J.; Choi, Y.K.; Kim, K.W. SSeCKS regulates angiogenesis and tight junction formation in blood-brain barrier. *Nat. Med.* **2003**, *9*, 900–906. [CrossRef] [PubMed]

56. Reed, M.J.; Damodarasamy, M.; Banks, W.A. The extracellular matrix of the blood–brain barrier: Structural and functional roles in health, aging, and Alzheimer's disease. *Tissue Barriers* **2019**, *7*, 1651157. [CrossRef] [PubMed]

57. Yao, Y.; Chen, Z.L.; Norris, E.H.; Strickland, S. Astrocytic laminin regulates pericyte differentiation and maintains blood brain barrier integrity. *Nat. Commun.* **2014**, *5*, 413. [CrossRef]

58. Daneman, R.; Zhou, L.; Kebede, A.A.; Barres, B.A. Pericytes are required for blood§€" brain barrier integrity during embryogenesis. *Nature* **2010**, *468*, 562–566. [CrossRef]

59. Liebner, S.; Corada, M.; Bangsow, T.; Babbage, J.; Taddei, A.; Czupalla, C.J.; Reis, M.; Felici, A.; Wolburg, H.; Fruttiger, M.; et al. Wnt/β-catenin signaling controls development of the blood—Brain barrier. *J. Cell Biol.* **2008**, *183*, 409–417. [CrossRef]

60. Stenman, J.M.; Rajagopal, J.; Carroll, T.J.; Ishibashi, M.; McMahon, J.; McMahon, A.P. Canonical Wnt Signaling Regulates Organ-Specific Assembly and Differentiation of CNS Vasculature. *Science* **2008**, *322*, 1247–1250.

61. Obermeier, B.; Daneman, R.; Ransohoff, R.M. Development, maintenance and disruption of the blood-brain barrier. *Nat. Med.* **2013**, *19*, 1584–1596. [CrossRef]

62. Chow, B.W.; Gu, C. Gradual Suppression of Transcytosis Governs Functional Blood-Retinal Barrier Formation. *Neuron* **2017**, *93*, 1325–1333.e3. [CrossRef] [PubMed]

63. Lindahl, P.; Johansson, B.R.; Levéen, P.; Betsholtz, C. Pericyte loss and microaneurysm formation in PDGF-B-deficient mice. *Science* **1997**, *277*, 242–245. [CrossRef] [PubMed]

64. Armulik, A.; Genové, G.; Mäe, M.; Nisancioglu, M.H.; Wallgard, E.; Niaudet, C.; He, L.; Norlin, J.; Lindblom, P.; Strittmatter, K.; et al. Pericytes regulate the blood-brain barrier. *Nature* **2010**, *468*, 557–561. [CrossRef]

65. Zhao, Z.; Nelson, A.R.; Betsholtz, C.; Zlokovic, B.V. Establishment and Dysfunction of the Blood-Brain Barrier. *Cell* **2015**, *163*, 1064–1078. [CrossRef] [PubMed]

66. Teichert, M.; Milde, L.; Holm, A.; Stanicek, L.; Gengenbacher, N.; Savant, S.; Ruckdeschel, T.; Hasanov, Z.; Srivastava, K.; Hu, J.; et al. Pericyte-expressed Tie2 controls angiogenesis and vessel maturation. *Nat. Commun.* **2017**, *8*, 16106. [CrossRef] [PubMed]

67. Armulik, A.; Abramsson, A.; Betsholtz, C. Endothelial/pericyte interactions. *Circ. Res.* **2005**, *97*, 512–523. [CrossRef] [PubMed]

68. Park, D.Y.; Lee, J.; Kim, J.; Kim, K.; Hong, S.; Han, S.; Kubota, Y.; Augustin, H.G.; Ding, L.; Kim, J.W.; et al. Plastic roles of pericytes in the blood-retinal barrier. *Nat. Commun.* **2017**, *8*, 15296. [CrossRef]

69. Jong Lee, H.; Young Koh, G. Shear stress activates Tie2 receptor tyrosine kinase in human endothelial cells. *Biochem. Biophys. Res. Commun.* **2003**, *304*, 399–404. [CrossRef]

70. McMillin, M.A.; Frampton, G.A.; Seiwell, A.P.; Patel, N.S.; Jacobs, A.N.; DeMorrow, S. TGFβ1 exacerbates blood-brain barrier permeability in a mouse model of hepatic encephalopathy via upregulation of MMP9 and downregulation of claudin-5. *Lab. Investig.* **2015**, *95*, 903–913. [CrossRef]

71. Behzadian, M.A.; Wang, X.L.; Windsor, L.J.; Ghaly, N.; Caldwell, R.B. TGF-β increases retinal endothelial cell permeability by increasing MMP-9: Possible role of glial cells in endothelial barrier function. *Investig. Ophthalmol. Vis. Sci.* **2001**, *42*, 853–859.

72. Braunger, B.M.; Leimbeck, S.V.; Schlecht, A.; Volz, C.; Jägle, H.; Tamm, E.R. Deletion of ocular transforming growth factor β signaling mimics essential characteristics of diabetic retinopathy. *Am. J. Pathol.* **2015**, *185*, 1749–1768. [CrossRef] [PubMed]

73. Paredes, I.; Himmels, P.; Ruiz de Almodóvar, C. Neurovascular Communication during CNS Development. *Dev. Cell* **2018**, *45*, 10–32. [CrossRef]

74. Robertson, J.M. Astrocytes and the evolution of the human brain. *Med. Hypotheses* **2014**, *82*, 236–239. [CrossRef] [PubMed]

75. Yao, H.; Wang, T.; Deng, J.; Liu, D.; Li, X.; Deng, J. The development of blood-retinal barrier during the interaction of astrocytes with vascular wall cells. *Neural Regen. Res.* **2014**, *9*, 1047–1054. [CrossRef]

76. Bell, R.D.; Winkler, E.A.; Singh, I.; Sagare, A.P.; Deane, R.; Wu, Z.; Holtzman, D.M.; Betsholtz, C.; Armulik, A.; Sallstrom, J.; et al. Apolipoprotein e controls cerebrovascular integrity via cyclophilin A. *Nature* **2012**, *485*, 512–516. [CrossRef]

77. Tait, M.J.; Saadoun, S.; Bell, B.A.; Papadopoulos, M.C. Water movements in the brain: Role of aquaporins. *Trends Neurosci.* **2008**, *31*, 37–43. [CrossRef] [PubMed]

78. Geraldes, P.; Hiraoka-Yamamoto, J.; Matsumoto, M.; Clermont, A.; Leitges, M.; Marette, A.; Aiello, L.P.; Kern, T.S.; King, G.L. Activation of PKC-and SHP-1 by hyperglycemia causes vascular cell apoptosis and diabetic retinopathy. *Nat. Med.* **2009**, *15*, 1298–1306. [CrossRef]

79. Cai, J.; Ruan, Q.; Chen, Z.J.; Han, S. Connection of pericyte-angiopoietin-Tie-2 system in diabetic retinopathy: Friend or foe? *Future Med. Chem.* **2012**, *4*, 2163–2176. [CrossRef]

80. Hu, J.; Dziumbla, S.; Lin, J.; Bibli, S.I.; Zukunft, S.; De Mos, J.; Awwad, K.; Frömel, T.; Jungmann, A.; Devraj, K.; et al. Inhibition of soluble epoxide hydrolase prevents diabetic retinopathy. *Nature* **2017**, *552*, 248–252. [CrossRef]

81. Sweeney, M.D.; Ayyadurai, S.; Zlokovic, B.V. Pericytes of the neurovascular unit: Key functions and signaling pathways. *Nat. Neurosci.* **2016**, *19*, 771–783. [CrossRef]

82. Geranmayeh, M.H.; Rahbarghazi, R.; Farhoudi, M. Targeting pericytes for neurovascular regeneration. *Cell Commun. Signal.* **2019**, *17*, 26. [CrossRef] [PubMed]

83. Procter, T.V.; Williams, A.; Montagne, A. Interplay between Brain Pericytes and Endothelial Cells in Dementia. *Am. J. Pathol.* **2021**, *191*, 1917–1931. [CrossRef] [PubMed]

84. Sengillo, J.D.; Winkler, E.A.; Walker, C.T.; Sullivan, J.S.; Johnson, M.; Zlokovic, B.V. Deficiency in mural vascular cells coincides with blood-brain barrier disruption in alzheimer's disease. *Brain Pathol.* **2013**, *23*, 303–310. [CrossRef] [PubMed]

85. Sweeney, M.D.; Zhao, Z.; Montagne, A.; Nelson, A.R.; Zlokovic, B.V. Blood-brain barrier: From physiology to disease and back. *Physiol. Rev.* **2019**, *99*, 21–78. [CrossRef]

86. Veszelka, S.; Kittel, Á.; Deli, M.A. Tools for modelling blood-brain barrier penetrability. *Solubility Deliv. ADME Probl. Drugs Drug Candidates* **2011**, 166–188. [CrossRef]

87. Montagne, A.; Barnes, S.R.; Sweeney, M.D.; Halliday, M.R.; Sagare, A.P.; Zhao, Z.; Toga, A.W.; Jacobs, R.E.; Liu, C.Y.; Amezcua, L.; et al. Blood Brain barrier breakdown in the aging human hippocampus. *Neuron* **2015**, *85*, 296–302. [CrossRef] [PubMed]

88. Nation, D.A.; Sweeney, M.D.; Montagne, A.; Sagare, A.P.; D'Orazio, L.M.; Pachicano, M.; Sepehrband, F.; Nelson, A.R.; Buennagel, D.P.; Harrington, M.G.; et al. Blood–brain barrier breakdown is an early biomarker of human cognitive dysfunction. *Nat. Med.* **2019**, *25*, 270–276. [CrossRef]

89. Wang, D.; Chen, F.; Han, Z.; Yin, Z.; Ge, X.; Lei, P. Relationship Between Amyloid-β Deposition and Blood-Brain Barrier Dysfunction in Alzheimer's Disease. *Front. Cell. Neurosci.* **2021**, *15*, 695479. [CrossRef]

90. Yamazaki, Y.; Shinohara, M.; Shinohara, M.; Yamazaki, A.; Murray, M.E.; Liesinger, A.M.; Heckman, M.G.; Lesser, E.R.; Parisi, J.E.; Petersen, R.C.; et al. Selective loss of cortical endothelial tight junction proteins during Alzheimer's disease progression. *Brain* **2019**, *142*, 1077–1092. [CrossRef]

91. Wang, Y.; Cho, C.; Williams, J.; Smallwood, P.M.; Zhang, C.; Junge, H.J.; Nathans, J. Correction: Interplay of the Norrin and Wnt7a/Wnt7b signaling systems in blood–brain barrier and blood–retina barrier development and maintenance. *Proc. Natl. Acad. Sci. USA* **2019**, *116*, 3934. [CrossRef]

92. Gaston, J.D.; Bischel, L.L.; Fitzgerald, L.A.; Cusick, K.D.; Ringeisen, B.R.; Pirlo, R.K. Gene Expression Changes in Long-Term in Vitro Human Blood-Brain Barrier Models and Their Dependence on a Transwell Scaffold Materia. *J. Healthc. Eng.* **2017**, *2017*. [CrossRef]

93. Cecchelli, R.; Berezowski, V.; Lundquist, S.; Culot, M.; Renftel, M.; Dehouck, M.P.; Fenart, L. Modelling of the blood—Brain barrier in drug discovery and development. *Nat. Rev. Drug Discov.* **2007**, *6*, 650–661. [CrossRef] [PubMed]

94. Kim, S.; Lee, H.; Chung, M.; Jeon, N.L. Engineering of functional, perfusable 3D microvascular networks on a chip. *Lab Chip* **2013**, *13*, 1489–1500. [CrossRef] [PubMed]

95. Whisler, J.A.; Chen, M.B.; Kamm, R.D. Control of Perfusable Microvascular Network Morphology Using a Multiculture Microfluidic System. *Tissue Eng. Part C Methods* **2014**, *20*, 543–552. [CrossRef] [PubMed]

96. Wang, X.; Phan, D.T.T.; Sobrino, A.; George, S.C.; Hughes, C.C.W.; Lee, A.P. Engineering anastomosis between living capillary networks and endothelial cell-lined microfluidic channels. *Lab Chip* **2016**, *16*, 282–290. [CrossRef] [PubMed]

97. Offeddu, G.S.; Possenti, L.; Loessberg-Zahl, J.T.; Zunino, P.; Roberts, J.; Han, X.; Hickman, D.; Knutson, C.G.; Kamm, R.D. Application of Transmural Flow Across In Vitro Microvasculature Enables Direct Sampling of Interstitial Therapeutic Molecule Distribution. *Small* **2019**, *15*, 1902393. [CrossRef] [PubMed]

98. Shi, Y.; Inoue, H.; Wu, J.C.; Yamanaka, S. Induced pluripotent stem cell technology: A decade of progress. *Nat. Rev. Drug Discov.* **2017**, *16*, 115–130. [CrossRef] [PubMed]

99. Delsing, L.; Herland, A.; Falk, A.; Hicks, R.; Synnergren, J.; Zetterberg, H. Models of the blood-brain barrier using iPSC-derived cells. *Mol. Cell. Neurosci.* **2020**, *107*, 103533. [CrossRef]

100. Lippmann, E.S.; Azarin, S.M.; Kay, J.E.; Nessler, R.A.; Wilson, H.K.; Al-Ahmad, A.; Palecek, S.P.; Shusta, E.V. Derivation of blood-brain barrier endothelial cells from human pluripotent stem cells. *Nat. Biotechnol.* **2012**, *30*, 783–791. [CrossRef]

101. Lu, T.M.; Houghton, S.; Magdeldin, T.; Barcia Durán, J.G.; Minotti, A.P.; Snead, A.; Sproul, A.; Nguyen, D.H.T.; Xiang, J.; Fine, H.A.; et al. Pluripotent stem cell-derived epithelium misidentified as brain microvascular endothelium requires ETS factors to acquire vascular fate. *Proc. Natl. Acad. Sci. USA* **2021**, *118*, e2016950118. [CrossRef]

102. Soldner, F.; Jaenisch, R. Stem Cells, Genome Editing, and the Path to Translational Medicine. *Cell* **2018**, *175*, 615–632. [CrossRef] [PubMed]

103. Song, H.W.; Foreman, K.L.; Gastfriend, B.D.; Kuo, J.S.; Palecek, S.P.; Shusta, E.V. Transcriptomic comparison of human and mouse brain microvessels. *Sci. Rep.* **2020**, *10*, 12358. [CrossRef] [PubMed]

104. Gardner, T.W.; Lieth, E.; Khin, S.A.; Barber, A.J.; Bonsall, D.J.; Lesher, T.; Rice, K.; Brennan, W.A. Astrocytes increase barrier properties and ZO-1 expression in retinal vascular endothelial cells. *Investig. Ophthalmol. Vis. Sci.* **1997**, *38*, 2423–2427.

105. Bryan, B.A.; D'Amore, P.A. Chapter 16 Pericyte Isolation and Use in Endothelial/Pericyte Coculture Models. *Methods Enzymol.* **2008**, *443*, 315–331. [CrossRef]

106. Wisniewska-Kruk, J.; Hoeben, K.A.; Vogels, I.M.C.; Gaillard, P.J.; Van Noorden, C.J.F.; Schlingemann, R.O.; Klaassen, I. A novel co-culture model of the blood-retinal barrier based on primary retinal endothelial cells, pericytes and astrocytes. *Exp. Eye Res.* **2012**, *96*, 181–190. [CrossRef]

107. Sánchez-Palencia, D.M.; Bigger-Allen, A.; Saint-Geniez, M.; Arboleda-Velásquez, J.F.; D'Amore, P.A. Coculture Assays for Endothelial Cells-Mural Cells Interactions. In *Tumor Angiogenesis Assays: Methods and Protocols*; Ribatti, D., Ed.; Springer: New York, NY, USA, 2016; pp. 35–47.

108. Eyre, J.J.; Williams, R.L.; Levis, H.J. A human retinal microvascular endothelial-pericyte co-culture model to study diabetic retinopathy in vitro. *Exp. Eye Res.* **2020**, *201*, 108293. [CrossRef]

109. Dehouck, M.P.; Méresse, S.; Delorme, P.; Fruchart, J.C.; Cecchelli, R. An easier, reproducible, and mass-production method to study the blood-brain barrier in vitro. *J. Neurochem.* **1990**, *54*, 1798–1801. [CrossRef]

110. Vatine, G.D.; Al-Ahmad, A.; Barriga, B.K.; Svendsen, S.; Salim, A.; Garcia, L.; Garcia, V.J.; Ho, R.; Yucer, N.; Qian, T.; et al. Modeling Psychomotor Retardation using iPSCs from MCT8-Deficient Patients Indicates a Prominent Role for the Blood-Brain Barrier. *Cell Stem Cell* **2017**, *20*, 831–843.e5. [CrossRef]

111. Stebbins, M.J.; Gastfriend, B.D.; Canfield, S.G.; Lee, M.S.; Richards, D.; Faubion, M.G.; Li, W.J.; Daneman, R.; Palecek, S.P.; Shusta, E.V. Human pluripotent stem cell–derived brain pericyte–like cells induce blood-brain barrier properties. *Sci. Adv.* **2019**, *5*, eaau7375. [CrossRef]

112. Stone, N.L.; England, T.J.; O'Sullivan, S.E. A novel transwell blood brain barrier model using primary human cells. *Front. Cell. Neurosci.* **2019**, *13*, 230. [CrossRef]

113. Srinivasan, B.; Kolli, A.R.; Esch, M.B.; Abaci, H.E.; Shuler, M.L.; Hickman, J.J. TEER Measurement Techniques for In Vitro Barrier Model Systems. *J. Lab. Autom.* **2015**, *20*, 107–126. [CrossRef] [PubMed]

114. Delsing, L.; Dönnes, P.; Sánchez, J.; Clausen, M.; Voulgaris, D.; Falk, A.; Herland, A.; Brolén, G.; Zetterberg, H.; Hicks, R.; et al. Barrier Properties and Transcriptome Expression in Human iPSC-Derived Models of the Blood–Brain Barrier. *Stem Cells* **2018**, *36*, 1816–1827. [CrossRef] [PubMed]

115. Bischel, L.L.; Coneski, P.N.; Lundin, J.G.; Wu, P.K.; Giller, C.B.; Wynne, J.; Ringeisen, B.R.; Pirlo, R.K. Electrospun gelatin biopapers as substrate for in vitro bilayer models of blood-brain barrier tissue. *J. Biomed. Mater. Res. Part A* **2016**, *104*, 901–909. [CrossRef] [PubMed]

116. Zhang, B.; Korolj, A.; Lai, B.F.L.; Radisic, M. Advances in organ-on-a-chip engineering. *Nat. Rev. Mater.* **2018**, *3*, 257–278. [CrossRef]

117. Yeste, J.; García-Ramírez, M.; Illa, X.; Guimerà, A.; Hernández, C.; Simó, R.; Villa, R. A compartmentalized microfluidic chip with crisscross microgrooves and electrophysiological electrodes for modeling the blood-retinal barrier. *Lab Chip* **2018**, *18*, 95–105. [CrossRef] [PubMed]

118. Urich, E.; Patsch, C.; Aigner, S.; Graf, M.; Iacone, R.; Freskgård, P.O. Multicellular self-assembled spheroidal model of the blood brain barrier. *Sci. Rep.* **2013**, *3*, 1500. [CrossRef]

119. Cho, C.F.; Wolfe, J.M.; Fadzen, C.M.; Calligaris, D.; Hornburg, K.; Chiocca, E.A.; Agar, N.Y.R.; Pentelute, B.L.; Lawler, S.E. Blood-brain-barrier spheroids as an in vitro screening platform for brain-penetrating agents. *Nat. Commun.* **2017**, *8*, 15623. [CrossRef]

120. Nzou, G.; Wicks, R.T.; Wicks, E.E.; Seale, S.A.; Sane, C.H.; Chen, A.; Murphy, S.V.; Jackson, J.D.; Atala, A.J. Human cortex spheroid with a functional blood brain barrier for high-throughput neurotoxicity screening and disease modeling. *Sci. Rep.* **2018**, *8*, 7413. [CrossRef]

121. Nzou, G.; Wicks, R.T.; VanOstrand, N.R.; Mekky, G.A.; Seale, S.A.; EL-Taibany, A.; Wicks, E.E.; Nechtman, C.M.; Marotte, E.J.; Makani, V.S.; et al. Multicellular 3D Neurovascular Unit Model for Assessing Hypoxia and Neuroinflammation Induced Blood-Brain Barrier Dysfunction. *Sci. Rep.* **2020**, *10*, 9766. [CrossRef]

122. Sokolova, V.; Mekky, G.; van der Meer, S.B.; Seeds, M.C.; Atala, A.J.; Epple, M. Transport of ultrasmall gold nanoparticles (2 nm) across the blood–brain barrier in a six-cell brain spheroid model. *Sci. Rep.* **2020**, *10*, 18033. [CrossRef]

123. Simonneau, C.; Duschmalé, M.; Gavrilov, A.; Brandenberg, N.; Ceroni, C.; Lassalle, E.; Knoetgen, H.; Niewoehner, J.; Villaseñor, R. Investigating receptor-mediated antibody transcytosis using Blood-Brain Barrier organoid arrays. *Fluids Barriers CNS* **2021**, *18*, 43. [CrossRef] [PubMed]

124. Eilenberger, C.; Rothbauer, M.; Selinger, F.; Gerhartl, A.; Jordan, C.; Harasek, M.; Schädl, B.; Grillari, J.; Weghuber, J.; Neuhaus, W.; et al. A Microfluidic Multisize Spheroid Array for Multiparametric Screening of Anticancer Drugs and Blood–Brain Barrier Transport Properties. *Adv. Sci.* **2021**, *8*, 2004856. [CrossRef]

125. Kumarasamy, M.; Sosnik, A. Heterocellular spheroids of the neurovascular blood-brain barrier as a platform for personalized nanoneuromedicine. *iScience* **2021**, *24*, 102183. [CrossRef] [PubMed]
126. Brown, J.A.; Pensabene, V.; Markov, D.A.; Allwardt, V.; Diana Neely, M.; Shi, M.; Britt, C.M.; Hoilett, O.S.; Yang, Q.; Brewer, B.M.; et al. Recreating blood-brain barrier physiology and structure on chip: A novel neurovascular microfluidic bioreactor. *Biomicrofluidics* **2015**, *9*, 054124. [CrossRef] [PubMed]
127. Adriani, G.; Ma, D.; Pavesi, A.; Kamm, R.D.; Goh, E.L.K. A 3D neurovascular microfluidic model consisting of neurons, astrocytes and cerebral endothelial cells as a blood-brain barrier. *Lab Chip* **2017**, *17*, 448–459. [CrossRef]
128. Shin, Y.; Choi, S.H.; Kim, E.; Bylykbashi, E.; Kim, J.A.; Chung, S.; Kim, D.Y.; Kamm, R.D.; Tanzi, R.E. Blood–Brain Barrier Dysfunction in a 3D In Vitro Model of Alzheimer's Disease. *Adv. Sci.* **2019**, *6*, 1900962. [CrossRef]
129. Ragelle, H.; Dernick, K.; Khemais, S.; Keppler, C.; Cousin, L.; Farouz, Y.; Louche, C.; Fauser, S.; Kustermann, S.; Tibbitt, M.W.; et al. Human Retinal Microvasculature-on-a-Chip for Drug Discovery. *Adv. Healthc. Mater.* **2020**, *9*, 2001531. [CrossRef]
130. Herland, A.; Van Der Meer, A.D.; FitzGerald, E.A.; Park, T.E.; Sleeboom, J.J.F.; Ingber, D.E. Distinct contributions of astrocytes and pericytes to neuroinflammation identified in a 3D human blood-brain barrier on a chip. *PLoS ONE* **2016**, *11*, e0150360. [CrossRef]
131. Linville, R.M.; DeStefano, J.G.; Sklar, M.B.; Xu, Z.; Farrell, A.M.; Bogorad, M.I.; Chu, C.; Walczak, P.; Cheng, L.; Mahairaki, V.; et al. Human iPSC-derived blood-brain barrier microvessels: Validation of barrier function and endothelial cell behavior. *Biomaterials* **2019**, *190–191*, 24–37. [CrossRef]
132. Polacheck, W.J.; Kutys, M.L.; Tefft, J.B.; Chen, C.S. Microfabricated blood vessels for modeling the vascular transport barrier. *Nat. Protoc.* **2019**, *14*, 1425–1454. [CrossRef]
133. Lee, S.R.S.; Chung, M.; Lee, S.R.S.; Jeon, N.L. 3D brain angiogenesis model to reconstitute functional human blood–brain barrier in vitro. *Biotechnol. Bioeng.* **2020**, *117*, 748–762. [CrossRef]
134. Bang, S.; Lee, S.R.; Ko, J.; Son, K.; Tahk, D.; Ahn, J.; Im, C.; Jeon, N.L. A Low Permeability Microfluidic Blood-Brain Barrier Platform with Direct Contact between Perfusable Vascular Network and Astrocytes. *Sci. Rep.* **2017**, *7*, 8083. [CrossRef] [PubMed]
135. Blanchard, J.W.; Bula, M.; Davila-Velderrain, J.; Akay, L.A.; Zhu, L.; Frank, A.; Victor, M.B.; Bonner, J.M.; Mathys, H.; Lin, Y.T.; et al. Reconstruction of the human blood–brain barrier in vitro reveals a pathogenic mechanism of APOE4 in pericytes. *Nat. Med.* **2020**, *26*, 952–963. [CrossRef] [PubMed]
136. Hajal, C.; Shin, Y.; Li, L.; Serrano, J.C.; Jacks, T.; Kamm, R.D. The CCL2-CCR2 astrocyte-cancer cell axis in tumor extravasation at the brain. *Sci. Adv.* **2021**, *7*, eabg8139. [CrossRef] [PubMed]
137. Offeddu, G.S.; Serrano, J.C.; Chen, S.W.; Shelton, S.E.; Shin, Y.; Floryan, M.; Kamm, R.D. Microheart: A microfluidic pump for functional vascular culture in microphysiological systems. *J. Biomech.* **2021**, *119*, 110330. [CrossRef] [PubMed]
138. Booth, R.; Kim, H. Characterization of a microfluidic in vitro model of the blood-brain barrier (µBBB). *Lab Chip* **2012**, *12*, 1784–1792. [CrossRef]
139. Griep, L.M.; Wolbers, F.; De Wagenaar, B.; Ter Braak, P.M.; Weksler, B.B.; Romero, I.A.; Couraud, P.O.; Vermes, I.; Van Der Meer, A.D.; Van Den Berg, A. BBB on CHIP: Microfluidic platform to mechanically and biochemically modulate blood-brain barrier function. *Biomed. Microdevices* **2013**, *15*, 145–150. [CrossRef]
140. Xu, H.; Li, Z.; Yu, Y.; Sizdahkhani, S.; Ho, W.S.; Yin, F.; Wang, L.; Zhu, G.; Zhang, M.; Jiang, L.; et al. A dynamic in vivo-like organotypic blood-brain barrier model to probe metastatic brain tumors. *Sci. Rep.* **2016**, *6*, 36670. [CrossRef]
141. Sidhaye, J.; Knoblich, J.A. Brain organoids: An ensemble of bioassays to investigate human neurodevelopment and disease. *Cell Death Differ.* **2021**, *28*, 52–67. [CrossRef]
142. Zhang, S.; Wan, Z.; Kamm, R.D. Vascularized organoids on a chip: Strategies for engineering organoids with functional vasculature. *Lab Chip* **2021**, *21*, 473–488. [CrossRef]
143. Mansour, A.A.; Gonçalves, J.T.; Bloyd, C.W.; Li, H.; Fernandes, S.; Quang, D.; Johnston, S.; Parylak, S.L.; Jin, X.; Gage, F.H. An in vivo model of functional and vascularized human brain organoids. *Nat. Biotechnol.* **2018**, *36*, 432–441. [CrossRef] [PubMed]
144. Wimmer, R.A.; Leopoldi, A.; Aichinger, M.; Wick, N.; Hantusch, B.; Novatchkova, M.; Taubenschmid, J.; Hämmerle, M.; Esk, C.; Bagley, J.A.; et al. Human blood vessel organoids as a model of diabetic vasculopathy. *Nature* **2019**, *565*, 505–510. [CrossRef]
145. Bhalerao, A.; Sivandzade, F.; Archie, S.R.; Chowdhury, E.A.; Noorani, B.; Cucullo, L. In vitro modeling of the neurovascular unit: Advances in the field. *Fluids Barriers CNS* **2020**, *17*, 22. [CrossRef] [PubMed]
146. Cakir, B.; Xiang, Y.; Tanaka, Y.; Kural, M.H.; Parent, M.; Kang, Y.J.; Chapeton, K.; Patterson, B.; Yuan, Y.; He, C.S.; et al. Engineering of human brain organoids with a functional vascular-like system. *Nat. Methods* **2019**, *16*, 1169–1175. [CrossRef] [PubMed]
147. Cowan, C.S.; Renner, M.; De Gennaro, M.; Gross-Scherf, B.; Goldblum, D.; Hou, Y.; Munz, M.; Rodrigues, T.M.; Krol, J.; Szikra, T.; et al. Cell Types of the Human Retina and Its Organoids at Single-Cell Resolution. *Cell* **2020**, *182*, 1623–1640.e34. [CrossRef]
148. O'Hara-Wright, M.; Gonzalez-Cordero, A. Retinal organoids: A window into human retinal development. *Development* **2020**, *147*, dev189746. [CrossRef]
149. Paul, S.M.; Mytelka, D.S.; Dunwiddie, C.T.; Persinger, C.C.; Munos, B.H.; Lindborg, S.R.; Schacht, A.L. How to improve RD productivity: The pharmaceutical industry's grand challenge. *Nat. Rev. Drug Discov.* **2010**, *9*, 203–214. [CrossRef]
150. Bichsel, C.A.; Hall, S.R.R.; Schmid, R.A.; Guenat, O.T.; Geiser, T. Primary Human Lung Pericytes Support and Stabilize in Vitro Perfusable Microvessels. *Tissue Eng. Part A* **2015**, *21*, 2166–2176. [CrossRef]
151. Escartin, C.; Galea, E.; Lakatos, A.; O'Callaghan, J.P.; Petzold, G.C.; Serrano-Pozo, A.; Steinhäuser, C.; Volterra, A.; Carmignoto, G.; Agarwal, A.; et al. Reactive astrocyte nomenclature, definitions, and future directions. *Nat. Neurosci.* **2021**, *24*, 312–325. [CrossRef]

152. Sharma, A.; Valle, M.L.; Beveridge, C.; Liu, Y.; Sharma, S. Unraveling the role of genetics in the pathogenesis of diabetic retinopathy. *Eye* **2019**, *33*, 534–541. [CrossRef] [PubMed]
153. Sergejeva, O.; Botov, R.; Liutkevičiene, R.; Kriaučiuniene, L. Genetic factors associated with the development of age-related macular degeneration. *Medicina* **2016**, *52*, 79–88. [CrossRef] [PubMed]
154. Maurissen, T.L.; Woltjen, K. Synergistic gene editing in human iPS cells via cell cycle and DNA repair modulation. *Nat. Commun.* **2020**, *11*, 2876. [CrossRef] [PubMed]
155. Soldner, F.; Laganière, J.; Cheng, A.W.; Hockemeyer, D.; Gao, Q.; Alagappan, R.; Khurana, V.; Golbe, L.I.; Myers, R.H.; Lindquist, S.; et al. Generation of isogenic pluripotent stem cells differing exclusively at two early onset parkinson point mutations. *Cell* **2011**, *146*, 318–331. [CrossRef] [PubMed]
156. Xu, C.L.; Park, K.S.; Tsang, S.H. CRISPR/Cas9 genome surgery for retinal diseases. *Drug Discov. Today Technol.* **2018**, *28*, 23–32. [CrossRef] [PubMed]

Journal of
Personalized Medicine

Article

Correlation of Volume of Macular Edema with Retinal Tomography Features in Diabetic Retinopathy Eyes

Santosh Gopi Krishna Gadde [1], Arpita Kshirsagar [2], Neha Anegondi [2], Thirumalesh B. Mochi [1], Stephane Heymans [3,4], Arkasubhra Ghosh [5,*] and Abhijit Sinha Roy [2,*]

1 Department of Retina, Narayana Nethralaya Eye Hospital, Bangalore 560099, India; drsantoshgk@gmail.com (S.G.K.G.); thirumaleshmb@gmail.com (T.B.M.)
2 Imaging, Biomechanics and Mathematical Modelling Solutions Lab, Narayana Nethralaya Foundation, Bangalore 560099, India; arpitakshirsagar@gmail.com (A.K.); nanegondi@gmail.com (N.A.)
3 Department of Cardiology, CARIM School for Cardiovascular Diseases, Maastricht University, Universiteitssingel 50, 6229 ER Maastricht, The Netherlands; stephane.heymans@mumc.nl
4 Centre for Molecular and Vascular Biology, Department of Cardiovascular Sciences, KU Leuven, Herestraat 49, Bus 911, 3000 Leuven, Belgium
5 GROW Research Laboratory, Narayana Nethralaya Foundation, Bangalore 560099, India
* Correspondence: arkasubhra@narayananethralaya.com (A.G.); asroy@narayananethralaya.com (A.S.R.)

Citation: Gadde, S.G.K.; Kshirsagar, A.; Anegondi, N.; Mochi, T.B.; Heymans, S.; Ghosh, A.; Roy, A.S. Correlation of Volume of Macular Edema with Retinal Tomography Features in Diabetic Retinopathy Eyes. *J. Pers. Med.* **2021**, *11*, 1337. https://doi.org/10.3390/jpm11121337

Academic Editors: Jeffrey M. Sundstrom, Peter D. Westenskow and Andreas Ebneter

Received: 18 October 2021
Accepted: 1 December 2021
Published: 9 December 2021

Publisher's Note: MDPI stays neutral with regard to jurisdictional claims in published maps and institutional affiliations.

Abstract: Optical coherence tomography (OCT) enables the detection of macular edema, a significant pathological outcome of diabetic retinopathy (DR). The aim of the study was to correlate edema volume with the severity of diabetic retinopathy and response to treatment with intravitreal injections (compared to baseline). Diabetic retinopathy (DR; $n = 181$) eyes were imaged with OCT (Heidelberg Engineering, Germany). They were grouped as responders (a decrease in thickness after intravitreal injection of Bevacizumab), non-responders (persistent edema or reduced decrease in thickness), recurrent (recurrence of edema after injection), and treatment naïve (no change in edema at follow-up without any injection). The post-treatment imaging of eyes was included for all groups, except for the treatment naïve group. All eyes underwent a 9×6 mm raster scan to measure the edema volume (EV). Central foveal thickness (CFT), central foveal volume (CFV), and total retinal volume (TRV) were obtained from the early treatment diabetic retinopathy study (ETDRS) map. The median EV increased with DR severity, with PDR having the greatest EV (4.01 mm^3). This correlated positively with TRV ($p < 0.001$). Median CFV and CFT were the greatest in severe NPDR. Median EV was the greatest in the recurrent eyes (4.675 mm^3) and lowest (1.6 mm^3) in the treatment naïve group. Responders and non-responders groups had median values of 3.65 and 3.93 mm^3, respectively. This trend was not observed with CFV, CFT, and TRV. A linear regression yielded threshold values of CFV (~ 0.3 mm^3), CFT (~ 386 µm), and TRV (~ 9.06 mm^3), above which EV may be detected by the current scanner. In this study, EV provided a better distinction between the response groups when compared to retinal tomography parameters. The EV increased with disease severity. Thus, EV can be a more precise parameter to identify subclinical edema and aid in better treatment planning.

Keywords: diabetic retinopathy; retina; edema; OCT; tomography

1. Introduction

Diabetic retinopathy (DR) is the main complication of both type one and type two diabetes and is one of the leading causes of blindness [1]. Diabetic macular edema (DME) results from retinal microvascular changes that compromise the blood-retinal barrier [2]. The earliest signs of leakage have been observed in mild, non-proliferative DR (NPDR), and increase with higher grades of DR (moderate and severe), until the end stage, where it reaches the proliferative stage (PDR) [3]. The development of optical coherence tomography (OCT) has allowed the imaging of DME with high resolution tomography. Since the classification of DR is primarily based on fundus image features[3], there is a clinical need of the improved quantification of DME. Several studies in the recent past have attempted the quantification of DME zones in 2-D OCT images. These studies can be broadly classified into two types: (a) studies where the clinical evaluation of DME led to improved understanding of the disease and response to therapy [4–6]; (b) studies where methods were developed to automatically detect DME and to classify the disease accordingly [7–10]. However, none of these studies quantified the morphological featured of edema in terms of 3D tomography, i.e., the volume of the edema after treatment, and its correlation with other clinical imaging features. Therefore, this study analyzed the distribution of edema volume in a population of DME patients, who were either being treated to resolve their edema or were yet to be treated. Subsequently, the volume of their edema was correlated with current clinical imaging features, namely, central foveal volume (CFV), central foveal thickness (CFT), and total retinal volume (TRV), to establish the proportionate change in the volume of an edema for a specific change in a clinical feature after treatment. These clinical features were derived from early treatment diabetic retinopathy sectors (ETDRS) [11].

2. Methods

This was a retrospective, observational, cross-sectional study of the eyes of diabetic patients who were diagnosed with retinopathy. The study was approved by the Narayana Nethralaya institutional Ethics Committee, Bangalore, India. All methods were performed in accordance with the relevant guidelines and regulations of the hospital, as set by the Ethics Committee. Medical records of the patients were retrospectively reviewed, and the need for patient consent was waived by the Ethics Committee. Only those eyes having had B-scans, which were acquired with the same scan protocol, were chosen for retrospective analyses. The sample size was 181 patients, in the age range from 38 to 79 years. Among these, 19 patients did not have edema. The remaining patients (n = 162) were sub-divided into the following response sub-groups:

(a)   Responders—decrease in CFT by 100 μm or more after either the first or second intravitreal injection of steroid or anti-VEGF; (n = 60)

(b)   Non-responders—either persisting macular edema or having an increase in CFT by 100 μm or CFT $\leq$ 100 μm from previous OCT scans, after 3 consecutive intravitreal injections; (n = 63)

(c)   Recurrent—return of macular edema with an increase in thickness greater than 100 μm, when compared to the last visit, after an injection free period of 2 months; (n = 26)

(d)   Treatment naïve—no visible signs of change in macular edema from consecutive scans at regular follow-ups, or eyes without any prescribed treatment. (n = 29)

In responder, non-responder, and recurrent eyes, the post-treatment scans were analyzed. In the treatment naïve, there was no treatment involved by definition.

In case of anti-VEGF injections, patients other than the treatment naïve group had undergone intravitreous injections of bevacizumab (Roche, Basel, Switzerland). Recent studies have suggested the classification of responders and non-responders based on more than three repeat intravitreous injections [12,13]. In the Indian sub-continent, treatment with repeat injections is limited, due to the high costs of the injections. Thus, it becomes necessary to evaluate the treatment outcomes earlier than the protocols followed in Western populations. We aimed to investigate the distribution of tomographic and edema volume in the aforementioned subgroups after intravitreous injection and compare them with treatment naïve patients. Furthermore, only those patients with an identical central macular raster scan pattern, and without significant media haze based on visual examination by retina specialists (S.G, T.M), were included in the study.

Exclusion criteria included the presence of vitreous hemorrhage, coexistent uveitis, having had ocular surgery in the last 6 weeks, any intravitreal injections for other causes, presence of vein occlusions, age related macular degeneration, pseudophakic cystoid macular edema, and endophthalmitis. All the OCT scans were performed with a Heidelberg Spectralis™ (Heidelberg Engineering, Heidelberg, Germany). This device had an A-scan rate of 40,000 lines per second. The images were exported as zipped files (*.e2e format) and converted to video files (*.avi) using the Heidelberg Eye. The frames in the video files were extracted as images (*.png) for postprocessing. CFV and CFT were obtained from the central sector of the ETDRS thickness map. TRV was obtained by adding the volumes from all the sectors of the ETDRS map.

To calculate the volume of the edema, a 9 × 6 mm scan of the central retina, centered approximately at the fovea, was acquired. A total of 25 uniformly spaced B-scans were acquired within the scan area. Only those scans were used, where all the 25 B-scans had a maximum noise level of 90 or below. The B-scans were exported from the OCT device as 8-bit gray scale images for analyses. Each scan was further processed to quantify the edema zones. Since the edema was 3D in shape, each B-scan was essentially a 2D cross-sectional image of a 3D volume. Each B-scan was initially resolved with a Wiener filter (window size 5 × 5), which preserved the high frequency sections of the B-scan. Since the amount of noise can vary among the images, as well as signal strength, signal to noise ratio (SNR) balancing was used to further resolve the edema zones. The evident noise N in the image was computed as the mean pixel value within a window in the upper left portion of the image. The signal S was calculated as the mean pixel value within a window located from the rightmost image, where the signal value was high. The noise and signal values were chosen after a trial and error method, wherein images with different signal strength and noise values were considered. The values for N and S were averaged across the B scan images. The images were then SNR balanced using the equation

$$I_f = (I_0 - N)/(S - N)$$

where I_0 is the initial pixel value and I_f is the final pixel value [14]. This equation was applied to each B scan. Then, a median filter was applied to these SNR balanced images (window size of 15 × 15) to remove salt and pepper noise. A morphological operation was performed to isolate the edema zones with a minimum number of connected components [15]. The anterior and posterior boundary of the retina was segmented in each B-scan to restrict the quantification only to the region of interest (ROI). The anterior boundary (inner limiting membrane) was defined as the layer having the first horizontal gradient change from the top. Similarly, the posterior boundary (retinal pigment epithelium) was defined as the first gradient change in the horizontal direction from the bottom. All the dark pixels (intensity~0) in the ROI were identified as edema. The segmented edema regions were overlaid on the original image. The area occupied by the pixels was calculated and converted to millimeters squared (image pixel resolution was ~3.77 pixel/mm).

In this study, all or some of the 2D B-scans for a given eye captured the cross-sections of edema regions, if they existed at the location of the scan. Then, the area of the captured cross-section of the edema was calculated from the image. These areas were calculated for all 25 of the B-scans. Since the spatial separation between the B-scans was known (25 μm), the areas were simply integrated numerically using the Trapezoidal method of integration. If A_i represented the area of the edema region in the ith B-scan, then the edema volume was computed as:

$$\text{Volume of edema} = (h/2.0) \times \left(2\sum_{i=2}^{24} A_i + A_1 + A_{25} \right)$$

where $h = 25$ μm. This provided an estimate of the total volume of the edema within the scan area of 9 × 6 mm. Since volume is essentially the sum of all the areas, the difference in volume computed by areas that were derived from manual segmentation vs. an algorithm would be simply the sum of differences between the individual areas derived from the manual vs. algorithm segmentation of the edema cross-sections. All the above methods were implemented using MATLAB v7.10 (MathWorks Inc., Natick, MA, USA).

Statistical Analyses

The normality of distribution was checked with the Kolmogorov-Smirnov test. All continuous variables were reported as median along with 95% of CI interval. Additionally, a manual segmentation of the edema zones was performed by an experienced retina specialist for a sub-set of the total number of B-scans. The agreement between the manual and automated segmentation of the area of the edema zones in the B-scans was analyzed with ICC. This agreement between the areas of fluid-filled zones in the B-scans was also equal to the agreement between corresponding volumes of the fluid filled zones. The variables analyzed were the corrected distance visual acuity (CDVA in LogMAR), volume of the edema (mm^3), CFV (mm^3), CFT (μm), and TRV (mm^3). The variables were analyzed between the ETDRS grades and also between the response sub-groups.

The Kruskal Wallis test was used for the pairwise comparison of subgroups, using the Conover post hoc test. A *p*-value of less than 0.05 was considered statistically significant. The statistical analyses were performed with MedCalc v18.5 (MedCalc Inc., Ostend, Belgium).

3. Results

3.1. Segmentation and Edema Volume Calculation

The manual segmentation of edema zones was performed in 371 B-scans chosen randomly from the patient scans. The areas of segmented zones were calculated from both the manual method and from the algorithm. The intra-class correlation (ICC) between the manual and automated segmentation of areas was 0.91 (95% confidence interval ADDIN EN.CITE [1] 0.89–0.93). Among the 181 eyes, 19 were devoid of edema and were isolated as a separate group. The CFV, CFT, and TRV of these eyes were 0.31 (0.28–0.34) mm^3, 395 (342.4–439.2) μm and 10.1 (9.24–12.20) mm^3, respectively. Among the eyes with edema, there were 7 with mild NPDR, 44 with moderate NPDR, 53 with severe NPDR and 74 with PDR.

3.2. Tomographic Features vs. DR Grades

Table 1 summarizes the salient tomographic features of the DR eyes stratified on the basis of severity. Corrected distance visual acuity (CDVA), TRV, CFV, and CFT were similar between the grades of DR eyes ($p > 0.05$). However, edema volume increased with an increasing severity of the grade of DR ($p < 0.001$). Among the response groups, there were 60 responder eyes, 63 non-responder eyes, 26 recurrent eyes, and 29 treatment naïve eyes.

Table 1. Median with range of indices for the ETDRS grades of diabetic retinopathy.

	Mild NPDR	Mod NPDR	Severe NPDR	PDR	*p*-Value
Age (years)	66 (61.4 to 64.8)	64 (64.00 to 66.00)	62 (61.00 to 65.76)	62.5 (57.33 to 65.00)	0.5
CDVA (LogMAR)	0.3 (0.36 to 0.52)	0.18 (0.16 to 0.30)	0.3 (0.18 to 0.48)	0.477 (0.33 to 0.48)	0.02
Edema Volume (mm^3)	2.86 (0.47 to 3.84)	2.60 (2.17 to 3.69)	3.85 (3.34 to 4.70)	4.011 (3.31 to 4.76)	0.17
Total Retinal Volume (mm^3)	9.05 (8.40 to 9.79)	9.52 (9.0 to 10.15)	11.82 (10.63 to 12.4)	11.16 (10.75 to 11.80)	<0.001
Central Foveal Volume (mm^3)	0.33 (0.28 to 0.35)	0.34 (0.32 to 0.38)	0.42 (0.31 to 0.48)	0.34 (0.31 to 0.36)	0.42
Central Foveal Thickness (μm)	416 (359.24 to 443.11)	434.5 (402.8 to 471.93)	461 (400.03 to 536.78)	435 (398.27 to 452.04)	0.58

CDVA—Corrected distance visual acuity, NPDR—Non-proliferative diabetic retinopathy, PDR—Proliferative diabetic retinopathy.

3.3. Tomographic Features vs. Treatment Response

Table 2 summarizes the salient tomographic features of the DR eyes stratified on the basis of response to treatment. Only CDVA was better in the responder and treatment naïve groups when compared to the non-responder and recurrent groups ($p = 0.05$). The remaining indices were similar between the groups ($p > 0.05$). The volume of edema was greatest in the recurrent group (median of 4.675 mm^3). The linear correlation between CDVA and the edema volume was not significant ($p > 0.05$), irrespective of either grade of DR or response to treatment.

Table 2. Median with range of indices for the DR eyes based on response to treatment.

	Non-Responder	Recurrent	Responder	Treatment Naïve	*p*-Value
Age (years)	64 (61.00 to 65.24)	65 (61.00 to 69.05)	64 (59.46 to 66.00)	62 (55.00 to 66.00)	0.56
CDVA (LogMAR)	0.48 (0.42 to 0.60)	0.6 (0.18 to 1.09)	0.3 (0.18 to 0.48)	0.477 (0.18 to 0.48)	0.05
Edema Volume (mm^3)	3.93 (3.31 to 4.49)	4.675 (2.83 to 5.41)	3.65 (2.69 to 3.96)	1.604 (0.37 to 3.82)	0.11
Total Retinal Volume (mm^3)	10.59 (9.82 to 11.39)	11.64 (10.81 to 12.42)	10.74 (10.28 to 11.78)	10.23 9 (32 to 11.94)	0.47
Central Foveal Volume (mm^3)	0.34 (0.32 to 0.36)	0.395 (0.35 to 0.46)	0.33 (0.280 to 0.40)	0.34 (0.31 to 0.37)	0.11
Central Foveal Thickness (μm)	431 (401.69 to 459.80)	494.5 (439.55 to 577.51)	415 (364.07 to 470.43)	438 (397.51 to 457.57)	0.18

3.4. Correlation among Features

The CFT and edema volume were significantly correlated (r = 0.40, *p* < 0.001) as shown in the linear regression data (Figure 1). Similarly, the linear regression data of CFV and TRV were significantly correlated with the edema volume (r = 0.41 and 0.65 respectively, *p* < 0.001 (Figures 2 and 3)). Interestingly, the estimated zero edema volume CFT (Figure 1) was 386.06 μm, which was very close to the median CFT of the 19 eyes with an absence of edema. Similarly, the zero edema volume CFV (Figure 2) was 0.30 mm^3, which was very close to the median CFV of the same 19 eyes. In case of TRV, the difference between the two was slightly greater (9.06 mm^3 from Figure 3 versus 10.1 mm^3 in the 19 eyes). Figure 4 shows a sample B-scan (A), after SNR balancing, (B) with the segmented edema regions overlaid on (A).

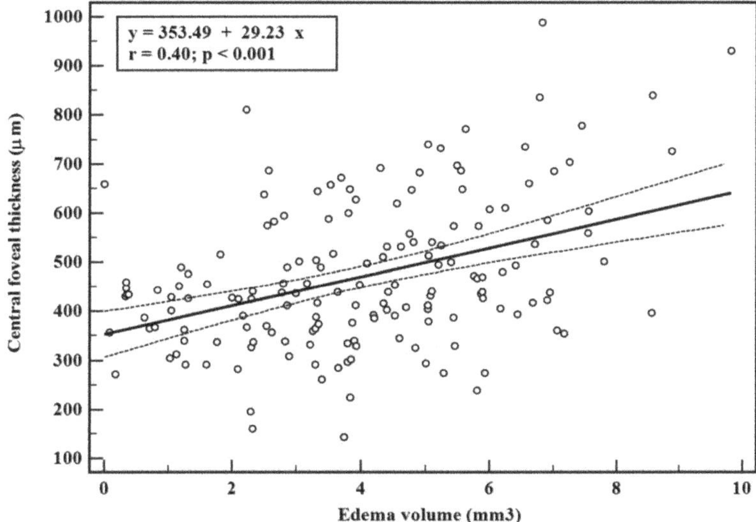

Figure 1. Linear regression with 95% confidence interval between central foveal thickness and edema volume. Only eyes with an edema volume greater than zero were included in the regression. Thickness was derived from the central sector of the ETDRS map.

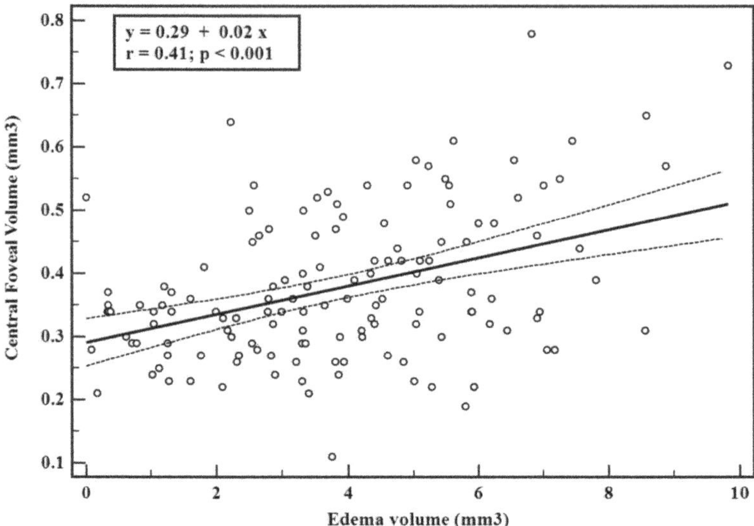

Figure 2. Linear regression with 95% confidence interval between central foveal volume and edema volume. Only eyes with an edema volume greater than zero were included in the regression. Volume was derived from the central sector of the ETDRS map.

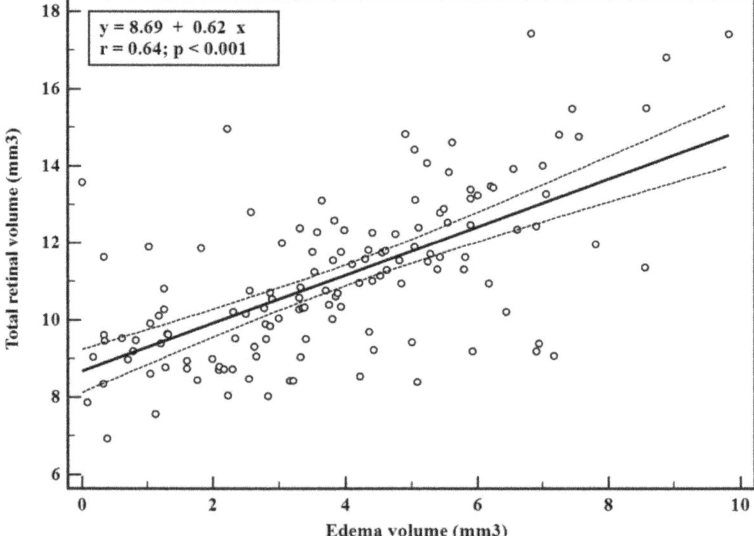

Figure 3. Linear regression with 95% confidence interval between total retinal volume and edema volume. Only eyes with an edema volume greater than zero were included in the regression. Total retinal volume was derived by adding the volumes of all the sectors of the ETDRS map.

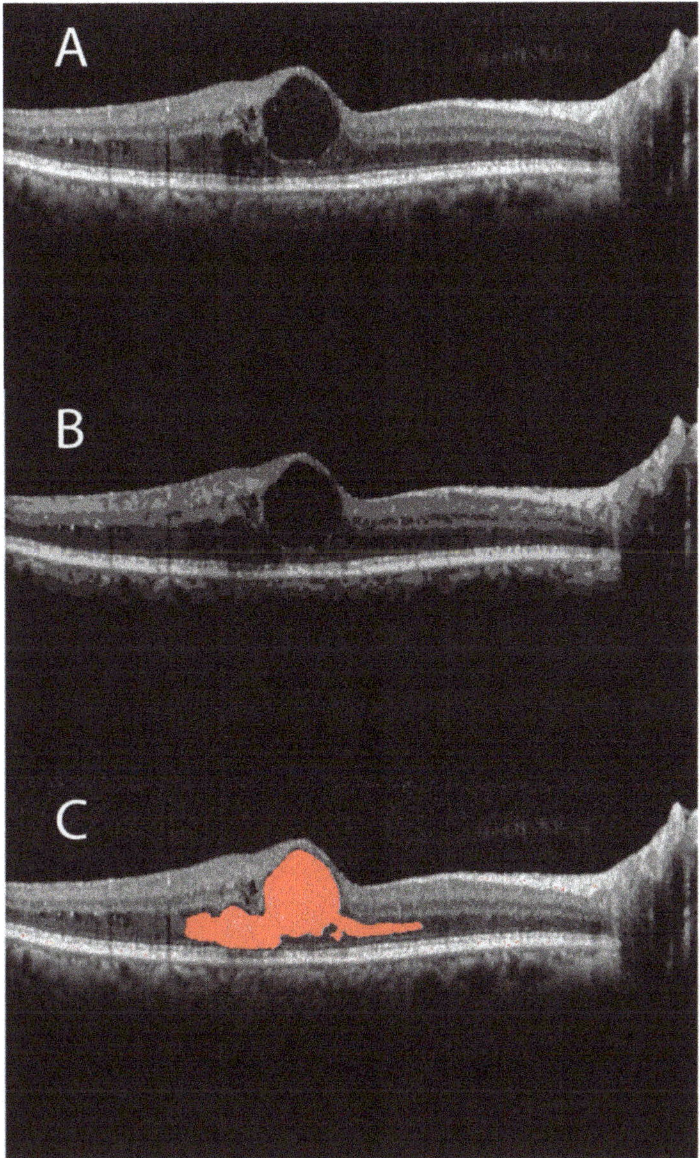

Figure 4. (**A**) A sample B-scan of a patient (categorized as recurrent and with an ETDRS grading as PDR) showing fluid filled regions; (**B**) the same B-scan is shown after signal to noise ratio balancing; (**C**) segmented fluid region is highlighted in red.

4. Discussion

The identification and quantification of edema in DR eyes is a subject of immense interest. With the advent of OCT, the high resolution mapping of edema zones is possible, along with the segmentation of the retinal layers. Increases in retinal thickness due to edema is a primary endpoint in the assessment of edema in DR eyes [4–6,16]. The same can be used to assess the efficacy of treatments and progression of the disease as well. Recent studies have focused on the automated classification of DR, by detection of edema in diabetic patients [7–9]. However, none of these studies assessed the volume of the edema as a primary endpoint. In this study, the volume of the edema was quantified using serial B-scans and numerical integration to obtain a representative measure of the fluid volume. These were then compared with the clinical classification of DR and its response to treatment, to assess whether current classifications accurately capture change in edema volume, if any.

The calculation of edema volume over a large scan is an important feature of the retina in DR, since the edema may not be localized in the foveal region alone and the region of maximal edema could be away from the fovea. Thus, quantification of the volume of the edema over a larger scan area makes clinical sense. In this study, CFV and TRV were derived from the ETDRS maps with diameters of 1 mm and 3 mm, respectively. In other words, TRV was calculated for an *en face* area of 28.3 mm^2 (= $\pi \times 3 \times 3$) using 12 radial scans. However, the edema volume was calculated for an *en face* area of 54 mm^2 (= 9×6) using 25 rectangular scans. Despite this difference, the linear correlations predicted nearly the same CFT, CFV, and TRV at zero edema volume as the 19 eyes with an absence of edema. An absence of edema can also be seen as the absence of large pockets of edema, which are clearly visible to the naked eye. It may also be treated as "dispersed" edema. Therefore, it appeared that progression of DR led to swelling of the retinal tissues before fluid-filled spaces actually formed in the retina. Detection of fluid-filled volume is also limited by the axial and lateral resolution of the OCT scanner, as volumes which are smaller than the resolution cannot be detected. Thus, it may be concluded that a threshold value of CFT~386 µm, CFV~0.3 mm^3, and TRV~9.06 mm^3 (Figures 1–3) needs to be reached before edema could be detectable by the OCT scanner that was used in this study. These observations need further validation with larger sample sizes and also with different populations. The linear equations may also be used estimate edema volume from known measurements of CFT, CFV, and TRV from ETDRS maps, thereby serving as predictive models.

Past studies have established that the volume of edema should increase with an increasing severity of DR [3,17]. PDR had the greatest edema volume. However, the ETDRS grading scale does not consider the presence or absence of edema in the retina. Thus, several analyzed eyes in the study had no edema. A separate grading scale, based on the tomography of the retina and edema volume, may be useful to further monitor the progression of disease and response to treatment. Among the response groups, the edema volume was the greatest in the non-responders when compared to baseline, followed by the responders and recurrent eyes. Treatment naïve eyes had the least edema volume. These trends make logical sense but weren't quantified previously. Further, no clear trends were observed with TRV, CFV, and CFT between the response groups. This implies that the quantification of edema volume may be a better marker to assess response to treatment than CFT, CFV, and TRV. These observations also require further validation.

Deep learning was used to quantify the correlation of retinal thickness and fluid volumes with post-treatment CDVA in a relatively large sample size of eyes (n = 629) [18]. Voxel imaging was used in this study [18]. The agreement between the model prediction and the actual CDVA was moderate (R^2 = 0.21) [18], but better than the correlation between edema volume and CDVA from our study. Another study [19] used deep learning to quantify the macular fluids in AMD, DME, and RVO eyes. Here, volumetric data from Spectralis were used to detect intraretinal cystoid fluid and sub-retinal fluid (SRF), with an AUC of 0.97 and 0.87, respectively. One more study [20] used deep learning to associate anti-VEGF treatment with visual acuity (BCVA), intra-retinal fluid (IRF) and SRF. It was concluded that presence of SRF was associated with low baseline BCVA. Thus, future combinations of deep learning, a larger sample size, and our method of edema volume analysis may improve the correlations presented in this study. The type of treatment and the rate of change of fluid volume during treatment are also important determinants of residual fluid volume in a patient. Combining this with deep learning may further improve the prediction of visual acuity and/or volume of fluid at the end of the treatment. In another study, edema volume was quantified in age-related macular degeneration eyes using volumetric imaging (voxel size of $30 \times 30 \times 1.95$ μm^3) [21]. The measured fluid volumes ranged from 0 to 2 mm^3 [21]. Another study used a voxel size of $11.7 \times 46.9 \times 2.7$ μm to image 10 patients, who underwent anti-VEGF injections over a year [20]. Detected volumes decreased over time and ranged from 0 to 1.8 mm^3 [22]. Thus, volumetric imaging was the primary protocol of interest in earlier studies. Voxel imaging was generally performed over a smaller *en face* area of 6×6 mm and took a longer time to acquire [22,23]. In comparison, our method is relatively faster and covered a larger *en face* area of 9×6 mm. In the future, agreement between volume quantification from voxel imaging and our method needs to be assessed for the same *en face* area.

A limitation of this study was that different types of edemas were not analyzed separately, and this needs to be addressed in future studies. A study on the correlation of BCVA with inner and outer retinal thickness found that inner retinal thickness negatively correlated with visual acuity, while outer retinal thickness was positively correlated, or had visual acuity gain, in response to the intravitreal injection of ranibizumab [18]. They speculated that the recovery of the outer retina in response to treatment. In future studies, assessing differential correlations between thickness changes in the inner and outer retinal layers, with changes in edema volume following treatment, would be of more value. However, it is clear that the spatial distribution of edema volume was largely heterogeneous over the scan area, since it correlated with TRV but not with CFT and CFV. Both CFT and CFV were specific to the center retina, i.e., the 1 mm diameter circle of the ETDRS map. Since this is a single center, single population study, it is a limitation which we will address in future studies with diverse patient populations in larger cohorts to validate our observations. In summary, the volume of edema was quantified in response groups with DR and was observed to be significantly correlated with retinal tomography features along with disease severity. It was also observed to be an improved differentiator between the response groups.

Author Contributions: Conceptualization, S.G.K.G., A.G. and A.S.R.; Methodology, S.G.K.G. and A.S.R.; Software, A.K. and N.A.; Validation, A.K., N.A., T.B.M. and A.S.R.; Formal Analysis, A.K., N.A. and A.S.R.; Investigation, S.G.K.G. and T.B.M.; Resources, A.S.R.; Data Curation, N.A.; Writing—Original Draft Preparation, A.K. and N.A.; Writing—Review & Editing, S.G.K.G., T.B.M., A.G. and A.S.R.; Visualization, A.S.R.; Supervision, A.G. and A.S.R.; Project Administration, S.H., A.G. and A.S.R.; Funding Acquisition, None. All authors have read and agreed to the published version of the manuscript.

Funding: This research received no external funding.

Institutional Review Board Statement: The study was conducted according to the guidelines of the Declaration of Helsinki, and approved by the Institutional Review Board of Narayana Nethralaya eye hospital (Ref No: C/2015/03/04 and 11/03/2015).

Informed Consent Statement: Patient consent was waived due to retrospective nature of the study.

Data Availability Statement: Data is not available.

Acknowledgments: Authors wish to thank Niveditha Govindsamy for assistance in revising the manuscript.

Conflicts of Interest: The authors declare no conflict of interest.

References

1. Nentwich, M.M.; Ulbig, M.W. Diabetic retinopathy-ocular complications of diabetes mellitus. *World J. Diabetes* **2015**, *6*, 489–499. [CrossRef] [PubMed]
2. Ferris, F.L.; Patz, A. Macular edema. A complication of diabetic retinopathy. *Surv. Ophthalmol.* **1984**, *28*, 452–461. [CrossRef]
3. Wu, L.; Fernandez-Loaiza, P.; Sauma, J.; Hernandez-Bogantes, E.; Masis, M. Classification of diabetic retinopathy and diabetic macular edema. *World J. Diabetes* **2013**, *4*, 290–294. [CrossRef] [PubMed]
4. Strøm, C.; Sander, B.; Larsen, N.; Larsen, M.; Lund-Andersen, H. Diabetic macular edema assessed with optical coherence tomography and stereo fundus photography. *Investig. Ophthalmol. Vis. Sci.* **2002**, *43*, 241–245.

5. Sánchez-Tocino, H.; Alvarez-Vidal, A.; Maldonado, M.J.; Moreno-Montañés, J.; García-Layana, A. Retinal thickness study with optical coherence tomography in patients with diabetes. *Investig. Ophthalmol. Vis. Sci.* **2002**, *43*, 1588–1594.
6. Hee, M.R.; Puliafito, C.A.; Duker, J.S.; Reichel, E.; Coker, J.G.; Wilkins, J.R.; Schuman, J.S.; Swanson, E.A.; Fujimoto, J.G. Topography of diabetic macular edema with optical coherence tomography. *Ophthalmology* **1998**, *105*, 360–370. [CrossRef]
7. Cabrera Fernández, D.; Salinas, H.M.; Puliafito, C.A. Automated detection of retinal layer structures on optical coherence tomography images. *Opt. Express* **2005**, *13*, 10200–10216. [CrossRef]
8. Helmy, Y.M.; Atta Allah, H.R. Optical coherence tomography classification of diabetic cystoid macular edema. *Clin. Ophthalmol.* **2013**, *7*, 1731–1737.
9. Srinivasan, P.P.; Kim, L.A.; Mettu, P.S.; Cousins, S.W.; Comer, G.M.; Izatt, J.A.; Farsiu, S. Fully automated detection of diabetic macular edema and dry age-related macular degeneration from optical coherence tomography images. *Biomed. Opt. Express* **2014**, *5*, 3568–3577. [CrossRef]
10. Hassan, B.; Raja, G.; Hassan, T.; Usman Akram, M. Structure tensor based automated detection of macular edema and central serous retinopathy using optical coherence tomography images. *J. Opt. Soc. Am. A Opt. Image Sci. Vis.* **2016**, *33*, 455–463. [CrossRef]
11. Chan, A.; Duker, J.S.; Ko, T.H.; Fujimoto, J.G.; Schuman, J.S. Normal macular thickness measurements in healthy eyes using Stratus optical coherence tomography. *Arch. Ophthalmol.* **2006**, *124*, 193–198. [CrossRef]
12. Bressler, S.B.; Ayala, A.R.; Bressler, N.M.; Melia, M.; Qin, H.; Ferris, F.L., III; Flaxel, C.J.; Friedman, S.M.; Glassman, A.R.; Jampol, L.M.; et al. Persistent macular thickening after ranibizumab treatment for diabetic macular edema with vision impairment. *JAMA Ophthalmol.* **2016**, *134*, 278–285. [CrossRef]
13. Bressler, N.M.; Beaulieu, W.T.; Glassman, A.R.; Blinder, K.J.; Bressler, S.B.; Jampol, L.M.; Melia, M.; Wells, J.A., III. Persistent macular thickening following intravitreous aflibercept, bevacizumab, or ranibizumab for central-involved diabetic macular edema with vision impairment: A secondary analysis of a randomized clinical trial. *JAMA Ophthalmol.* **2018**, *136*, 257–269. [CrossRef]
14. Wilkins, G.R.; Houghton, O.M.; Oldenburg, A.L. Automated segmentation of intraretinal cystoid fluid in optical coherence tomography. *IEEE Trans. Biomed. Eng.* **2012**, *59*, 1109–1114. [CrossRef]
15. Soille, P. *Morphological Image Analysis: Principles and Applications*; Springer: Berlin/Heidelberg, Germany, 2003; pp. 1–368.
16. Das, R.; Spence, G.; Hogg, R.E.; Stevenson, M.; Chakravarthy, U. Disorganization of Inner Retina and Outer Retinal Morphology in Diabetic Macular Edema. *JAMA Ophthalmol.* **2018**, *136*, 202–208. [CrossRef]
17. Wilkinson, C.P.; Ferris, F.L., III; Klein, R.E.; Lee, P.P.; Agardh, C.D.; Davis, M.; Dills, D.; Kampik, A.; Pararajasegaram, R.; Verdaguer, J.T.; et al. Proposed international clinical diabetic retinopathy and diabetic macular edema disease severity scales. *Ophthalmology* **2003**, *110*, 1677–1682. [CrossRef]
18. Ebneter, A.; Wolf, S.; Abhishek, J.; Zinkernagel, M.S. Retinal layer response to ranibizumab during treatment of diabetic macular edema. *Retina* **2016**, *36*, 1314–1323. [CrossRef]
19. Schlegl, T.; Waldstein, S.M.; Bogunovic, H.; Endstraßer, F.; Sadeghipour, A.; Philip, A.M.; Podkowinski, D.; Gerendas, B.S.; Langs, G.; Schmidt-Erfurth, U. Fully Automated Detection and Quantification of Macular Fluid in OCT Using Deep Learning. *Ophthalmology* **2018**, *125*, 549–558. [CrossRef]
20. Roberts, P.K.; Vogl, W.D.; Gerendas, B.S.; Glassman, A.R.; Bogunovic, H.; Jampol, L.M.; Schmidt-Erfurth, U.M. Quantification of Fluid Resolution and Visual Acuity Gain in Patients With Diabetic Macular Edema Using Deep Learning: A Post Hoc Analysis of a Randomized Clinical Trial. *JAMA Ophthalmol.* **2020**, *138*, 945–953. [CrossRef]
21. Gerendas, B.S.; Bogunovic, H.; Sadeghipour, A.; Schlegl, T.; Langs, G.; Waldstein, S.M.; Schmidt-Erfurth, U. Computational image analysis for prognosis determination in DME. *Vis. Res.* **2017**, *139*, 204–210. [CrossRef]
22. Chen, X.; Niemeijer, M.; Zhang, L.; Lee, K.; Abramoff, M.D.; Sonka, M. Three-dimensional segmentation of fluid-associated abnormalities in retinal OCT: Probability constrained graph-search-graph-cut. *IEEE Trans. Med. Imaging* **2012**, *31*, 1521–1531. [CrossRef] [PubMed]
23. Xu, X.; Lee, K.; Zhang, L.; Sonka, M.; Abramoff, M.D. Stratified sampling voxel classification for segmentation of intraretinal and sub retinal fluid in longitudinal clinical OCT data. *IEEE Trans. Med. Imaging* **2015**, *34*, 1616–1623. [CrossRef] [PubMed]

Journal of
Personalized
Medicine

MDPI

Article

miRNA Levels as a Biomarker for Anti-VEGF Response in Patients with Diabetic Macular Edema

Maartje J. C. Vader [1], Yasmin I. Habani [1], Reinier O. Schlingemann [1,2] and Ingeborg Klaassen [1,*]

[1] Ocular Angiogenesis Group, Department of Ophthalmology, Amsterdam Cardiovascular Sciences, Amsterdam Neuroscience, Amsterdam UMC, University of Amsterdam, Meibergdreef 9, 1105 AZ Amsterdam, The Netherlands; m.j.vader@amsterdamumc.nl (M.J.C.V.); y.i.habani@amsterdamumc.nl (Y.I.H.); r.o.schlingemann@amsterdamumc.nl (R.O.S.)

[2] Department of Ophthalmology, University of Lausanne, Jules-Gonin Eye Hospital, Fondation Asile des Aveugles, Avenue de France 15, 1004 Lausanne, Switzerland

* Correspondence: i.klaassen@amsterdamumc.nl

Abstract: Background: The aim of this study was to investigate whether miRNA levels in the circulation could serve as a predictive biomarker for responsiveness to anti-vascular endothelial growth factor (VEGF) therapy in patients with diabetic macular edema. Methods: Whole blood samples were collected at baseline from 135 patients who were included in the BRDME study, a randomized controlled comparative trial of monthly bevacizumab or ranibizumab treatment for 6 months in patients with diabetic macular edema (Trialregister.nl, NTR3247). Best corrected visual acuity letter score (BCVA) and retinal central area thickness (CAT) were measured monthly during the 6-month follow-up. Levels of selected miRNAs were quantified. Results: Following linear regression analysis, the levels of four miRNAs were negatively associated with baseline CAT. Multivariable regression analysis confirmed this association for miR-181a. No associations with changes in CAT after 3 or 6 months of anti-VEGF treatment were found. In addition, no associations with miRNA levels with baseline BCVA or change in BCVA after 3 or 6 months of anti-VEGF treatment were found. Conclusions: Circulating miR-181a levels were negatively associated with CAT at baseline. However, no associations between miRNA levels and the response to anti-VEGF therapy were found.

Keywords: diabetic macular edema; visual acuity; central area thickness; biomarker; microRNA; anti-VEGF

Citation: Vader, M.J.C.; Habani, Y.I.; Schlingemann, R.O.; Klaassen, I. miRNA Levels as a Biomarker for Anti-VEGF Response in Patients with Diabetic Macular Edema. *J. Pers. Med.* **2021**, *11*, 1297. https://doi.org/10.3390/jpm11121297

Academic Editors: Peter D. Westenskow and Andreas Ebneter

Received: 6 October 2021
Accepted: 29 November 2021
Published: 4 December 2021

Publisher's Note: MDPI stays neutral with regard to jurisdictional claims in published maps and institutional affiliations.

1. Introduction

Patients with diabetic macular edema (DME) are commonly treated with intravitreal injections of anti-vascular endothelial growth factor (VEGF) agents. Unfortunately, a substantial number of patients exhibit a suboptimal response to anti-VEGF agents regarding the gain in visual acuity and might benefit from alternative therapeutics options, such as intravitreal corticosteroid injections or additional photocoagulation therapy [1–3]. To prevent long-term vision loss, it is important to identify this patient group at an early stage to ensure that they receive the proper treatment regimen. However, predictive biomarkers for anti-VEGF treatment response are lacking.

DME is the most important cause of visual impairment in patients with diabetic retinopathy. VEGF, the main known mediator of DME, is a protein secreted by retinal cells during hypoxia which is responsible for disruption of the blood–retinal barrier, resulting in fluid and protein leakage into the retina, causing DME, and eventually, visual impairment [4]. DME patients are therefore treated with intravitreal anti-VEGF agents, of which bevacizumab, ranibizumab and aflibercept are commonly used in clinical practice.

However, up to 40% of DME patients treated with anti-VEGF agents exhibit a suboptimal response with minimal visual acuity gain or are non-responders. These patients would be better off switching to another anti-VEGF agent, or should initially be treated with alternative treatments [1–3]. Considering the varying response to anti-VEGF agents and the different treatment possibilities, treating clinicians need to know in advance whether a patient will be a responder or a non-responder (Figure 1). However, to date, no sufficient predictive biomarkers have been reported that are able to distinguish responders from non-responders.

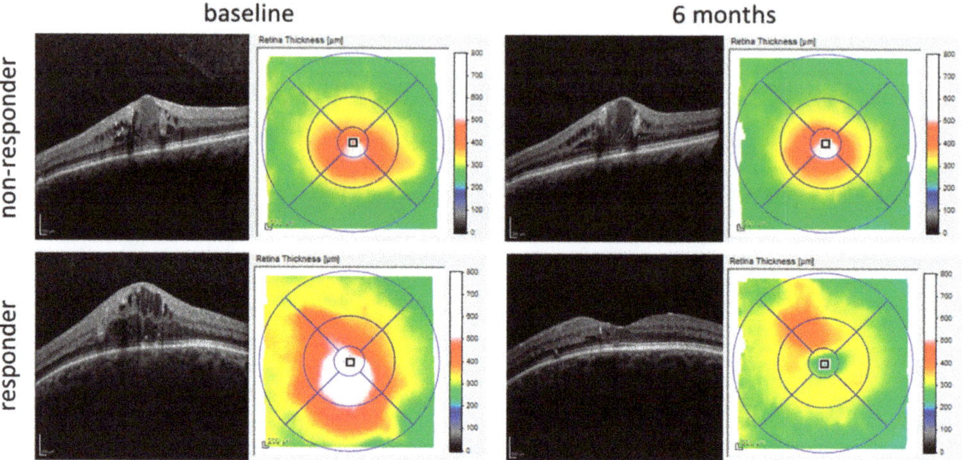

Figure 1. OCT images of DME patients obtained with Heidelberg Spectralis at baseline and after 6 months of anti-VEGF treatment. The central area thickness was unchanged after treatment in the non-responder patient, whereas the central area thickness was clearly decreased and macular edema was almost resolved in a responder patient.

Research on the role of microRNAs (miRNAs) in the pathogenesis of diabetes, diabetic retinopathy and diabetic macular edema is emerging. miRNAs are highly conserved, approximately 22 nucleotides long, noncoding RNAs that regulate gene expression. miRNAs are expressed in all human cells and are engaged in several physiological processes [5]. Distinct expression levels of specified miRNAs are known to be involved in the development of non-proliferative diabetic retinopathy and in the progression to proliferative diabetic retinopathy [5–9]. Only a few papers have reported a role for miRNAs in the development of DME [10–12]. However, it is unknown whether these miRNA expression profiles could also predict DME development or even predict responsiveness to anti-VEGF therapy.

The use of microRNA levels in circulation to predict therapy efficacy has previously been established in rheumatoid arthritis patients treated with anti-TNFα [13], and in the prediction of the response to radiotherapy in head and neck squamous cell carcinoma patients [14]. In the current study, we investigated whether miRNA levels in the circulation at baseline could serve as a biomarker for the responsiveness to anti-VEGF therapy in patients with diabetic macular edema. We selected seven miRNAs: miR-21, miR-181c and miR-1197, based on their reported roles as serum biomarkers for proliferative diabetic retinopathy (PDR) [8]; miR-181a and miR-222, for their roles in retinal neovascularization [15,16]; miR-133b, for its role in wound healing and the regulation of connective tissue growth factor (CTGF) [17]; and miR-7, for its role as a negative mediator of angiogenesis [18]. Using linear regression, we analyzed possible correlations between miRNA levels and the best corrected visual acuity letter score (BCVA) as well as central area thickness (CAT) at baseline and its change after 3- and 6-month follow-up.

2. Materials and Methods

2.1. Study Population

Patient material for miRNA analyses was obtained from participants of the BRDME study [19]. The BRDME study is a randomized, double-masked, multicenter clinical trial, conducted to generate conclusive evidence on the non-inferiority of bevacizumab to ranibizumab in the treatment of patients with DME (Trialregister.nl, NTR3247). The study was approved by the Institutional Review Board of the Amsterdam University Medical Centers, location AMC, and performed in accordance with the Declaration of Helsinki.

In the period from June 2012 to February 2018, 170 participants were enrolled in the BRDME study. Patients were eligible for participation if they presented vision loss due to DME and may benefit from treatment with anti-VEGF agents, were over 18 years of age, diagnosed with type 1 or type 2 diabetes mellitus with a glycosylated hemoglobin (HbA1c) of less than 12%, a CAT of >325 μm, and a BCVA between 24 and 79 letters. Over a period of 6 months, patients received monthly injections of either 1.25 mg bevacizumab (n = 86) or 0.5 mg ranibizumab (n = 84) for a follow-up time of 6 months.

The BCVA of the studied eye and CAT of the studied eye were measured monthly. BCVA was examined using the standardized Early Treatment Diabetic Retinopathy Study chart and registered as the number of letters read by the participant. CAT is defined as the 1 mm central retinal thickness area, as described in the ETDRS. Only at screening and after 6 months was a more comprehensive ophthalmic examination performed. Data on CAT were obtained from optical coherence tomography (OCT) scans, and different OCT devices were used depending on the available system at the participating center (Zeiss Cirrus, Heidelberg Spectralis or Topcon). CAT values obtained from Zeiss Cirrus and Topcon devices were converted to Heidelberg Spectralis outcomes, according to the conversion table derived by Giani et al. [20]. Blood samples for DNA and RNA extraction were collected during the screening visit, before administration of the first injection. The diagnosis of DME was confirmed by an independent reading center, the Belfast Reading Center, Belfast, United Kingdom, part of the Network of Ophthalmic Reading Centers of the United Kingdom. A summary of the clinical design and data analysis is given in Figure 2.

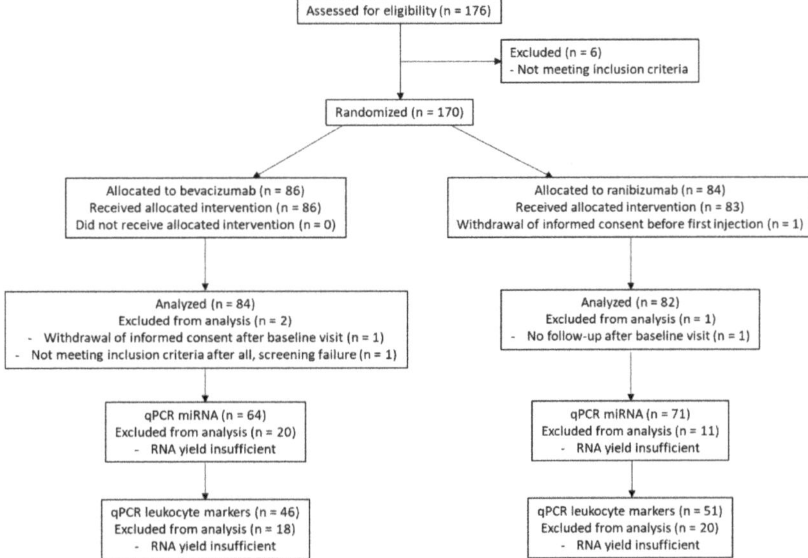

Figure 2. Flow chart summarizing the clinical design and data analysis.

2.2. Outcome Measurements

The primary outcomes included the association between miRNA expression levels and the change in BCVA from baseline to 3 and 6 months, and the association between miRNA levels and change in CAT from baseline to 3 and 6 months. In addition, differences in miRNA expression levels between responder groups were examined. Participants were appointed to one of the following groups based on changes in CAT: rapid response, defined as a reduction of >50% in CAT after 3 months of treatment; slow response, reduction in CAT between 10% and 50% at 3 months, but a reduction of >50% at 6 months; and no response, a less than 25% reduction in CAT at both 3 and 6 months.

2.3. miRNA Profiling

Whole blood samples of 2.5 mL were collected from 135 participants, which were drawn into PAXgene RNA tubes; eventually, RNA was extracted using the PAXgene blood RNA kit (Qiagen, Hilden, Germany), following the manufacturer's instructions. miRNA expression profiles were quantified using real-time quantitative reverse transcription polymerase chain reaction (qRT-PCR). A 1 µg aliquot of total RNA of each sample was reverse-transcribed into cDNA using the miScript II RT Kit (Qiagen). The samples were diluted 20 times with nuclease-free water. Real-time qPCR was performed using a CFX96 real-time PCR detection system (Bio-Rad Laboratories, Hercules, CA, USA). A master mix was prepared for each primer set, consisting of a 5 µL iQ SYBR green supermix (Bio-Rad) and 1 µL primer mix. Additionally, 4 µL of cDNA (diluted 1:80) in 6 µL master mix was amplified using the following PCR protocol: 95 °C for 10 min and 60 °C for 1 min, followed by 44 cycles of 95 °C for 10 s and 60 °C for 1 min, followed by 95 °C for 30 s and a melting program (60–95 °C). Calculating the relative gene expression, the equation: $R = E^{-Ct}$ was used, where E is the mean efficiency of all samples as determined by LinReg [21], and Ct is the threshold for the gene that was determined during qPCR. The following miRNAs were analyzed in our study cohort: miR-7, miR-21, miR-133b, miR-181a, miR-181c, miR-222 and miR-1197. These were selected based on reported associations with angiogenesis, diabetes or diabetic retinopathy [5,6,8]. Primers were purchased from Exiqon. All reactions were performed in quadruplicates. Gene expression data were normalized using UnisP6.

2.4. mRNA Expression of Leukocyte Markers

The mRNA levels of the following leukocyte markers were analyzed: PTPRC/CD45 (pan-leukocyte marker), CD4 (CD4$^+$ T-cells), CD8A (CD8$^+$ T-cells), CD19 (B-cells) and CD14 (monocytes). A 1 µg aliquot of total RNA of each sample was treated with DNase-I (amplification grade; Invitrogen, Waltham, MA, USA) and reverse-transcribed into cDNA using the Maxima First Strand cDNA Synthesis Kit (Thermo Scientific, Waltham, MA, USA). Real-time quantitative PCR (RT-qPCR) was performed on 20× diluted cDNA samples using a CFX96 real-time PCR detection system (Bio-Rad Laboratories) and the specificity of primers was confirmed as described previously [22]. Primer details are available upon request. Gene expression data are expressed as absolute amounts. All reactions were performed in triplicates.

2.5. Statistical Analysis

Linear regression analysis was performed to evaluate the relationship between miRNA levels and baseline BCVA, changes in BCVA, baseline CAT and changes in CAT from baseline to 3 and 6 months. miRNA and mRNA data were log10-transformed to obtain a normal distribution. Multivariable regression analysis was used to confirm associations found by univariable regression analysis.

For responder and non-responder analyses, a threshold of 283 µm was subtracted from OCT values, which equaled the mean normal retinal thickness, as measured with Heidelberg Spectralis, because all OCT measurements were converted to Heidelberg Spectralis values [20]. Relative changes in CAT were used for the classification of rapid responders, slow responders and non-responders, as described in Section 2.2. For responder analysis, miRNA levels were compared using the Kruskal–Wallis test with multiple comparisons.

Significance levels of <0.05 were considered as statistically significant in all analyses mentioned above.

3. Results

3.1. Patient Characteristics

The mean age of the patients in the study was 63.7 ± 11.4 years (±standard deviation); 85 were male and 51 were female. The majority of the participants (*n* = 119) were diagnosed with type 2 diabetes mellitus, whereas 16 participants were diagnosed with type 1 diabetes mellitus. The mean duration since the diagnosis of diabetes mellitus was 16.0 ± 11.1 years. The mean baseline BCVA score was 68.6 ± 9.8 letters, and participants had a mean CAT of 462.5 ± 101.3 μm. Hemoglobin A1c (HbA1c) levels of only 60 patients were documented, with a mean HbA1c of 7.78%.

3.2. miRNA Levels and Visual Acuity Outcomes

Following linear regression analysis, levels of miRNAs in the circulation at baseline were not associated with BCVA at baseline (Table 1). In addition, no association with miRNA levels or change in BCVA at 3 and 6 months was found.

Table 1. Linear regression analysis for log10-transformed miRNA levels and baseline BCVA and changes in BCVA at 3 and 6 months.

	miRNA	miR7	miR21	miR133b	miR181a	miR181c	miR222	miR1197
	Beta	0.132	0.152	−0.047	0.102	0.080	0.154	0.078
Baseline BCVA	SE	1.868	1.486	1.782	1.904	1.345	1.911	0.990
	p-value	0.128	0.078	0.586	0.240	0.354	0.074	0.371
	Beta	−0.135	−0.088	0.068	0.020	0.046	−0.041	−0.040
ΔBCVA 3 months	SE	1.299	1.042	1.238	1.331	0.938	1.344	0.690
	p-value	0.118	0.310	0.430	0.821	0.579	0.635	0.642
	Beta	−0.062	0.004	0.167	0.108	0.081	0.060	0.008
ΔBCVA 6 months	SE	1.437	1.149	1.343	1.453	1.027	1.475	0.758
	p-value	0.473	0.962	0.053	0.214	0.353	0.490	0.922

Beta, beta coefficient; SE, standard error.

3.3. miRNA Levels and Central Area Thickness

Following linear regression analysis, CAT at baseline was negatively associated with miRNA levels in circulation at baseline for miR-7, miR-181a, miR-222 and miR-1197 (Table 2, Figure 3). Multivariable linear regression analyses confirmed a significant association of miR-181a with baseline CAT. No associations of any miRNA levels and change in CAT at 3 or 6 months could be demonstrated (Table 2, Figure 3).

3.4. Responder Analyses

Following the criteria for responsiveness as described in the Methods section, 66 patients were defined as rapid responders, 18 patients as slow responders, and 29 patients as non-responders. Using the Kruskal–Wallis test, no significant differences were found in miRNA levels between these responder groups (Figure 4).

3.5. miRNA Levels and Leukocyte Markers

The majority of mRNA isolated from whole blood comes from leukocytes, and these cells play an important role in the pathogenesis of DME; therefore, we checked whether the expression of leukocyte markers is related to miRNA expression and VA and CAT outcomes. The following leukocyte markers were included in the analyses: CD4, CD8a, CD14, CD19 and CD45.

Regression analysis demonstrated a number of associations between miRNA levels and leukocyte markers (Figure 5). miR-21 was positively associated with the marker of CD4[+] T-cells CD4 and the monocyte marker CD14. miR-133b was negatively associated with the B-cell marker CD19 and the pan-leukocyte marker CD45. miR-222 was positively associated with CD14 and miR-1197 was negatively associated with CD45. Expression levels of leukocyte markers were not associated with BCVA or CAT at baseline or with CAT changes at 3 or 6 months.

Table 2. Univariable and multivariable linear regression analysis for log10-transformed miRNA levels and baseline CAT and changes in CAT at 3 and 6 months.

		miRNA	miR7	miR21	miR133b	miR181a	miR181c	miR222	miR1197
Baseline CAT	Univariable	Beta	−0.194	−0.154	−0.151	−0.298	−0.157	−0.217	−0.171
		SE	19.132	15.373	18.244	18.906	13.792	19.538	10.119
		p-value	0.024	0.074	0.080	0.000	0.070	0.011	0.048
	Multivariable	Beta	−0.083	0.013	−0.002	−0.348	0.070	0.097	−0.106
		SE	30.418	27.553	22.193	33.824	18.543	43.481	10.780
		p-value	0.540	0.932	0.986	0.020	0.546	0.608	0.246
ΔCAT 3 months	Univariable	Beta	0.138	0.058	0.001	0.054	−0.028	0.025	0.104
		SE	21.117	16.982	20.178	21.619	15.260	21.876	11.167
		p-value	0.111	0.506	0.993	0.535	0.747	0.774	0.228
	Multivariable	Beta	0.273	0.034	−0.074	0.168	−0.129	−0.259	0.113
		SE	34.108	30.896	24.886	37.928	20.793	48.757	12.088
		p-value	0.051	0.827	0.493	0.270	0.278	0.182	0.226
ΔCAT 6 months	Univariable	Beta	0.120	0.053	0.010	0.137	0.018	0.066	0.144
		SE	21.871	17.553	20.850	22.160	15.773	22.563	11.482
		p-value	0.164	0.541	0.913	0.112	0.839	0.446	0.096
	Multivariable	Beta	0.191	−0.009	−0.091	0.286	−0.122	−0.218	0.142
		SE	35.111	31.804	25.617	39.043	21.404	50.190	12.444
		p-value	0.169	0.957	0.394	0.061	0.302	0.259	0.128

Beta, beta coefficient; SE, standard error.

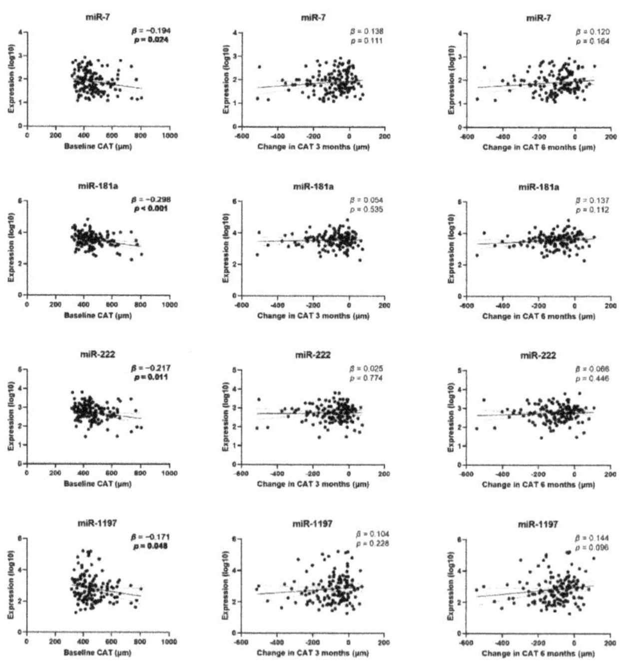

Figure 3. Association between circulating miRNAs and baseline CAT or change in CAT at 3 and 6 months. The scatter plots show the simple linear regression line with 95% confidence bands. Beta (*β*) coefficients and *p*-value (*p*) are given. *p*-values < 0.05 are indicated in bold.

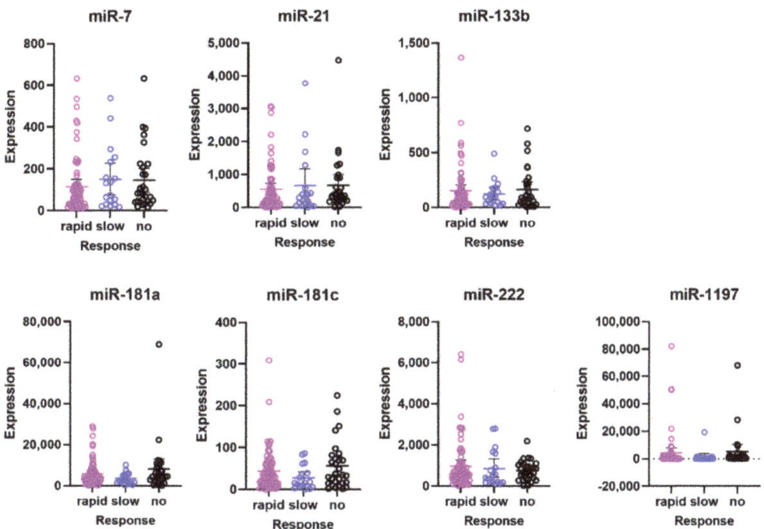

Figure 4. Expression of circulating miRNAs at baseline in different responder groups. Rapid responders exhibited a reduction of >50% in CAT after 3 months of treatment, slow responders exhibited a reduction in CAT between 10% and 50% at 3 months but a reduction of >50% at 6 months, and patients with no response exhibited a less than 25% reduction in CAT at both 3 and 6 months.

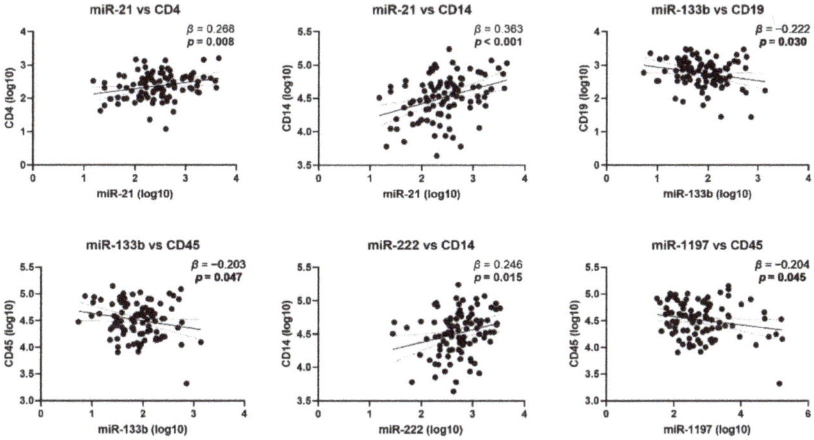

Figure 5. Association between circulating miRNAs and leukocyte markers at baseline. The scatter plots show the simple linear regression line with 95% confidence bands. Beta (β) coefficients and *p*-values (*p*) are given. *p*-values < 0.05 are indicated in bold.

4. Discussion

The use of microRNA levels as biomarkers for disease progression or in the prediction for therapy response is emerging [13,14]. In the present study, we compared the expression levels of selected miRNAs with BCVA and central area thickness (CAT) at baseline and after 3 and 6 months of anti-VEGF therapy. Four miRNAs, miR-7, miR-181a, miR-222 and miR-1197, exhibited a negative association with baseline CAT, as analyzed with univariable regression analysis. For miR-181a, this association could be confirmed with the use of multivariable regression analysis. No associations were found between miRNA levels and changes in CAT or BCVA after therapy, which means that these miRNAs are not suitable as predictors for therapy efficacy. We could also not demonstrate an association between miRNA levels and baseline BCVA. Therefore, expression in the circulation of the studied miRNAs does not seem to be related to the DME response to anti-VEGF.

The main positive finding of this study is that levels at baseline of miR-181a are negatively associated with baseline CAT, suggesting that miR-181a levels decrease when CAT increases. This finding seems in line with a recent study which reported that miR-181a is enriched in the normal mouse retina, but reduced in the retinas of mice in an oxygen-induced retinopathy (OIR) model [15]. The downregulation of miR-181a was accompanied by the upregulation of its target gene endothelial cell-specific molecule 1 (endocan or ESM-1), which has previously been identified to be highly enriched in retinal endothelial tip cells [23], and which has been shown to play an important role in the regulation of angiogenesis [24]. The intraocular injection of miR-181a resulted in the suppression of retinal neovascularization in the OIR model [15].

In addition, extracellular microvesicles (EVs) derived from apoptotic human T lymphocytes (LMPs) were found to have anti-angiogenic properties in the OIR model, and miR-181a was shown to be one of the most abundant miRNAs in these LMPs [25]. This confirms an anti-angiogenic role for miR-181a and suggests LMPs as a source of this miRNA, which may also explain our findings in the circulation.

To explore this further, we investigated possible associations between mRNA levels of leukocyte markers and miRNA levels. We were unable to find a relationship between miR-181a and the abundance of T-cells with markers for CD4 and CD8A, however, suggesting that leukocytes may not be the source of miR-181a, or that the used method is not appropriate. However, associations with leukocyte markers could be established for miR-21, miR-133b, miR-222 and miR-1197, suggesting that leukocytes may be a source for these miRNAs. Other sources of miRNAs may be diverse, including passive leakage from inflamed or injured cells or platelets, active secretion via cell-derived membrane vesicles such as microparticles, exosomes, shedding vesicles, and apoptotic bodies, and active secretion by a protein–miRNA with lipoproteins (e.g., high-density lipoprotein: HDL) and Argonaute protein (e.g., Ago2) [26].

We used whole serum samples; therefore, we could not distinguish between miRNAs from these different sources, although miRNAs derived from leukocytes were expected to be the most abundant in our samples. It remains to be determined which sources of circulating miRNAs can best be used as biomarkers for the diagnosis, prevention, and treatment of disease, but thus far, research is in favor of exosomal samples compared to whole serum samples [26]. Future experiments with exosomal-derived miRNAs from serum may discover better associations with therapy efficacy in DME than we found in the current study.

In addition to the lack of an association of miRNA levels with changes in CAT or BCVA after therapy, we could also find no differences between the different responder groups. In our study, we used selected miRNAs. Possibly, by using a discovery-based method and analyzing all possible miRNAs available, future studies will reveal other miRNAs that do associate with therapy efficacy.

Another limitation of this study is that we only had blood samples collected at the beginning of the study, before the start of anti-VEGF treatment. The expression of miRNAs can change rapidly in response to changes in the microenvironment; therefore, a single measurement is only a snapshot and may not truly reflect the dynamic regulation of miRNAs in relation to changes in BCVA and CAT after therapy. To better understand the role of these miRNAs in the response to anti-VEGF treatment, blood samples should also be examined during and after treatment in future studies.

5. Conclusions

Our findings demonstrated no role for the miRNAs studied in predicting response to anti-VEGF therapy. However, we did observe an association between miR-181a levels and

CAT at baseline. Future research is needed to further investigate the predictive value of miRNAs as biomarkers of anti-VEGF treatment response, and to fully understand the role of miRNAs in the pathophysiology of DME.

Author Contributions: Conceptualization, R.O.S. and I.K.; methodology, I.K.; formal analysis, M.J.C.V. and I.K.; investigation, Y.I.H.; writing—original draft preparation, M.J.C.V.; writing—review and editing, R.O.S. and I.K. All authors have read and agreed to the published version of the manuscript.

Funding: This research was funded by ZonMw, The Netherlands, Organisation for Health Research and Development Grant 171202019 and the following foundations: Landelijke Stichting voor Blinden en Slechtzienden, Algemene Nederlandse Vereniging Ter Voorkoming Van Blindheid, Stichting Blindenpenning, Stichting Oogfonds Nederland, which contributed through UitZicht (Grant UitZicht2016-31). Other grants were provided by Rotterdamse Stichting Blindenbelangen (Grant B20160059), Stichting Ooglijders, Stichting tot Verbetering van het Lot der Blinden, Stichting Blindenhulp. This study was published with the help of the Edward en Marianne Blaauw Fonds voor Oogheelkunde (Edward and Marianne Blaauw Fund for Ophthalmology).

Institutional Review Board Statement: This study was approved by the Institutional Review Board of the Amsterdam University Medical Centers, location AMC, and performed in accordance with the Declaration of Helsinki.

Informed Consent Statement: Informed consent was obtained from all subjects involved in the study.

Data Availability Statement: Raw data were generated at the Department of Biology, KU Leuven, and the Department of Ophthalmology, Amsterdam UMC. Derived data supporting the findings of this study are available from the corresponding author (I.K.) on request.

Conflicts of Interest: The authors declare no conflict of interest.

References

1. Chatziralli, I. Editorial-Suboptimal response to intravitreal anti-VEGF treatment for patients with diabetic macular edema: Is there any point in switching treatment? *Eur. Rev. Med. Pharmacol. Sci.* **2018**, *22*, 5047–5050.
2. Chen, Y.P.; Wu, A.L.; Chuang, C.C.; Chen, S.N. Factors influencing clinical outcomes in patients with diabetic macular edema treated with intravitreal ranibizumab: Comparison between responder and non-responder cases. *Sci. Rep.* **2019**, *9*, 10952. [CrossRef]
3. Gonzalez, V.H.; Campbell, J.; Holekamp, N.M.; Kiss, S.; Loewenstein, A.; Augustin, A.J.; Ma, J.; Ho, A.C.; Patel, V.; Whitcup, S.M.; et al. Early and Long-Term Responses to Anti-Vascular Endothelial Growth Factor Therapy in Diabetic Macular Edema: Analysis of Protocol I Data. *Am. J. Ophthalmol.* **2016**, *172*, 72–79. [CrossRef] [PubMed]
4. Witmer, A. Vascular endothelial growth factors and angiogenesis in eye disease. *Prog. Retin. Eye Res.* **2003**, *22*, 1–29. [CrossRef]
5. Mastropasqua, R.; Toto, L.; Cipollone, F.; Santovito, D.; Carpineto, P.; Mastropasqua, L. Role of microRNAs in the modulation of diabetic retinopathy. *Prog. Retin. Eye Res.* **2014**, *43*, 92–107. [CrossRef]
6. Gong, Q.; Su, G. Roles of miRNAs and long noncoding RNAs in the progression of diabetic retinopathy. *Biosci. Rep.* **2017**, *37*, BSR20171157. [CrossRef]
7. Liu, H.N.; Li, X.; Wu, N.; Tong, M.M.; Chen, S.; Zhu, S.S.; Qian, W.; Chen, X.L. Serum microRNA-221 as a biomarker for diabetic retinopathy in patients associated with type 2 diabetes. *Int. J. Ophthalmol.* **2018**, *11*, 1889–1894. [PubMed]
8. Qing, S.; Yuan, S.; Yun, C.; Hui, H.; Mao, P.; Wen, F.; Ding, Y.; Liu, Q. Serum miRNA biomarkers serve as a fingerprint for proliferative diabetic retinopathy. *Cell. Physiol. Biochem.* **2014**, *34*, 1733–1740. [CrossRef] [PubMed]
9. Friedrich, J.; Steel, D.H.W.; Schlingemann, R.O.; Koss, M.J.; Hammes, H.P.; Krenning, G.; Klaassen, I. microRNA Expression Profile in the Vitreous of Proliferative Diabetic Retinopathy Patients and Differences from Patients Treated with Anti-VEGF Therapy. *Transl. Vis. Sci. Technol.* **2020**, *9*, 16. [CrossRef]
10. Jiang, L.; Cao, H.; Deng, T.; Yang, M.; Meng, T.; Yang, H.; Luo, X. Serum exosomal miR-377-3p inhibits retinal pigment epithelium proliferation and offers a biomarker for diabetic macular edema. *J. Int. Med. Res.* **2021**, *49*, 3000605211002975. [CrossRef] [PubMed]
11. Grieco, G.E.; Sebastiani, G.; Eandi, C.M.; Neri, G.; Nigi, L.; Brusco, N.; D'Aurizio, R.; Posarelli, M.; Bacci, T.; De Benedetto, E.; et al. MicroRNA Expression in the Aqueous Humor of Patients with Diabetic Macular Edema. *Int. J. Mol. Sci.* **2020**, *21*, 7328. [CrossRef]
12. Cho, H.; Hwang, M.; Hong, E.H.; Yu, H.; Park, H.H.; Koh, S.H.; Shin, Y.U. Micro RNAs in the aqueous humour of patients with diabetic macular oedema. *Clin. Exp. Ophthalmol.* **2020**, *48*, 624–635. [CrossRef] [PubMed]

13. Castro-Villegas, C.; Pérez-Sánchez, C.; Escudero, A.; Filipescu, I.; Verdu, M.; Ruiz-Limón, P.; Aguirre, M.A.; Jiménez-Gomez, Y.; Font, P.; Rodriguez-Ariza, A.; et al. Circulating miRNAs as potential biomarkers of therapy effectiveness in rheumatoid arthritis patients treated with anti-TNFα. *Arthritis Res. Ther.* **2015**, *17*, 49. [CrossRef]
14. Citron, F.; Segatto, I.; Musco, L.; Pellarin, I.; Rampioni Vinciguerra, G.L.; Franchin, G.; Fanetti, G.; Miccichè, F.; Giacomarra, V.; Lupato, V.; et al. miR-9 modulates and predicts the response to radiotherapy and EGFR inhibition in HNSCC. *EMBO Mol. Med.* **2021**, *13*, e12872. [CrossRef] [PubMed]
15. Chen, X.; Yao, Y.; Yuan, F.; Xie, B. Overexpression of miR-181a-5p inhibits retinal neovascularization through endocan and the ERK1/2 signaling pathway. *J. Cell. Physiol.* **2020**, *235*, 9323–9335. [CrossRef]
16. Dentelli, P.; Rosso, A.; Orso, F.; Olgasi, C.; Taverna, D.; Brizzi, M.F. microRNA-222 controls neovascularization by regulating signal transducer and activator of transcription 5A expression. *Arter. Thromb. Vasc. Biol.* **2010**, *30*, 1562–1568. [CrossRef]
17. Robinson, P.M.; Chuang, T.D.; Sriram, S.; Pi, L.; Luo, X.P.; Petersen, B.E.; Schultz, G.S. MicroRNA signature in wound healing following excimer laser ablation: Role of miR-133b on TGFβ1, CTGF, SMA, and COL1A1 expression levels in rabbit corneal fibroblasts. *Investig. Ophthalmol. Vis. Sci.* **2013**, *54*, 6944–6951. [CrossRef] [PubMed]
18. Babae, N.; Bourajjaj, M.; Liu, Y.; Van Beijnum, J.R.; Cerisoli, F.; Scaria, P.V.; Verheul, M.; Van Berkel, M.P.; Pieters, E.H.; Van Haastert, R.J.; et al. Systemic miRNA-7 delivery inhibits tumor angiogenesis and growth in murine xenograft glioblastoma. *Oncotarget* **2014**, *5*, 6687–6700. [CrossRef] [PubMed]
19. Vader, M.J.C.; Schauwvlieghe, A.M.E.; Verbraak, F.D.; Dijkman, G.; Hooymans, J.M.M.; Los, L.I.; Zwinderman, A.H.; Peto, T.; Hoyng, C.B.; van Leeuwen, R.; et al. Comparing the Efficacy of Bevacizumab and Ranibizumab in Patients with Diabetic Macular Edema (BRDME): The BRDME Study, a Randomized Trial. *Ophthalmol. Retin.* **2020**, *4*, 777–788. [CrossRef]
20. Giani, A.; Cigada, M.; Choudhry, N.; Deiro, A.P.; Oldani, M.; Pellegrini, M.; Invernizzi, A.; Duca, P.; Miller, J.W.; Staurenghi, G. Reproducibility of retinal thickness measurements on normal and pathologic eyes by different optical coherence tomography instruments. *Am. J. Ophthalmol.* **2010**, *150*, 815–824. [CrossRef]
21. Ruijter, J.M.; Ramakers, C.; Hoogaars, W.M.; Karlen, Y.; Bakker, O.; van den Hoff, M.J.; Moorman, A.F. Amplification efficiency: Linking baseline and bias in the analysis of quantitative PCR data. *Nucleic Acids Res.* **2009**, *37*, e45. [CrossRef]
22. Klaassen, I.; Hughes, J.M.; Vogels, I.M.; Schalkwijk, C.G.; Van Noorden, C.J.; Schlingemann, R.O. Altered expression of genes related to blood-retina barrier disruption in streptozotocin-induced diabetes. *Exp. Eye Res.* **2009**, *89*, 4–15. [CrossRef] [PubMed]
23. del Toro, R.; Prahst, C.; Mathivet, T.; Siegfried, G.; Kaminker, J.S.; Larrivee, B.; Breant, C.; Duarte, A.; Takakura, N.; Fukamizu, A.; et al. Identification and functional analysis of endothelial tip cell-enriched genes. *Blood* **2010**, *116*, 4025–4033. [CrossRef]
24. Rocha, S.F.; Schiller, M.; Jing, D.; Li, H.; Butz, S.; Vestweber, D.; Biljes, D.; Drexler, H.C.; Nieminen-Kelhä, M.; Vajkoczy, P.; et al. Esm1 modulates endothelial tip cell behavior and vascular permeability by enhancing VEGF bioavailability. *Circ. Res.* **2014**, *115*, 581–590. [CrossRef]
25. Yang, C.; Tahiri, H.; Cai, C.; Gu, M.; Gagnon, C.; Hardy, P. microRNA-181a inhibits ocular neovascularization by interfering with vascular endothelial growth factor expression. *Cardiovasc. Ther.* **2018**, *36*, e12329. [CrossRef] [PubMed]
26. Kamal, N.N.M.; Shahidan, W.N.S. Non-Exosomal and Exosomal Circulatory MicroRNAs: Which Are More Valid as Biomarkers? *Front. Pharmacol.* **2019**, *10*, 1500. [CrossRef] [PubMed]

Journal of
Personalized
Medicine

Article

Semi-Quantitative Multiplex Profiling of the Complement System Identifies Associations of Complement Proteins with Genetic Variants and Metabolites in Age-Related Macular Degeneration

I. Erkin Acar [1,†], Esther Willems [2,3,4,†], Eveline Kersten [1], Jenneke Keizer-Garritsen [4], Else Kragt [4], Bjorn Bakker [1], Tessel E. Galesloot [5], Carel B. Hoyng [1], Sascha Fauser [6], Alain J. van Gool [4], Yara T. E. Lechanteur [1], Elod Koertvely [6], Everson Nogoceke [6], Jolein Gloerich [4], Marien I. de Jonge [2,3], Laura Lorés-Motta [1,6,†] and Anneke I. den Hollander [1,7,*,†]

1   Department of Ophthalmology, Donders Institute for Brain, Cognition and Behaviour, Radboud University Medical Center, 6525 GA Nijmegen, The Netherlands; erkin.acar@radboudumc.nl (I.E.A.); Eveline.Kersten@radboudumc.nl (E.K.); Bjorn.Bakker@radboudumc.nl (B.B.); Carel.Hoyng@radboudumc.nl (C.B.H.); yara.lechanteur@radboudumc.nl (Y.T.E.L.); lauraloresmotta@protonmail.com (L.L.-M.)
2   Laboratory of Medical Immunology, Department of Laboratory Medicine, Radboud Institute for Molecular Life Sciences, Radboud University Medical Center, 6525 GA Nijmegen, The Netherlands; Esther.Willems1@radboudumc.nl (E.W.), marien.dejonge@radboudumc.nl (M.I.d.J.)
3   Radboud Center for Infectious Diseases, Radboud University Medical Center, 6525 GA Nijmegen, The Netherlands
4   Translational Metabolic Laboratory, Department of Laboratory Medicine, Radboud Institute for Molecular Life Sciences, Radboud University Medical Center, 6525 GA Nijmegen, The Netherlands; Jenneke.Keizer-Garritsen@radboudumc.nl (J.K.-G.); Else.Kragt@gmail.com (E.K.); Alain.vanGool@radboudumc.nl (A.J.v.G.); Jolein.Gloerich@radboudumc.nl (J.G.)
5   Department for Health Evidence, Radboud Institute for Health Sciences, Radboud University Medical Center, 6525 GA Nijmegen, The Netherlands; tessel.galesloot@radboudumc.nl
6   Roche Pharma Research and Early Development, Roche Innovation Center Basel, 124 Grenzacherstrasse, 4070 Basel, Switzerland; sascha.fauser@roche.com (S.F.); elod.koertvely@roche.com (E.K.); everson.nogoceke@roche.com (E.N.)
7   Department of Human Genetics, Donders Institute for Brain, Cognition and Behaviour, Radboud University Medical Center, 6525 GA Nijmegen, The Netherlands
*   Correspondence: Anneke.denHollander@radboudumc.nl
†   Contributed equally.

Citation: Acar, I.E.; Willems, E.; Kersten, E.; Keizer-Garritsen, J.; Kragt, E.; Bakker, B.; Galesloot, T.E.; Hoyng, C.B.; Fauser, S.; van Gool, A.J.; et al. Semi-Quantitative Multiplex Profiling of the Complement System Identifies Associations of Complement Proteins with Genetic Variants and Metabolites in Age-Related Macular Degeneration. *J. Pers. Med.* **2021**, *11*, 1256. https://doi.org/10.3390/jpm11121256

Academic Editors: Peter D. Westenskow and Andreas Ebneter

Received: 20 October 2021
Accepted: 19 November 2021
Published: 25 November 2021

Publisher's Note: MDPI stays neutral with regard to jurisdictional claims in published maps and institutional affiliations.

Abstract: Age-related macular degeneration (AMD) is a major cause of vision loss among the elderly in the Western world. The complement system has been identified as one of the main AMD disease pathways. We performed a comprehensive expression analysis of 32 complement proteins in plasma samples of 255 AMD patients and 221 control individuals using mass spectrometry-based semi-quantitative multiplex profiling. We detected significant associations of complement protein levels with age, sex and body-mass index (BMI), and potential associations of C-reactive protein, factor H related-2 (FHR-2) and collectin-11 with AMD. In addition, we confirmed previously described associations and identified new associations of AMD variants with complement levels. New associations include increased C4 levels for rs181705462 at the *C2/CFB* locus, decreased vitronectin (VTN) levels for rs11080055 at the *TMEM97/VTN* locus and decreased factor I levels for rs10033900 at the *CFI* locus. Finally, we detected significant associations between AMD-associated metabolites and complement proteins in plasma. The most significant complement-metabolite associations included increased high density lipoprotein (HDL) subparticle levels with decreased C3, factor H (FH) and VTN levels. The results of our study indicate that demographic factors, genetic variants and circulating metabolites are associated with complement protein components. We suggest that these factors should be considered to design personalized treatment approaches and to increase the success of clinical trials targeting the complement system.

Keywords: age-related macular degeneration; AMD; complement system; semi-quantitative multiplex profilin; mass spectrometry; C4; vitronectin; factor I; genetic variants; metabolites; HDL

1. Introduction

Age-related macular degeneration (AMD) is a major cause of vision loss among the elderly in the Western world. Due to increased ageing of the population, the number of individuals affected by AMD worldwide is expected to rise from 200 million currently to approximately 288 million by 2040 [1]. Early forms of AMD are characterized by the accumulation of drusen between the retinal pigment epithelium (RPE) and Bruch's membrane. As the disease progresses, the drusen expand and coalesce, leading to two types of advanced stages. Geographic atrophy (GA) is defined by atrophy of photoreceptors, RPE and choriocapillaris in the macula, leading to progressive and irreversible loss of central vision. Neovascular AMD (nAMD) is characterized by the abnormal ingrowth of new blood vessels from the choriocapillaris into the sub-RPE or subretinal space. These aberrant vessels are fragile and tend to leak fluid and blood, causing rapid and severe vision loss. Neovascularization in nAMD can effectively be treated with intravitreal injections of anti-vascular endothelial growth factor (VEGF), but no approved treatment is currently available to prevent degeneration in GA.

AMD is a multifactorial disease caused by a combination of environmental and genetic risk factors. Increasing age, female sex, smoking and a high body-mass index (BMI) are associated with an increased risk for AMD, while use of antioxidants decreases the risk of AMD disease progression [2–5]. Genome-wide association studies have identified the complement system, extracellular matrix remodelling and lipid metabolism as main AMD disease pathways [6]. The strongest independent genetic associations were reported for the common p.Tyr402His (rs1061170) variant in the complement factor H (*CFH*) gene [7–10] and for a common variant at the *ARMS2/HTRA1* locus [11–13]. Since then, eight additional genetic variants at the *CFH* locus have been independently associated with AMD. In addition, four variants in or near the complement factor B (*CFB*) and complement component 2 (*C2*) genes, three variants in or near the complement component 3 (*C3*) gene, two variants in or near the complement factor I (*CFI*) gene, one variant in the complement component 9 (*C9*) gene and one variant near the vitronectin (*VTN*) gene have been associated with AMD [6,14,15]. These variants include intergenic, intronic and coding variants that either confer an increased or decreased risk for AMD. Coding variants in these genes include the variants p.Gly119Arg (rs141853578) in *CFI*, p.Arg102Gly (rs2230199) and p.Lys155Gln (rs147859257) in *C3*, and p.Pro167Ser (rs34882957) in *C9*, which are associated with an increased risk of AMD. Two coding variants in the *CFB* gene, p.Leu9His (rs4151667) and p.Arg32Gln (rs641153), are associated with a decreased risk for AMD. In addition to these variants that are individually associated with AMD, an aggregated effect of rare coding variants in the *CFH*, *CFI*, *C3* and *C9* genes has also been reported to associate with AMD using gene-based approaches [6,16–19].

An important role of the complement system in AMD already emerged from earlier studies demonstrating that complement components are constituents of drusen [8,14,20–22]. In addition, altered concentrations of complement components have been detected in the systemic circulation of AMD patients. Several have studies demonstrated that complement activation markers such as C3a, C3d and C5a are elevated in AMD patients compared to controls [23–25]. Moreover, several genetic variants in or near genes of the complement system are associated with altered concentrations of complement protein components in the systemic circulation of AMD patients. The AMD-protective p.Leu9His (rs4151667) variant in the *CFB* gene is associated with reduced factor B (FB) levels [25,26], while the AMD-risk variant p.Gly119Arg (rs141853578) in the *CFI* gene is associated with reduced factor I (FI) levels [27]. The AMD-risk variant p.Pro167Ser (rs34882957) in the *C9* gene was initially associated with elevated C9 levels [28], but a more recent study rather reported decreased C9 levels in carriers of the p.Pro167Ser variant [29]. Common variants at the *CFH* locus were recently shown to be strongly associated with levels of the factor H-related proteins FHR-1, FHR-2, FHR-3 and FHR-4, which are encoded by the *CFHR* genes located at the *CFH* locus downstream of the *CFH* gene. The AMD risk-conferring variant rs570618[T] is associated with increased FHR-1, FHR-2, FHR-3 and FHR-4 levels, while the AMD-protective variant rs10922109[A] is associated with decreased FHR-1, FHR-2, FHR-3 and FHR-4 levels [30–32]. Nevertheless, for many of the AMD-associated genetic variants in or near genes of the complement system, the effect on complement protein levels remains unknown.

A recent study observed strong associations of systemic complement activation measurements (defined by the C3d/C3 ratio) with AMD-associated metabolites, including lipoprotein subfractions (large and very large high-density lipoproteins (L- and XL-HDL), very-low-density lipoproteins (VLDL)), other lipids/apolipoproteins (remnant-C, apolipoprotein B (ApoB), triglycerides), fatty acids (monounsaturated fatty acids (MUFA), saturated fatty acids (SFA), total fatty acids (TotFA)), and amino acids (leucine (Leu), isoleucine (Ile) and alanine (Ala)) [33]. Increased L-and XL-HDL levels were associated with increased complement activation, and both HDL levels and complement activation were increased in AMD patients compared to controls. On the other hand, decreased VLDL, remnant-C, ApoB, triglycerides, MUFA, SFA, TotFA, Ile, Leu and Ala levels were associated with increased complement activation, and these metabolites were decreased in AMD compared to controls. These associations may indicate biological interactions between the main AMD disease pathways: lipid metabolism and complement activation. Associations between other complement components and AMD-associated metabolites have not yet been studied in the context of AMD.

In this study, we performed a comprehensive analysis of 32 complement proteins in plasma samples of AMD patients and controls using mass spectrometry-based semi-quantitative multiplex profiling [34]. First, we aimed to determine associations of complement proteins with demographic factors and with AMD. Next, we aimed to determine the association of genetic variants with levels of complement proteins in plasma. Finally, we aimed to determine associations between AMD-associated metabolites and complement proteins in plasma.

2. Results

Using a multiplex selected reaction monitoring (SRM) liquid chromatography–mass spectrometry (MS) assay [34], levels of 64 peptides from 32 proteins of the complement system were measured in 476 plasma samples (details of the measured peptides are provided in Supplementary Table S1). These samples belonged to 221 control individuals and 255 AMD patients selected from the EUGENDA database (descriptive statistics of the study cohort are provided in Supplementary Table S2). Genotyping and metabolomics data was available for the majority of the included individuals.

2.1. Complement Peptide Levels Are Associated with Age, Sex, and BMI

First, we evaluated whether complement peptide levels vary with age, sex, smoking and BMI. Pearson correlation analyses were performed on the entire cohort using rank-based inverse normal transformed peptide levels. After correcting for the false discovery rate (FDR), 12 peptides were significantly correlated with age (Table 1, Supplementary Table S3). Peptides originating from factor D (FD), C9 and C7 were positively correlated with age, whereas peptides originating from MASP1/3, FCN3, MASP1, C8G (two peptides), C8A (two peptides), C4BPA and clusterin were negatively correlated with age (Table 1). The most significant correlation was found for factor D (r = 0.25, p-value = 2.50×10^{-8}, P_{FDR} = 1.60×10^{-6}, Table 1).

Table 1. Association of demographic factors with complement peptide levels. Pearson correlation was computed for each demographic factor, and the significantly associated protein peptides are shown. Full list of Pearson correlations to age, sex, smoking and BMI can be found in Supplementary Tables S3–S6.

Demographic	Peptide ID	Protein	Pearson's *r*	*p*-Value	*p*-Value$_{FDR}$
	39	FD	0.25	2.50×10^{-8}	1.60×10^{-6}
	46	MASP1/3	−0.19	2.78×10^{-5}	8.89×10^{-4}
	52	FCN3	−0.18	9.94×10^{-5}	2.12×10^{-3}
	35	C9	0.14	1.65×10^{-3}	2.64×10^{-2}
	32	C8G	−0.14	2.80×10^{-3}	2.94×10^{-2}
Age	31	C8G	−0.14	2.85×10^{-3}	2.94×10^{-2}
	48	MASP1	−0.13	3.21×10^{-3}	2.94×10^{-2}
	25	C7	0.13	3.87×10^{-3}	3.10×10^{-2}
	27	C8A	−0.13	4.94×10^{-3}	3.16×10^{-2}
	26	C8A	−0.13	4.64×10^{-3}	3.16×10^{-2}
	55	C4BPA	−0.12	7.62×10^{-3}	4.07×10^{-2}
	60	Clusterin	−0.12	7.07×10^{-3}	4.07×10^{-2}
	23	C6	0.19	3.17×10^{-5}	2.03×10^{-3}
	50	MBL2	−0.18	9.60×10^{-5}	3.07×10^{-3}
	51	MBL2	−0.17	2.06×10^{-4}	3.29×10^{-3}
	57	C4BPB	0.17	1.69×10^{-4}	3.29×10^{-3}
	58	IC1	0.16	6.38×10^{-4}	8.17×10^{-3}
Sex	59	IC1	0.15	7.97×10^{-4}	8.50×10^{-3}
	43	FI	0.14	2.61×10^{-3}	1.86×10^{-2}
	61	Clusterin	0.14	2.47×10^{-3}	1.86×10^{-2}
	60	Clusterin	0.14	2.58×10^{-3}	1.86×10^{-2}
	39	FD	−0.14	2.95×10^{-3}	1.89×10^{-2}
	27	C8A	0.13	5.85×10^{-3}	3.40×10^{-2}
	44	CRP	0.28	9.49×10^{-9}	6.07×10^{-7}
	12	C3	0.24	1.20×10^{-6}	3.85×10^{-5}
	41	FH	0.21	1.05×10^{-5}	2.25×10^{-4}
	43	FI	0.20	5.55×10^{-5}	8.88×10^{-4}
BMI	42	FH	0.19	1.27×10^{-4}	1.63×10^{-3}
	40	FH	0.18	1.77×10^{-4}	1.89×10^{-3}
	63	Vitronectin	0.17	4.03×10^{-4}	3.69×10^{-3}
	13	C3	0.14	5.40×10^{-3}	4.32×10^{-2}
	38	FB	0.13	6.53×10^{-3}	4.64×10^{-2}

Pearson correlation identified 11 peptides that were significantly correlated with sex (Table 1, Supplementary Table S4). Peptides originating from C6, C4BPB, IC1, FI, C8A and clusterin were higher in females, while peptides originating from MBL2 and FD were lower in males. The most significant association with sex was noted for peptide 23 originating from the C6 protein ($r = 0.19$, *p*-value $= 3.17 \times 10^{-5}$, $P_{FDR} = 2.03 \times 10^{-3}$). No significant correlations were identified for smoking status with complement peptides (Supplementary Table S5). However, nine peptides, originating from CRP, C3, FH, FI, vitronectin, and FB, were all positively correlated with increased BMI (Table 1, Supplementary Table S6). The most significantly BMI-associated peptide originated from the CRP protein ($r = 0.28$, *p*-value $= 9.49 \times 10^{-9}$, $P_{FDR} = 6.07 \times 10^{-7}$). All correlation coefficients were between the absolute value of 0 and 0.3, which suggests weak correlations. A schematic overview of correlations between complement peptide levels and demographics is provided in Supplementary Figure S1.

2.2. Association Analysis of Complement Peptide Levels with AMD

To determine associations of complement peptide levels with AMD, we performed an association analysis of rank-based inverse normal transformed peptide levels with

AMD status by means of linear regression analysis adjusted for age, sex, BMI, smoking and triglycerides, since triglyceride levels were previously associated with complement activation levels [35]. After correction for multiple testing using FDR, no peptides remained significantly associated with AMD status (Supplementary Table S7). However, peptides originating from CRP, FHR-2 and collectin-11 had a *p*-value < 0.05 prior to FDR correction (Table 2) and they were elevated in AMD patients compared to controls (Figure 1A–C).

Table 2. Suggestive associations of complement peptide levels with age-related macular degeneration. Beta values show the effect estimates from the regression analysis results. AMD status was the dependent variable in the regression model, while peptide levels, age, sex, BMI, smoking status and triglyceride levels were added as independent variables.

Peptide id	Protein	B	SE	*p*-Value	*p*-Value$_{\text{FDR}}$
64	FHR-2	0.234	0.094	0.013	0.517
44	CRP	0.226	0.094	0.016	0.517
54	Collectin 11	0.187	0.094	0.046	0.913

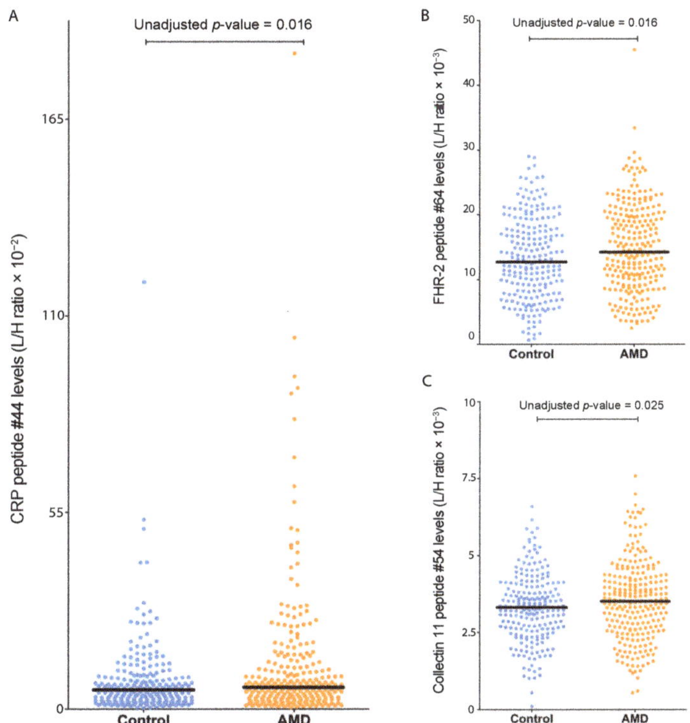

Figure 1. Suggestive association of CRP, FHR-2 and collectin-11 peptide levels with AMD. Showing the distribution of peptide levels (in L/H ratio = endogenous peptide/heavy labelled standard) in blood plasma from controls and AMD patients for (**A**) CRP peptide #44 levels (ESDTSYVSLK), (**B**) FHR-2 peptide #64 levels (TGDIVEFVCK), and (**C**) collectin-11 peptide #54 levels (VFIGINDLEK). *p*-values represent the results from the regression model that is adjusted for age, gender, BMI, smoking status and triglyceride levels.

2.3. Complement Peptide Levels Are Associated with AMD Variants

Next, we aimed to determine associations of complement peptide levels with AMD-associated genetic variants. The list of all the variants included in our study can be found in

Supplementary Table S7. We initially focused on cis-acting effects of the genetic variants on complement peptide levels, thus we studied associations of genetic variants with proteins encoded by genes located in or near these AMD loci (Table 3, Supplementary Tables S9–S19). Association analyses of complement peptides with AMD-associated variants as identified by the IAMDGC GWAS [6] were carried out adjusting for age, sex and AMD status.

Table 3. Cis-acting effects of AMD-associated genetic variants on complement peptide levels. Complement protein levels in the locus of each known AMD variant were tested for association with each known AMD variant separately. Alleles were chosen for the risk increasing effect.

Peptide Id	Protein	AMD Variant *	Allele	AF	Locus Gene(s) ^	B	SE	p-Value	p-Value$_{FDR}$	OR AMD #	p-Value AMD #
64	FHR-2	rs61818925	G	0.639	*CFH (CFHR3/CFHR1)*	0.570	0.075	2.84×10^{-13}	1.13×10^{-12}	1.667	6.0×10^{-165}
64	FHR-2	rs570618	T	0.442	*CFH*	0.560	0.074	3.24×10^{-13}	1.13×10^{-12}	2.380	2.0×10^{-590}
64	FHR-2	rs10922109	C	0.617	*CFH*	0.540	0.078	2.20×10^{-11}	5.13×10^{-11}	2.632	9.6×10^{-618}
64	FHR-2	rs148553336	T	0.996	*CFH*	1.320	0.585	0.025	0.044	3.448	8.6×10^{-26}
41	FH	rs148553336	T	0.996	*CFH*	1.669	0.573	0.004	0.084	3.448	8.6×10^{-26}
40	FH	rs148553336	T	0.996	*CFH*	1.399	0.576	0.016	0.112	3.448	8.6×10^{-26}
42	FH	rs148553336	T	0.996	*CFH*	1.403	0.581	0.016	0.112	3.448	8.6×10^{-26}
41	FH	rs61818925	G	0.639	*CFH (CFHR3/CFHR1)*	0.179	0.079	0.024	0.126	1.667	6.0×10^{-165}
42	FH	rs10922109	C	0.617	*CFH*	−0.164	0.082	0.047	0.197	2.632	9.6×10^{-618}
17	C4	rs181705462	T	0.009	*C2/CFB/SKIV2L*	1.713	0.401	2.55×10^{-5}	0.001	1.550	3.1×10^{-10}
15	C4	rs181705462	T	0.009	*C2/CFB/SKIV2L*	1.485	0.406	2.91×10^{-4}	0.003	1.550	3.1×10^{-10}
16	C4	rs181705462	T	0.009	*C2/CFB/SKIV2L*	1.468	0.409	3.73×10^{-4}	0.003	1.550	3.1×10^{-10}
14	C4	rs181705462	T	0.009	*C2/CFB/SKIV2L*	1.318	0.410	0.001	0.006	1.550	3.1×10^{-10}
63	Vitronectin	rs11080055	A	0.522	*TMEM97/VTN*	0.206	0.072	0.005	0.010	1.099	1.0×10^{-8}

AF = Allele frequency, * AMD associated variants from the IAMDGC GWAS on AMD (Fritsche et al., 2015) [6], ^ Nearest genes to the variant, # Results from the primary IAMDGC GWAS in Fritsche et al., 2015 [6].

2.3.1. FHR-2 Peptide Levels Are Associated with AMD Variants at the CFH Locus

The multiplex complement assay contains three peptides from FH and one peptide from FHR-2, which are both encoded by genes located at the *CFH* locus (Supplementary Table S1). Eight genetic variants at the *CFH* locus were previously associated with AMD (rs570618, rs121913059, rs187328863, rs35292876, rs191281603, rs10922109, rs148553336 and rs61818925) [6]. Association analyses were performed for these eight variants with FH and FHR-2 peptide levels. After correction for multiple testing using FDR, no peptides from FH were significantly associated with genetic variants at the *CFH* locus, while strong associations were identified for the FHR-2 peptide with rs61818925, rs570618, rs10922109 and rs148553336 (Table 3, Supplementary Tables S9 and S10).

The AMD variant rs61818925 was strongly associated with higher FHR-2 peptide levels (B = 0.570, SE = 0.075, p-value = 2.84×10^{-13}, $P_{FDR} = 1.13 \times 10^{-12}$, Table 3). The median FHR-2 peptide levels were significantly lower in individuals carrying the TT genotype compared to those carrying the GG genotype for rs61818925 (p-value = 2.0×10^{-13}, Figure 2A). The AMD variant rs10922109 was associated with higher FHR-2 peptide levels (B = 0.540, SE = 0.078, p-value = 2.20×10^{-11}, $P_{FDR} = 5.13 \times 10^{-11}$). The median FHR-2 peptide levels were significantly lower in individuals carrying the homozygous AA genotype compared to those carrying the homozygous CC genotype for rs10922109 (p-value = 9.5×10^{-11}, Figure 2B). The AMD variant rs148553336 was associated with higher FHR-2 peptide levels (B = 1.320, SE = 0.585, p-value = 0.025, $P_{FDR} = 0.04$), with significantly lower median FHR-2 peptide levels in individuals carrying the minor TC genotype compared to those carrying the homozygous TT genotype for rs148553336 (p-value = 0.026, Figure 2C). The AMD variant rs570618 was associated with higher FHR-2 peptide levels (B = 0.560, SE = 0.074, p-value = 3.24×10^{-13}, $P_{FDR} = 1.13 \times 10^{-12}$). Individuals carrying the homozygous GG genotype had significantly lower median FHR-2 peptide levels compared to those carrying the homozygous TT genotype for rs570618 (p-value = 2.0×10^{-10}, Figure 2D).

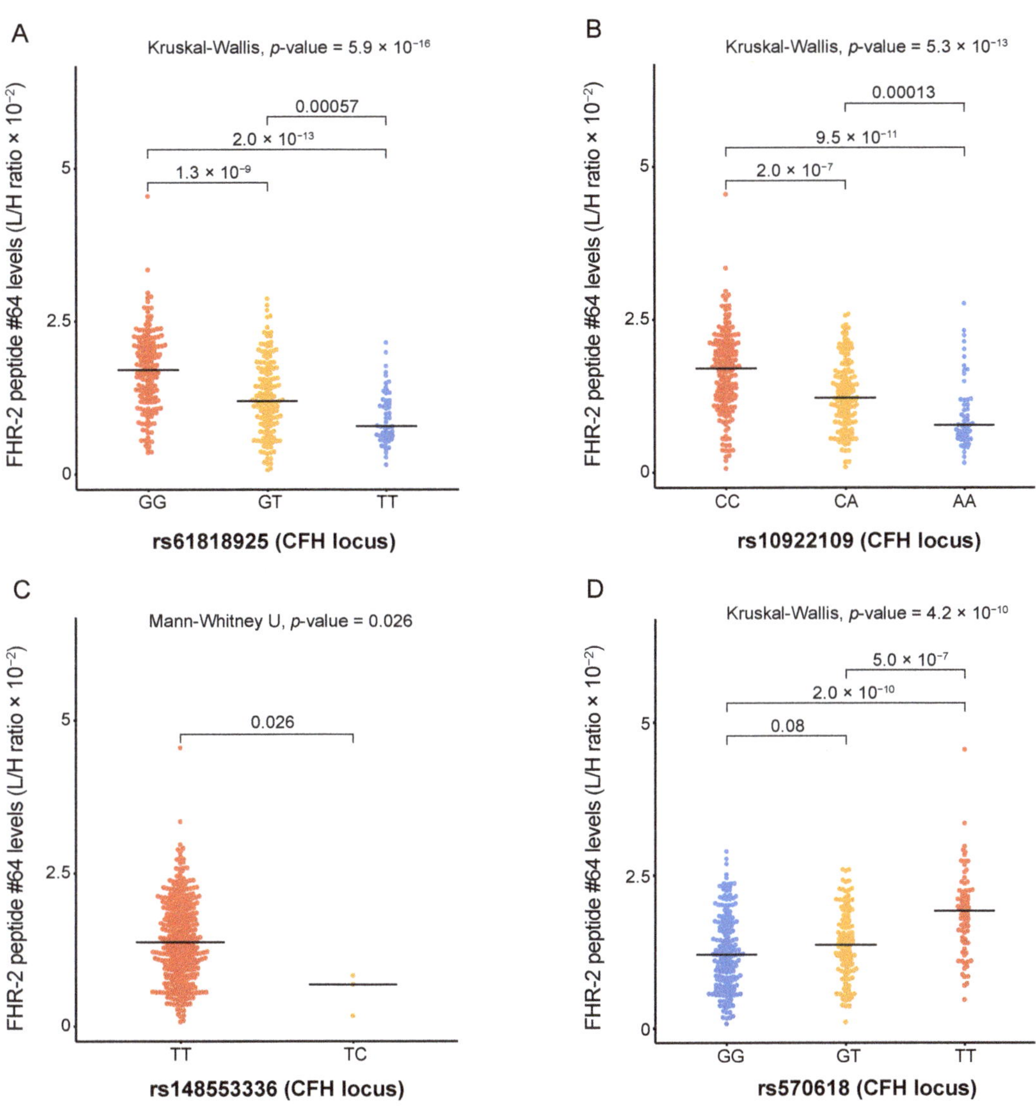

Figure 2. Factor-H related-2 (FHR-2) peptide levels show significant differences between genotype groups of AMD-associated variants at the CFH locus. Red indicates homozygous AMD-risk increasing genotype, while blue shows the homozygous AMD protective genotype and yellow indicates the heterozygous genotype. The Kruskal–Wallis test was included to test if there was a difference between the three genotype groups, where possible. Medians of two genotype groups were compared with the Mann–Whitney-U test. Showing the distribution of peptide levels in light/heavy (L/H) ratio in blood plasma for (**A**) FHR-2 peptide #64 (TGDIVEFVCK), stratified by rs61818925 genotype at the CFH locus; (**B**) FHR-2 peptide #64 (TGDIVEFVCK), stratified by rs10922109 genotype at the CFH locus; (**C**) FHR-2 peptide #64 (TGDIVEFVCK), stratified by rs148553336 genotype at the CFH locus; (**D**) FHR-2 peptide #64 (TGDIVEFVCK), stratified by rs570618 genotype at the CFH locus.

Associations of peptides from FH with genetic variants at the *CFH* locus did not reach significance after FDR correction, but three variants showed suggestive associations with FH peptide levels with a *p*-value <0.05 prior to FDR correction (Table 3; Supplementary

Table S10). The median FH level comparisons with *CFH* locus variants showed borderline significance between different genotypes and can be seen in Supplementary Figure S2A–C.

In addition to the eight AMD-associated variants at the *CFH* locus, a collective enrichment of rare coding variants in the *CFH* gene has been reported in AMD using gene-based approaches [6]. Two rare *CFH* variants were present in this study cohort in more than one individual: *CFH* p.Arg175Gln and *CFH* p.Ser193Leu (details of the number of carriers for rare variants are displayed in Supplementary Table S11). A Mann–Whitney U test was performed comparing FH levels in carriers versus non-carriers. Both coding variants in the *CFH* gene were not associated with altered FH peptide levels.

2.3.2. C4 Peptide Levels Are Associated with AMD Variants at the C2/CFB/SKIV2L Locus

Four genetic variants are independently associated with AMD at the *C2/CFB/SKIV2L* locus [6], which encompasses four complement genes: the *C2*, *CFB*, *C4A* and *C4B* genes. Notably, the rs181705462 variant is located closer to the *C4A* gene (3 kb) than to the *C2* (34 kb) and *CFB* (27 kb) genes (Figure 3A). The multiplex complement assay contains three peptides from FB, two peptides from C2, and four peptides from C4 (Supplementary Table S1). Association analyses were performed for the four AMD-associated variants at the *C2/CFB/SKIV2L* locus with FB, C2 and C4 peptide levels. After correction for multiple testing using FDR, no peptides from FB and C2 were associated with genetic variants at the *C2/CFB/SKIV2L* locus (Supplementary Tables S12 and S13), while significant associations were identified for all four C4 peptides with the rs181705462 variant located near the *C4A* gene (Table 3, Supplementary Table S14). The AMD variant rs181705462 was associated with higher C4 peptide levels for all four C4 peptides, of which C4 peptide 17 showed the strongest effect (B = 1.713, SE = 0.401, *p*-value = 2.55 $\times$ 10^{-5}, P_{FDR} = 0.001). The median C4 peptide levels were significantly higher in individuals carrying the GT genotype compared to those carrying the homozygous GG genotype for rs181705462 (*p*-value = 0.0031 for peptide 17; Figure 3B–E).

The rs116503776 variant at the *C2/CFB/SKIV2L* locus tags two coding variants in the *CFB* gene: p.Leu9His (rs4151667) and p.Arg32Gln (rs641153) [6] (Figure 3A). The p.Leu9His (rs4151667) variant was previously associated with reduced FB levels [25,26]. We therefore also analysed the median levels of FB in relation to the p.Leu9His (rs4151667) genotype, however, no significant difference was observed. Supplementary Figure S3A–C).

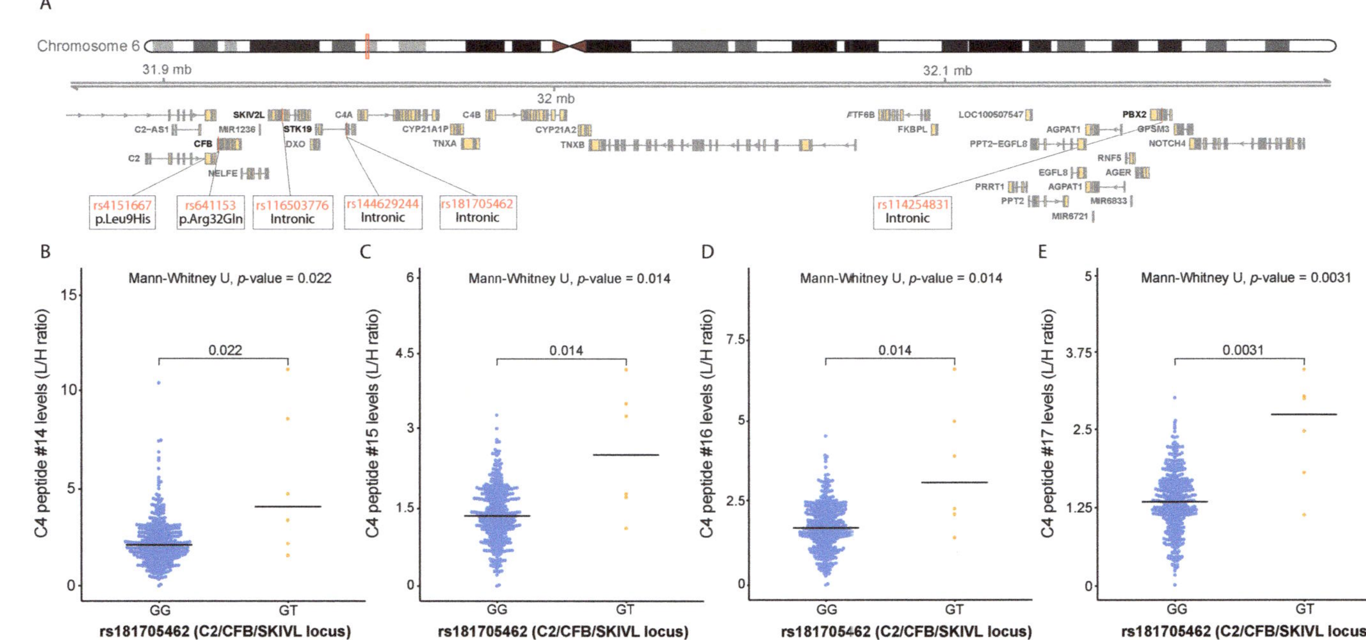

Figure 3. Complement C4 peptide levels show significant differences between genotype groups of AMD-associated variants at the C2/CFB/SKIV2L locus. Blue shows the homozygous AMD protective genotype and yellow indicates the heterozygous genotype. Medians of two genotype groups were compared with the Mann–Whitney-U test. (**A**) Overview of the C2/CFB/SKIV2L locus, showing the four top AMD-associated variants at this locus and two coding variants in CFB tagged by rs116503776, indicating their location in relation to the genes at this locus. Graphs showing the distribution of peptide levels in light/heavy (L/H) in blood plasma for (**B**) complement C4 peptide #14 (DFALLSLQVPLK), stratified by rs181705462 genotype at the C2/CFB/SKIV2L locus; (**C**) complement C4 peptide #15 (VDFTLSSER), stratified by rs181705462 genotype at the C2/CFB/SKIV2L locus; (**D**) complement C4 peptide #16 (VGDTLNLNLR), stratified by rs181705462 genotype at the C2/CFB/SKIV2L locus; (**E**) complement C4 peptide #17 (SHALQLNNR), stratified by rs181705462 genotype at the C2/CFB/SKIV2L locus.

2.3.3. VTN Peptide Levels Are Associated with AMD Variants at the TMEM97/VTN Locus

The multiplex complement assay contains two peptides originating from vitronectin (VTN). A common variant (rs11080055) at the *TMEM97/VTN* locus is independently associated with AMD [6]. The rs11080055 variant was associated with higher VTN peptide levels (for peptide 63: B = 0.206, SE = 0.072, *p*-value = 0.005, P_{FDR} = 0.010; Table 3, Supplementary Table S15). The median VTN peptide levels were lower in individuals carrying the homozygous CC genotype compared to those carrying the homozygous AA genotype for rs11080055 (for peptide 63: *p*-value = 0.00022; Figure 4A,B).

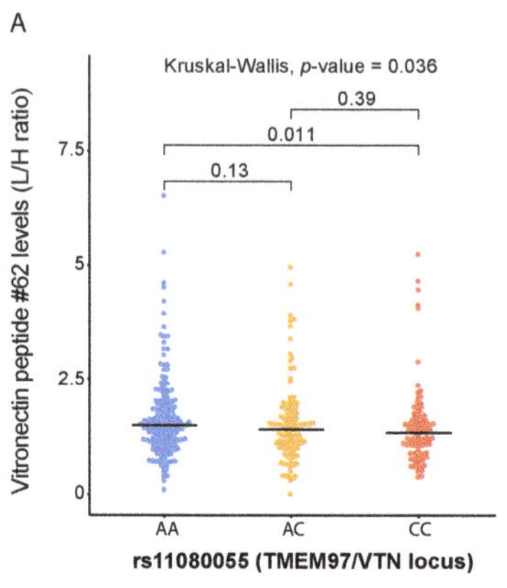

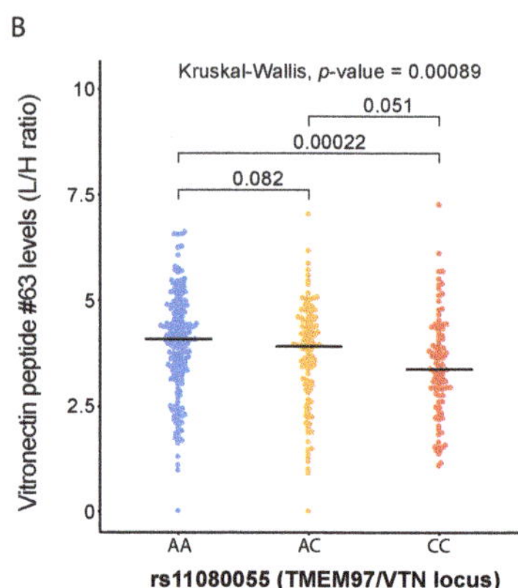

Figure 4. Vitronectin (VTN) levels show significant differences between genotype groups of AMD-associated variants at the TMEM97/VTN locus. Red indicates homozygous AMD-risk increasing genotype, while blue shows the homozygous AMD protective genotype, and yellow indicates the heterozygous genotype. The Kruskal–Wallis test was included to test if there was a difference between the three genotype groups, where possible. Medians of two genotype groups were compared with the Mann–Whitney-U test. Showing the distribution of peptide levels in light/heavy (L/H) in blood plasma for (**A**) VTN peptide #62 (DVWGIEGPIDAAFTR), stratified by rs11080055 genotype at the C2/CFB/SKIV2L locus; (**B**) VTN peptide #63 (FEDGVLDPDYPR), stratified by rs11080055 genotype at the C2/CFB/SKIV2L locus.

2.3.4. FI Peptide Levels Are Associated with AMD Variants at the CFI Locus

Two genetic variants at the *CFI* locus are associated with AMD: the common variant rs10033900 and the rare coding variant p.Gly119Arg (rs141853578) [6]. The target peptide for FI encompasses the p.Gly119 residue, thus the multiplex complement assay cannot be used to reliably analyse the effect of the p.Gly119Arg (rs141853578) variant on FI levels. The AMD risk-conferring variant rs10033900 was not significantly associated with FI peptide levels (Table 3, Supplementary Table S16). In addition to the two AMD-associated variants, a collective enrichment of rare coding variants in the *CFI* gene has been reported in AMD using gene-based approaches [6,16]. The following variants were present in this study cohort in more than one individual: *CFI* p.Leu131Arg, *CFI* p.Arg406His and *CFI* p.Pro553Ser (details of the number of carriers for rare variants are displayed in Supplementary Table S11). A Mann–Whitney U test was performed comparing carriers versus non-carriers for both rare and common variants which suggested significant difference *CFI* p.Leu131Arg variant (Supplementary Figure S4).

2.3.5. C9 Peptide Levels Are Associated with AMD Variants at the C9 Locus

A low-frequency coding variant in the *C9* gene, rs62358361, is associated with AMD [6]. The multiplex complement assay contains three peptides originated from C9; however, association analyses that were performed for C9 peptide levels with the *C9* rs62358361 variant showed no significant results (Supplementary Table S17). The coding variant p.Pro167Ser (rs34882957) was previously associated with altered C9 levels [28,29]. The Mann–Whitney U test comparing carriers versus non-carriers identified significant differences in levels of all three C9 peptides with *C9* genotype (p-value = 1.91×10^{-10} for peptide #34; Supplementary Table S11). C9 peptide levels are significantly lower in carriers of the *C9* p.Pro167Ser (rs34882957) variant compared to non-carriers (Figure 5A–C).

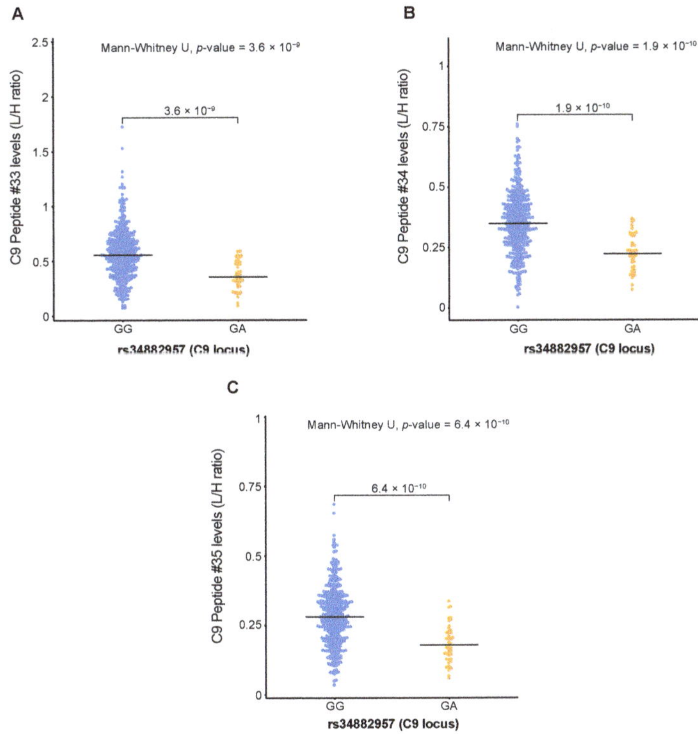

Figure 5. Complement C9 levels show significant differences between genotype groups of rare AMD risk variants at the C9 locus. Showing the distribution of peptide levels in light/heavy (L/H) in blood plasma for (**A**) peptide #33 (TSNFNAAISLK), (**B**) peptide #34 (VVEESELAR) and (**C**) peptide #35 (LSPIYNLVPVK), stratified by C9 p.Pro167Ser (rs34882957) genotype. Blue indicates homozygous AMD protective genotype, and yellow indicates the heterozygous genotype.

2.3.6. Association Analysis of C3 Peptide Levels with AMD Variants at the C3 Locus

Three variants in or near the *C3* gene were previously associated with AMD [6]. Two peptides from the C3 protein were included in the multiplex complement assay. Association analyses of C3 peptide levels with *C3* genotypes did not identify any significant associations, neither by using general linear models nor with a Mann–Whitney U test (Supplementary Tables S11 and S18).

2.3.7. Association Analysis of Complement Peptide Levels with AMD-Associated Genetic Variants at Other Loci

In addition to cis-acting effects, we also analysed the effects of AMD-associated genetic variants on complement peptide levels originated from proteins encoded by other loci. These associations did not reach statistical significance after FDR correction, but 173 suggestive associations with complement peptide levels with a p-value < 0.05 prior to FDR correction were identified (Supplementary Table S19). Within the complement system loci, there were three variants that had more than 10 suggestive associations with other complement protein peptides: the AMD risk-conferring variant rs12019136 at the *C3* locus showed suggestive associations with increased levels of 11 peptides, the protective AMD variant rs148553336 at the *CFH* locus showed suggestive associations with decreased levels of 14 peptides, and the AMD risk-conferring variant rs181705462 at the *C2/CFB/SKIV2L* locus showed suggestive associations with increased levels of 11 peptides.

2.4. Complement Peptides Are Associated with Lipoproteins and Other AMD Metabolites

In a large metabolomics study including 2267 AMD cases and 4266 controls, a total of 60 metabolites were found to be associated with AMD [33] (see Supplementary Table S20 with the associated metabolites and the effect on AMD). These metabolites included elevated levels of large and extra-large HDL subclasses, decreased levels of VLDL, citrate and amino acids. Moreover, 57 out to these AMD metabolites were associated with complement activation levels measured as the C3d/C3 ratio.

A total of 452 samples (239 cases, 213 controls) from the previous metabolomics study were also included in the current complement profiling study, which allowed us to assess whether complement peptides are associated with AMD metabolites. Association of complement peptide levels with AMD metabolites was performed adjusting for age, sex and AMD status. After FDR adjustment for multiple testing, a total of 327 significant associations were documented (Table 4, Supplementary Table S21). The strongest association was found for a C3 peptide with lower total cholesterol in very large HDL particles (B = -0.250, SE = 0.044, *p*-value = 2.79×10^{-8}, $P_{FDR} = 3.70 \times 10^{-5}$), Table 4. The top 20 associations were found for peptides in C3, FH, C9 and vitronectin. These peptides all associated with lower HDL subparticle levels except for C9, which was associated with higher phenylalanine levels (Table 4). These top 20 associations were also observed in a stratified analysis for disease status (Supplementary Table S21).

Table 4. Top 20 associations of complement peptide levels with AMD-associated metabolites. Each AMD-associated metabolite was checked for association with complement protein peptides in univariate regression analyses.

Metabolite	OR_{AMD} *	Peptide Id	Protein	B	SE	*p*-Value	*p*-Value$_{FDR}$
XL-HDL-C	1.082	12	C3	-0.250	0.044	2.79×10^{-8}	3.70×10^{-5}
XL-HDL-L	1.102	12	C3	-0.246	0.044	2.81×10^{-8}	3.70×10^{-5}
XL-HDL-P	1.105	12	C3	-0.249	0.044	2.89×10^{-8}	3.70×10^{-5}
XL-HDL-FC	1.092	12	C3	-0.240	0.044	6.61×10^{-8}	6.34×10^{-5}
XL-HDL-C	1.082	42	FH	-0.239	0.044	9.93×10^{-8}	6.71×10^{-5}
XL-HDL-P	1.105	42	FH	-0.237	0.044	1.05×10^{-7}	6.71×10^{-5}
XL-HDL-FC	1.092	42	FH	-0.233	0.044	1.48×10^{-7}	7.12×10^{-5}
XL-HDL-L	1.102	42	FH	-0.233	0.044	1.37×10^{-7}	7.12×10^{-5}
XL-HDL-PL	1.123	12	C3	-0.229	0.043	1.89×10^{-7}	8.06×10^{-5}
Phe	0.857	34	C9	0.227	0.044	2.97×10^{-7}	9.97×10^{-5}
L-HDL-P	1.092	12	C3	-0.218	0.042	2.65×10^{-7}	9.97×10^{-5}
L-HDL-CE	1.095	12	C3	-0.219	0.042	3.38×10^{-7}	9.97×10^{-5}
L-HDL-FC	1.097	12	C3	-0.224	0.043	3.14×10^{-7}	9.97×10^{-5}
L-HDL-C	1.095	12	C3	-0.218	0.042	3.90×10^{-7}	1.07×10^{-4}
XL-HDL-P	1.105	63	Vitronectin	-0.225	0.044	5.01×10^{-7}	1.13×10^{-4}
XL-HDL-PL	1.123	42	FH	-0.220	0.043	5.29×10^{-7}	1.13×10^{-4}
L-HDL-P	1.092	42	FH	-0.213	0.042	4.51×10^{-7}	1.13×10^{-4}
L-HDL-CE	1.095	42	FH	-0.214	0.042	5.27×10^{-7}	1.13×10^{-4}
L-HDL-FC	1.097	42	FH	-0.218	0.043	5.66×10^{-7}	1.14×10^{-4}
L-HDL-C	1.095	42	FH	-0.213	0.042	6.12×10^{-7}	1.18×10^{-4}

* Acar et al., 2020 [33].

Of the 327 significantly associated metabolites, those increased in AMD are HDL cholesterol, HDL-2 cholesterol and subparticles of large and extra-large HDL (Supplementary Tables S20 and S21). These HDL-related metabolites were all associated with decreased peptides levels of C3, FH, vitronectin, CRP, FCN3, FI, MASP3, FB, MASP1, C4BPB, C1QA, C8B, C2, FD and properdin (Supplementary Table S21). Metabolites decreased in AMD were VLDL subparticles, apoB, amino acids, citrate, several triglyceride measurements, fatty acid measurements and remnant cholesterol (Supplementary Tables S20 and S21). These metabolites were associated both with higher complement peptide levels and with lower complement peptide levels (Supplementary Table S21). VLDL subparticles were associated with higher peptide levels of FCN3, and lower peptide levels of C7 and C9. APOB was associated with higher levels of FCN3 and lower levels of C7.

The amino acid phenylalanine was associated with higher CRP, FB, C1QA and C8G, isoleucine and leucin were associated with higher FCN3 and alanine was associated with lower C9 and CRP peptide levels. Citrate was associate with lower CRP and C8B peptide levels. Serum triglyceride levels were associated with higher FCN3 levels and lower levels of C7 and C9, and triglyceride in HDL was associated with lower levels of C1QB, C1QC, C1R, C2, C6, C7, C8B, C8G and C9. Fatty acid measurements were associated with higher FCN3 peptide levels, and lower C1QC, C7 and C9 peptide levels. Finally, remnant cholesterol was associated with lower C7 levels (Supplementary Table S21).

3. Discussion

In this study we performed a comprehensive analysis of 32 complement proteins in plasma samples of AMD patients using mass spectrometry-based semi-quantitative multiplex profiling [34]. We detected significant associations of complement protein levels with age, sex and BMI, and identified potential associations of CRP, FHR-2 and collectin-11 with AMD. In addition, our proteogenomics analyses identified significant associations of genetic variants with levels of complement proteins in plasma, adjusted for AMD status, sex and age: FHR-2 peptide levels were associated with AMD variants at the *CFH* locus, C4 peptide levels were associated with an AMD variant at the *C2/CFB/SKIV2L* locus, VTN peptide levels were associated with an AMD variant at the *TMEM97/VTN* locus, FI peptide levels were associated with AMD variants at the *CFI* locus and C9 peptide levels were associated with an AMD variant at the *C9* locus. Finally, we detected significant associations between AMD-associated metabolites and complement proteins in plasma, also adjusted for AMD status, sex and age. The most significant complement-metabolite associations included HDL subparticle levels with decreased C3, FH and VTN levels. A schematic overview of the identified associations is provided in Figure 6.

We observed an increase of FD, C9 and C7 levels with age, whereas MASP1/3, FCN3, MASP1, C8G, C8A, C4BPA and clusterin decreased with age. A previous study also detected an increase of C9 levels with age [36], which is in agreement with our findings. However, that study reported increased C8 and decreased FD levels with age, which does not correspond to our findings. A potential explanation for the discrepancies between the studies is the difference in age range: the study by Gaya de Costa et al. included a population of healthy individuals with an age range of 20 to 69 years (mean age 45 years) [36], while the AMD case–control cohort described in this study has a mean age of 74 years (Supplementary Table S2). Additional studies specifically in the elderly population will gain a better understanding of the effect of ageing on complement component concentrations across the entire age spectrum. We also identified significant associations of sex and BMI with complement peptide levels. Peptides originating from C6, C4BPB, IC1, FI, C8A and clusterin were increased in females, while peptides originating from MBL2 and FD were decreased. A previous study also detected significant sex differences of complement activity and complement levels in a healthy population [36]. In that study lower concentrations of C3, properdin, MBL, ficolin-3 and terminal component levels were found in females, while FD concentrations were higher. The difference between sex-associated complement proteins could be caused by age difference as it was mentioned for age associations as well. In our current study, peptides originated from CRP, C3, FH, FI, vitronectin and FB, and were all positively correlated with increased BMI. This finding is in agreement with previous studies, which demonstrated that complement factors are expressed in adipose tissue and increased CRP, C3, FH and FB levels are positively associated with BMI [37,38].

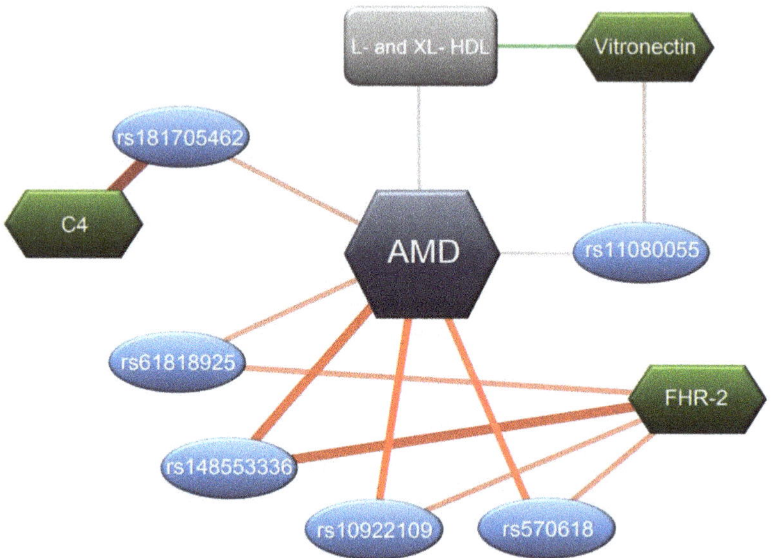

Figure 6. Schematic overview of associations of complement proteins with genetic variants and metabolites. AMD is highlighted in the center in red, complement proteins are highlighted in blue diamonds, genetic variants are highlighted in purple circles and metabolites are highlighted in yellow squares. Only the significantly associated proteins, genetic variants and metabolites are shown. Green colored connections show associations with beta estimates smaller than zero, which indicate associations with lower levels. Red colored connections show associations with beta estimates larger than zero, which generally indicate associations with higher levels. The color of the line is closer to gray as the beta estimate gets closer to zero. Association lines are represented thicker for the higher absolute beta estimates, and thinner for the lower absolute beta estimate.

None of the analysed levels of complement-derived peptides were significantly associated with AMD after FDR correction. However, the levels of three distinct peptides had a p-value < 0.05 prior to correction and are thus suggestive for association with AMD. In our analysis, the peptide from CRP was elevated in AMD patients compared to controls, which confirms previous studies reporting elevated CRP levels in multiple cohorts [39]. In addition, a peptide from the FHR-2 protein was elevated in AMD patients compared to controls, which is in agreement with our recent work reporting elevated FHR-2 levels, in addition to elevated FHR-1, FHR-3 and FHR-4 levels in AMD patients using ELISA measurements [30,31]. Finally, a peptide from collectin-11 was elevated in AMD patients compared to controls. A previous study demonstrated that binding of collectin-11 to retinal pigment epithelial cells induces complement activation [40]. Collectin-11 was shown to activate inflammatory responses through recognition of L-fucose, and fucosidase-treated RPE cells failed to activate complement. This suggests that collectin-11 is of relevance to the inflammatory status of RPE cells and may play a role in AMD pathogenesis. Further analysis of collectin-11 in larger AMD case–control cohorts is warranted to further confirm this association.

This study enabled us to systematically analyse the effect of genetic variants on concentrations of complement components. Our study confirmed previously described associations of FHR-2 levels with genetic variants at the *CFH* locus [31], and reduced

FB levels with the p.Leu9His (rs4151667) variant in the *CFB* gene [25,26]. The AMD-risk variant p.Pro167Ser (rs34882957) in the *C9* gene was initially associated with elevated C9 levels [28], but a more recent study rather reported decreased C9 levels in carriers of the p.Pro167Ser variant [29]. The current study identified decreased C9 levels in carriers of the p.Pro167Ser variant in the *C9* gene and gave consistent results for three independent peptides from C9, supporting the latter study that the C9 variant levels are genuinely lower.

Importantly, our study also identified new associations of AMD variants with complement levels. The AMD risk-conferring variant rs181705462 at the *C2/CFB/SKIV2L* locus was associated with higher C4 peptide levels. The rs181705462 variant is located closer to the *C4A* gene (3 kb) than to the *C2* (34 kb) and *CFB* (27 kb) genes, supporting that the variant may be a relevant determinant of C4 levels, yet the complexity of this locus does not allow us to draw strong conclusions. A previous study by Grassmann et al. reported a protective effect of copy number variation of the *C4A* gene on AMD [41]. The copy number variation was reported to be tagged by rs429608, while the *C4A* copy number variation is independent of rs181705462. This suggests that the elevated C4 peptide levels associated with rs181705462 are not driven by the *C4A* copy number variation. Moreover, the direction of effect in our study seems to be opposite to the study by Grassmann et al., as our study identified an association of increased AMD risk with higher C4 peptide levels, while the study by Grassmann et al. suggests a protective effect with increased *C4A* copy number and thus increased C4A protein levels. Our mass spectrometry-based assay was not able to differentiate between the C4A and C4B isoforms as the targeted peptides are identical in both isoforms (the isoforms differ in only several amino acids). Further studies are needed to determine which isoform is elevated and how this relates to a higher AMD risk conferred by the rs181705462 genotype.

Our study also demonstrates that the AMD variant rs11080055 at the *TMEM97/VTN* locus is associated with increased VTN peptide levels. This is in agreement with a recent study that examined the effect of rs704, a non-synonymous variant in the *VTN* gene [42]. The AMD risk-conferring variant rs704 is associated with increased VTN expression, supporting the involvement of altered VTN levels in AMD pathogenesis. VTN is an inhibitor of the terminal complement complex, and also serves various other functions, such as maintaining retinal integrity [42]. The rs704 variant induces collagen accumulation, and VTN is a major component of drusen and subretinal drusenoid deposits. It was also reported in the same study that rs704 causes a change in vitronectin protein, resulting in two isoforms. Our mass spectrometry assay was not designed to differentiate between these isoforms specifically. Therefore, we could not study the differences of these isoforms.

Additionally, we observed a suggestive association of the AMD risk-conferring variant rs10033900 at the *CFI* locus with reduced FI levels. A recent eQTL study reported associations of genetic variants (that are in high linkage disequilibrium with rs10033900) at the *CFI* locus with decreased gene expression of *CFI* in the retina [43]. Reduced FI levels have previously also been reported for the p.Gly119Arg variant and other rare coding variants in the *CFI* gene [27,44]. The effect of the p.Gly119Arg variant on FI levels could not reliably be analysed with the multiplex complement assay used in this study, as the target peptide for FI encompasses the p.Gly119 residue. In this study we confirmed that the rare coding *CFI* variant p.Leu131Arg is associated with lower FI levels, while no effect on FI levels was noted for the p.Arg406His and *CFI* p.Pro553Ser variants [44].

In addition to cis-acting effects, we also identified potential associations of complement peptide levels with genetic variants at other AMD loci, but none of these associations remained significant after FDR correction. Several variants showed suggestive associations with multiple complement components. For example, the protective AMD variant rs148553336[C] at the *CFH* locus showed suggestive associations with decreased FI, C3, FB and VTN peptide levels, in addition to decreased FHR-2 and FH associated with cis-acting effects at the *CFH* locus. The AMD risk-conferring variant rs181705462 at the *C2/CFB/SKIV2L* locus showed suggestive associations with increased FI, C9, C3 and FH peptide levels, in addition to increased C4 levels. A recent study using the same assay for semi-quantitative multiplex profiling of the complement system in patients with various complement deficiencies also identified potential trans-acting effects of mutations in complement genes [45]. This may be explained by functional coupling of complement factors within the pathways down- or upstream of the affected protein. Semi-quantitative multiplex profiling of the complement system in larger AMD case–control cohorts can further solidify trans-acting effects of genetic variants on a wide range of complement proteins.

The associations between genetic variants and the peptides of complement proteins were further checked in the QTLbase database (http://www.mulinlab.org/qtlbase/index.html, accessed on 12 November 2021) to assess novelty of our findings. In this database, rs570618 was reported to be associated with FH proteins. Only the effect estimates for hemopexin was reported for CFH, which was 0.09 (beta value). Given how small the effect estimate is, we would need higher statistical power to detect it in our study. For the variant rs10922109, an association with hemopexin was shown with the beta estimate of -0.13 within the database. We have shown a similar effect estimate for this variant (-0.16) which lost its statistical significance after multiple test correction. Lastly, the variant rs11080055 was shown as a pQTL; however, the effect estimate is not provided in the database.

Finally, we detected strong associations between metabolites and complement components. A recent study observed strong associations of systemic complement activation measurements (defined by the C3d/C3 ratio) with AMD-associated metabolites, including HDL and VLDL lipoprotein subfractions, other lipids/apolipoproteins (remnant-C, ApoB and triglycerides), fatty acids (MUFA, SFA and TotFA) and amino acids (Leu, Ile and Ala) [33]. In the current study, we extended these associations to a broad range of metabolites and complement components. Increased HDL levels were previously associated with an increased risk of AMD [33,46], and in this study we show that increased HDL levels are associated with decreased C3, FH, VTN, CRP, FCN3, FI, MASP3, FB, MASP1, C4BPB, C1QA, C8B, C2 and FD and properdin levels. Increased HDL levels have previously been associated with decreased FH and FB levels [37]; our study thus confirms and further extends these findings. In addition, our current study identified strong associations of VLDL subparticles, amino acids, citrate and fatty acids with altered complement component levels. Strong associations between AMD-associated metabolites and complement components may indicate biological interactions between the main AMD pathways, including the lipid metabolism and the complement system.

The results of this study have several implications for clinical translation. Multiple clinical trials, either targeting the central components C5 and C3 or complement regulators, have been completed or are currently ongoing [47]. Most of the candidate therapeutic complement compounds that have been tested so far have shown limited success [48]. Our current study suggests that several factors including age, sex, BMI, genetic variants and circulating metabolites may affect complement protein concentrations and may thus influence treatment success. The observed age-related differences of concentrations of complement components suggest that age should be considered in the design of therapeutics targeting the complement system, as different doses may be required among different age groups [36]. In addition, genetic variants may be considered for patient inclusion in clinical trials to increase the efficacy of the treatments that are being tested. Variants at the *CFH, CFI, C2/CFB, C9* and *VTN* loci are associated with altered complement component concentrations and could therefore be used to select patients for compounds targeting specific complement components. Two clinical trials supplementing FH (GEM103, Gemini Therapeutics) and *CFI* (GT005, Gyroscope Therapeutics) are the first to be selecting patients based on genotype (carrying risk variants in *CFH* and *CFI*, respectively) before inclusion in the trials. Our current study and previous work by others [17,25,26,29,30,44] suggest that genetic variants at the *CFH, CFI, C2/CFB, C9* and *VTN* loci can also be used to design personalised medicine approaches for AMD. Furthermore, the strong associations between circulating metabolites and complement components suggest that multiple disease pathways may need to be targeted for successful treatment of AMD.

Our study has several strengths and weaknesses. A strength of our study is that our semi-quantitative multiplex profiling assay is able to measure a broad range of complement components using a small sample volume [34]. The availability of detailed genetic data and metabolite measurements for our cohort enabled the integration of complement concentrations with various levels of data. Our current study included a fairly large sample size, consisting of 476 plasma samples. Although our sample size was sufficiently large to identify numerous significant associations, even larger studies are needed to follow-up on our identified suggestive associations that remained under the multiple testing adjusted threshold. Moreover, the genetic variants with low MAF (<5%) can be chance based findings due to their rarity and should not be considered to be conclusive with the number of samples included in this study. A limitation of the current design of our multiplex assay is that, due to peptide target design and the detection limit of our mass spectrometry assay, not all complement components can be assessed: FHR-2 could be analysed using the assay, but the other FHR proteins were below the detection limit. In addition, the target peptide for FI is specific for the p.Gly119 residue and can therefore not detect the effect of the *CFI* variant p.Gly119Arg on FI levels. Furthermore, the current assay cannot differentiate between the C4A and C4B isoforms, and further studies are needed to clarify which isoform (or whether the ratio between isoforms) is associated with the rs181705462 genotype. Finally, the current assay has been developed to measure complement component levels and does not allow the measurement of complement activation products.

In conclusion, we performed a comprehensive analysis of 32 complement proteins in plasma samples of AMD patients and controls using semi-quantitative multiplex profiling. Our study did not identify significantly associated peptides with AMD status; however, we identified numerous associations of complement components with age, sex, BMI, genetic variants and circulating metabolites. The results of our study suggest that these factors should be taken into account to design personalized treatment approaches and to increase the success of clinical trials targeting the complement system.

4. Materials and Methods

4.1. The Study Samples

For this study, EDTA plasma samples from 255 AMD patients and 221 control subjects were selected from the Dutch and German European Genetic Database (EUGENDA-Nijmegen and EUGENDA-Cologne) [49]. These EDTA plasma have been collected and stored (−80 °C within 1 h) according to the standard protocols. All participants provided written informed consent for clinical examination, epidemiological data collection and blood sampling for biochemical and genetic analyses. All studies were approved by the appropriate ethical committees (Commissie Mensgebonden Onderzoek [CMO] Arnhem-Nijmegen for EUGENDA-Nijmegen, Ethics Commission of Cologne University's Faculty of Medicine for EUGENDA-Cologne). Inclusion/exclusion criteria was based on the completeness of the dataset in terms of AMD grading, age > 55, and genetics and metabolomics data availability. Demographic factors of the study cohort are provided in Supplementary Table S2.

4.2. Genotyping

All individuals included in this study had previously been genotyped with a custom-modified Illumina HumanCoreExome array at the Centre for Inherited Disease Research (CIDR) and analysed within the IAMDGC GWAS (43,566 subjects; 16,144 advanced AMD cases and 17,832 controls of European ancestry in the primary analysis dataset) [6].

4.3. Metabolomic Analysis

Available metabolomic data from these patients were obtained from a previous study by Acar et al. [33]. In summary, the plasma samples were analysed by means of a high-throughput proton nuclear magnetic resonance (NMR) metabolomics platform (Nightingale Health, Ltd., Helsinki, Finland). After quality control, in total 146 high quality metabolites were selected, including amino acids, glycolysis measures, ketone bodies, inflammation-related measurements, fatty acids, and lipoprotein subclasses (Supplementary Table S20). Univariate logistic regression models were adjusted for age and sex, and were further combined into a random effects meta-analysis to determine the final association estimates.

4.4. Complement Targeted Multiplex Assay

Sample preparation and mass spectrometric analysis was performed as previously described [37]. Briefly, the plasma samples were reduced, alkylated, digested and stored at -80 °C. In prior to analysis, the digests were spiked with a mix of C-terminally $^{13}C^{15}N$ stable isotope labelled peptide standards (Thermo, JPT) for quantitation (L/H ratio = endogenous peptide/heavy labelled standard) and were desalted using Bond Elut OMIX tips (Agilent). Samples were analysed in 1-min target windows using the Waters Aquity MClass UPLC Xevo TQ-S, equipped with an ESI source and an iKeyTM (Waters peptide BEH C18, 130 Å, 1.7 μm, 150 μm × 100 mm). The peptides were eluted from the column using a gradient from 3 to 35% acetonitrile in 0.1% formic acid in 20 min at a flow rate of 2 μL/min. Raw data were analysed and exported using Skyline software v3.7 (MacCoss Lab, University of Washington, DC, USA [50]). The quality of the dataset was manually inspected to ensure correct peak detection and integration. Peaks that failed manual evaluation had poor technical quality indicators such as <0.75 dot product (ratio of the transitions as compared to the human plasma spectral library (Human_plasma_2012-08_all.splib.zip, build by H. Lam (2012), available at the PeptideAtlas website: http://www. peptideatlas.org/speclib/, last accessed 20 June 2020) or $<3\times$ signal-to-noise ratio. They were excluded and reported as 'invalid' or 'below limit of detection' (<LOD), respectively. This resulted in 64 high quality peptides targets for 32 different complement proteins (Supplementary Table S1).

4.5. Statistical Analysis

Statistical analyses were performed using SPSS (version 16.0; IBM) and standard build-in packages in R (version 3.6.3). Pearson correlations were calculated for demographic factors with Hmisc package in R, using Pearson parameter in rcorr function. Rank-based inverse normal transformation was used for the peptide levels. Univariate association analyses with the complement proteins as outcomes were performed using linear regression models adjusted for sex, age, smoking, BMI, and triglyceride levels. A post hoc multiple testing correction was performed to control the false discovery rate (FDR) using the Benjamini–Hochberg procedure to take the high correlation between complement peptides into account (Supplementary Figure S1). Genotype and metabolite associations were performed by univariate linear regression models adjusted by age, sex, and AMD status, with complement proteins as outcomes. Genotype associations were performed to determine the AMD risk increasing alleles. Differences between different genotype groups were explored using the Mann–Whitney U test, and Kruskal–Wallis test where possible. The threshold for statistical significance was defined as an FDR-corrected p-value (P_{FDR}) of less than 0.05. The p-values that were lower than 0.05, but were higher than 0.05 for P_{FDR}, were defined as suggestive associations.

Supplementary Materials: The following are available online at https://www.mdpi.com/article/10.3390/jpm11121256/s1, Figure S1: Correlation heatmap of demographic factors and peptides from proteins, Figure S2: Factor H (FH) peptide levels show significant differences between genotype groups of AMD-associated variants at the CFH locus, Figure S3: Factor B (FB) peptide levels show significant differences between genotype groups of AMD-associated variants at the C2/CFB/SKIV2L locus, Figure S4: Factor I (FI) levels show significant differences between genotype groups of rs10033900 and p.Leu131Arg rare variant at the CFI locus, Table S1. Peptides of the complement system analysed in the multiplex selected reaction monitoring liquid chromatography–mass spectrometry assay, Table S2. Demographic factors, clinical characteristics and other measurements of the study cohort, Table S3. Pearson correlation analyses of peptides in the complement pathway with age, Table S4. Pearson correlation analyses of peptides in the complement pathway with sex, Table S5. Pearson correlation analyses of peptides in the complement pathway with smoking status, Table S6. Pearson correlation analyses of peptides in the complement pathway with BMI, Table S7. Association of peptides in the complement pathway with age-related macular degeneration, Table S8. List of the genetic variants included in this study, Table S9. Association of FHR-2 peptides with *CFH* variants known to be associated with age-related macular degeneration, Table S10. Association of FH peptides with *CFH* variants known to be associated with age-related macular degeneration, Table S11. Rare variants in complement genes in the study cohort and Mann–Whitney U test results with peptide levels between rare variants carriers and non-carriers, Table S12. Association of FB peptides with variants at the *C2/CFB/SKIV2L* locus, known to be associated with age-related macular degeneration, Table S13. Association of C2 peptides with variants at the *C2/CFB/SKIV2L* locus, known to be associated with age-related macular degeneration, Table S14. Association of C4 peptides with variants at the *C2/CFB/SKIV2L* locus, known to be associated with age-related macular degeneration, Table S15. Association of vitronectin peptides with variants at the *TMEM97/VTN* locus, known to be associated with age-related macular degeneration, Table S16. Association of FI peptide with variant at the *CFI* locus, known to be associated with age-related macular degeneration, Table S17. Association of C9 peptides with a rare variant at the *C9* locus, known to be associated with age-related macular degeneration, Table S18. Association of C3 peptides with variants at *C3* locus, known to be associated with age-related macular degeneration, Table S19. Association of peptides from the complement system with variants at other AMD loci, Table S20. Metabolites included in the Nightingale platform and their association with AMD, Table S21. Association of peptides in the complement pathway with AMD-associated metabolites.

Author Contributions: Conceptualization, I.E.A., E.W., A.J.v.G., E.K. (Elod Koertvely), E.N., J.G., M.I.d.J., L.L.-M. and A.I.d.H.; formal analysis, I.E.A., E.W., J.G., M.I.d.J. and L.L.-M.; funding acquisition, E.K. (Eveline Kerstenand), E.K. (Elod Koertvely), E.N. and A.I.d.H.; investigation, I.E.A., E.W., J.K.-G., E.K. (Else Kragtand); methodology, E.W., J.G. and M.I.d.J.; resources, E.K. (Eveline Kerstenand), B.B., C.B.H., S.F., Y.T.E.L. and A.I.d.H.; supervision, E.K. (Elod Koertvely), E.N., J.G., M.I.d.J. and A.I.d.H.; visualization, I.E.A. and E.W.; writing—original draft, I.E.A., E.W., J.G., M.I.d.J., L.L.-M. and A.I.d.H.; writing—review and editing, I.E.A., E.W., E.K. (Eveline Kerstenand), J.K.-G., B.B., T.E.G., C.B.H., S.F., A.J.v.G., Y.T.E.L., E.K. (Elod Koertvely), E.K. (Else Kragtand), E.N., J.G., M.I.d.J., L.L.-M. and A.I.d.H. All authors have read and agreed to the published version of the manuscript.

Funding: This study was funded by the Roche Postdoctoral Fellowship program (RPF-ID 480 to Anneke I. den Hollander, Elod Kortvely, Everson Nogoceke), by the European Research Council under the European Union's Seventh Framework Programme (FP/2007-2013) (ERC Grant Agreement no. 310644 MACULA to Anneke I. den Hollander), and by the Bayer Ophthalmology Research Award (BORA, to Eveline Kersten).

Institutional Review Board Statement: The study was conducted according to the guidelines of the Declaration of Helsinki and approved by the Institutional Review Board Commissie Mensgebonden Onderzoek [CMO] Arnhem-Nijmegen for EUGENDA-Nijmegen and Ethics Commission of Cologne University's Faculty of Medicine for EUGENDA-Cologne (cod 2007/158 approved on 31 January 2008).

Informed Consent Statement: Informed consent was obtained from all subjects involved in the study.

Data Availability Statement: The complement peptide analysis Skyline and raw data files are available at the Panorama public repository: https://panoramaweb.org/CS_MRM_AMD.url (ProteomeXchange ID: PDX027689) (last accessed 20 June 2020).

Acknowledgments: We would like to thank all the patients and the healthy donors for donating their blood for these studies.

Conflicts of Interest: Sascha Fauser, Elod Koertvely and Everson Nogoceke are employees of F. Hoffmann-La Roche. I. Erkin Acar and Laura Lorés-Motta were employed with fellowships from F. Hoffmann-La Roche for part of the work. Anneke I. den Hollander is a consultant for Ionis Pharmaceuticals, Gyroscope Therapeutics, Gemini Therapeutics and F. Hoffmann-La Roche.

References

1. Wong, W.L.; Su, X.; Li, X.; Cheung, C.M.; Klein, R.; Cheng, C.Y.; Wong, T.Y. Global prevalence of age-related macular degeneration and disease burden projection for 2020 and 2040: A systematic review and meta-analysis. *Lancet Glob. Health* **2014**, *2*, e106–e116. [CrossRef]
2. Rudnicka, A.R.; Jarrar, Z.; Wormald, R.; Cook, D.G.; Fletcher, A.; Owen, C.G. Age and gender variations in age-related macular degeneration prevalence in populations of European ancestry: A meta-analysis. *Ophthalmology* **2012**, *119*, 571–580. [CrossRef] [PubMed]
3. Coleman, A.L.; Seitzman, R.L.; Cummings, S.R.; Yu, F.; Cauley, J.A.; Ensrud, K.E.; Stone, K.L.; Hochberg, M.C.; Pedula, K.L.; Thomas, E.L.; et al. The association of smoking and alcohol use with age-related macular degeneration in the oldest old: The Study of Osteoporotic Fractures. *Am. J. Ophthalmol.* **2010**, *149*, 160–169. [CrossRef] [PubMed]
4. Zhang, Q.Y.; Tie, L.J.; Wu, S.S.; Lv, P.L.; Huang, H.W.; Wang, W.Q.; Wang, H.; Ma, L. Overweight, Obesity, and Risk of Age-Related Macular Degeneration. *Investig. Ophthalmol. Vis. Sci.* **2016**, *57*, 1276–1283. [CrossRef] [PubMed]
5. Smith, W.; Assink, J.; Klein, R.; Mitchell, P.; Klaver, C.C.; Klein, B.E.; Hofman, A.; Jensen, S.; Wang, J.J.; de Jong, P.T. Risk factors for age-related macular degeneration: Pooled findings from three continents. *Ophthalmology* **2001**, *108*, 697–704. [CrossRef]
6. Fritsche, L.G.; Igl, W.; Bailey, J.N.; Grassmann, F.; Sengupta, S.; Bragg-Gresham, J.L.; Burdon, K.P.; Hebbring, S.J.; Wen, C.; Gorski, M.; et al. A large genome-wide association study of age-related macular degeneration highlights contributions of rare and common variants. *Nat. Genet.* **2016**, *48*, 134–143. [CrossRef] [PubMed]
7. Edwards, A.O.; Ritter, R., 3rd; Abel, K.J.; Manning, A.; Panhuysen, C.; Farrer, L.A. Complement factor H polymorphism and age-related macular degeneration. *Science* **2005**, *308*, 421–424. [CrossRef]
8. Hageman, G.S.; Anderson, D.H.; Johnson, L.V.; Hancox, L.S.; Taiber, A.J.; Hardisty, L.I.; Hageman, J.L.; Stockman, H.A.; Borchardt, J.D.; Gehrs, K.M.; et al. A common haplotype in the complement regulatory gene factor H (HF1/CFH) predisposes individuals to age-related macular degeneration. *Proc. Natl. Acad. Sci. USA* **2005**, *102*, 7227–7232. [CrossRef]
9. Haines, J.L.; Hauser, M.A.; Schmidt, S.; Scott, W.K.; Olson, L.M.; Gallins, P.; Spencer, K.L.; Kwan, S.Y.; Noureddine, M.; Gilbert, J.R.; et al. Complement factor H variant increases the risk of age-related macular degeneration. *Science* **2005**, *308*, 419–421. [CrossRef]
10. Klein, R.J.; Zeiss, C.; Chew, E.Y.; Tsai, J.Y.; Sackler, R.S.; Haynes, C.; Henning, A.K.; SanGiovanni, J.P.; Mane, S.M.; Mayne, S.T.; et al. Complement factor H polymorphism in age-related macular degeneration. *Science* **2005**, *308*, 385–389. [CrossRef]
11. Dewan, A.; Liu, M.; Hartman, S.; Zhang, S.S.; Liu, D.T.; Zhao, C.; Tam, P.O.; Chan, W.M.; Lam, D.S.; Snyder, M.; et al. HTRA1 promoter polymorphism in wet age-related macular degeneration. *Science* **2006**, *314*, 989–992. [CrossRef]
12. Rivera, A.; Fisher, S.A.; Fritsche, L.G.; Keilhauer, C.N.; Lichtner, P.; Meitinger, T.; Weber, B.H. Hypothetical LOC387715 is a second major susceptibility gene for age-related macular degeneration, contributing independently of complement factor H to disease risk. *Hum. Mol. Genet.* **2005**, *14*, 3227–3236. [CrossRef]
13. Yang, Z.; Camp, N.J.; Sun, H.; Tong, Z.; Gibbs, D.; Cameron, D.J.; Chen, H.; Zhao, Y.; Pearson, E.; Li, X.; et al. A variant of the HTRA1 gene increases susceptibility to age-related macular degeneration. *Science* **2006**, *314*, 992–993. [CrossRef]
14. Gold, B.; Merriam, J.E.; Zernant, J.; Hancox, L.S.; Taiber, A.J.; Gehrs, K.; Cramer, K.; Neel, J.; Bergeron, J.; Barile, G.R.; et al. Variation in factor B (BF) and complement component 2 (C2) genes is associated with age-related macular degeneration. *Nat. Genet.* **2006**, *38*, 458–462. [CrossRef]
15. Yates, J.R.; Sepp, T.; Matharu, B.K.; Khan, J.C.; Thurlby, D.A.; Shahid, H.; Clayton, D.G.; Hayward, C.; Morgan, J.; Wright, A.F.; et al. Complement C3 variant and the risk of age-related macular degeneration. *N. Engl. J. Med.* **2007**, *357*, 553–561. [CrossRef]
16. Seddon, J.M.; Yu, Y.; Miller, E.C.; Reynolds, R.; Tan, P.L.; Gowrisankar, S.; Goldstein, J.I.; Triebwasser, M.; Anderson, H.E.; Zerbib, J.; et al. Rare variants in CFI, C3 and C9 are associated with high risk of advanced age-related macular degeneration. *Nat. Genet.* **2013**, *45*, 1366–1370. [CrossRef]
17. Triebwasser, M.P.; Roberson, E.D.; Yu, Y.; Schramm, E.C.; Wagner, E.K.; Raychaudhuri, S.; Seddon, J.M.; Atkinson, J.P. Rare Variants in the Functional Domains of Complement Factor H Are Associated With Age-Related Macular Degeneration. *Investig. Ophthalmol. Vis. Sci.* **2015**, *56*, 6873–6878. [CrossRef]
18. Geerlings, M.J.; de Jong, E.K.; den Hollander, A.I. The complement system in age-related macular degeneration: A review of rare genetic variants and implications for personalized treatment. *Mol. Immunol.* **2017**, *84*, 65–76. [CrossRef]
19. de Breuk, A.; Acar, I.E.; Kersten, E.; Schijvenaars, M.; Colijn, J.M.; Haer-Wigman, L.; Bakker, B.; de Jong, S.; Meester-Smoor, M.A.; Verzijden, T.; et al. Development of a Genotype Assay for Age-Related Macular Degeneration: The EYE-RISK Consortium. *Ophthalmology* **2021**, *128*, 1604–1617. [CrossRef]
20. Van der Schaft, T.L.; Mooy, C.M.; de Bruijn, W.C.; de Jong, P.T. Early stages of age-related macular degeneration: An immunofluorescence and electron microscopy study. *Br. J. Ophthalmol.* **1993**, *77*, 657–661. [CrossRef]
21. Johnson, L.V.; Ozaki, S.; Staples, M.K.; Erickson, P.A.; Anderson, D.H. A potential role for immune complex pathogenesis in drusen formation. *Exp. Eye Res.* **2000**, *70*, 441–449. [CrossRef] [PubMed]
22. Mullins, R.F.; Russell, S.R.; Anderson, D.H.; Hageman, G.S. Drusen associated with aging and age-related macular degeneration contain proteins common to extracellular deposits associated with atherosclerosis, elastosis, amyloidosis, and dense deposit disease. *FASEB J.* **2000**, *14*, 835–846. [CrossRef] [PubMed]

23. Scholl, H.P.; Charbel Issa, P.; Walier, M.; Janzer, S.; Pollok-Kopp, B.; Borncke, F.; Fritsche, L.G.; Chong, N.V.; Fimmers, R.; Wienker, T.; et al. Systemic complement activation in age-related macular degeneration. *PLoS ONE* **2008**, *3*, e2593. [CrossRef]

24. Reynolds, R.; Hartnett, M.E.; Atkinson, J.P.; Giclas, P.C.; Rosner, B.; Seddon, J.M. Plasma complement components and activation fragments: Associations with age-related macular degeneration genotypes and phenotypes. *Investig. Ophthalmol. Vis. Sci.* **2009**, *50*, 5818–5827. [CrossRef]

25. Smailhodzic, D.; Klaver, C.C.; Klevering, B.J.; Boon, C.J.; Groenewoud, J.M.; Kirchhof, B.; Daha, M.R.; den Hollander, A.I.; Hoyng, C.B. Risk alleles in CFH and ARMS2 are independently associated with systemic complement activation in age-related macular degeneration. *Ophthalmology* **2012**, *119*, 339–346. [CrossRef]

26. Hecker, L.A.; Edwards, A.O.; Ryu, E.; Tosakulwong, N.; Baratz, K.H.; Brown, W.L.; Charbel Issa, P.; Scholl, H.P.; Pollok-Kopp, B.; Schmid-Kubista, K.E.; et al. Genetic control of the alternative pathway of complement in humans and age-related macular degeneration. *Hum. Mol. Genet.* **2010**, *19*, 209–215. [CrossRef]

27. Van de Ven, J.P.; Nilsson, S.C.; Tan, P.L.; Buitendijk, G.H.; Ristau, T.; Mohlin, F.C.; Nabuurs, S.B.; Schoenmaker-Koller, F.E.; Smailhodzic, D.; Campochiaro, P.A.; et al. A functional variant in the CFI gene confers a high risk of age-related macular degeneration. *Nat. Genet.* **2013**, *45*, 813–817. [CrossRef]

28. Geerlings, M.J.; Kremlitzka, M.; Bakker, B.; Nilsson, S.C.; Saksens, N.T.; Lechanteur, Y.T.; Pauper, M.; Corominas, J.; Fauser, S.; Hoyng, C.B.; et al. The Functional Effect of Rare Variants in Complement Genes on C3b Degradation in Patients With Age-Related Macular Degeneration. *JAMA Ophthalmol.* **2017**, *135*, 39–46. [CrossRef]

29. Mc Mahon, O.; Hallam, T.M.; Patel, S.; Harris, C.L.; Menny, A.; Zelek, W.M.; Widjajahakim, R.; Java, A.; Cox, T.; Tzoumas, N.; et al. The Rare C9 P167S Risk Variant for Age-related Macular Degeneration Increases Polymerization of the Terminal Component of the Complement Cascade. *Hum. Mol. Genet.* **2021**, *30*, 1188–1199. [CrossRef]

30. Cipriani, V.; Lores-Motta, L.; He, F.; Fathalla, D.; Tilakaratna, V.; McHarg, S.; Bayatti, N.; Acar, I.E.; Hoyng, C.B.; Fauser, S.; et al. Increased circulating levels of Factor H-Related Protein 4 are strongly associated with age-related macular degeneration. *Nat. Commun.* **2020**, *11*, 778. [CrossRef]

31. Lores-Motta, L.; van Beek, A.E.; Willems, E.; Zandstra, J.; van Mierlo, G.; Einhaus, A.; Mary, J.L.; Stucki, C.; Bakker, B.; Hoyng, C.B.; et al. Common haplotypes at the CFH locus and low-frequency variants in CFHR2 and CFHR5 associate with systemic FHR concentrations and age-related macular degeneration. *Am. J. Hum. Genet.* **2021**, *108*, 1367–1384. [CrossRef]

32. Cipriani, V.; Tierney, A.; Griffiths, J.R.; Zuber, V.; Sergouniotis, P.I.; Yates, J.R.W.; Moore, A.T.; Bishop, P.N.; Clark, S.J.; Unwin, R.D. Beyond factor H: The impact of genetic-risk variants for age-related macular degeneration on circulating factor-H-like 1 and factor-H-related protein concentrations. *Am. J. Hum. Genet.* **2021**, *108*, 1385–1400. [CrossRef]

33. Acar, I.E.; Lores-Motta, L.; Colijn, J.M.; Meester-Smoor, M.A.; Verzijden, T.; Cougnard-Gregoire, A.; Ajana, S.; Merle, B.M.J.; de Breuk, A.; Heesterbeek, T.J.; et al. Integrating Metabolomics, Genomics, and Disease Pathways in Age-Related Macular Degeneration: The EYE-RISK Consortium. *Ophthalmology* **2020**, *127*, 1693–1709. [CrossRef]

34. Willems, E.; Alkema, W.; Keizer-Garritsen, J.; Suppers, A.; van der Flier, M.; Philipsen, R.; van den Heuvel, L.P.; Volokhina, E.; van der Molen, R.G.; Herberg, J.A.; et al. Biosynthetic homeostasis and resilience of the complement system in health and infectious disease. *EBioMedicine* **2019**, *45*, 303–313. [CrossRef]

35. Lores-Motta, L.; Paun, C.C.; Corominas, J.; Pauper, M.; Geerlings, M.J.; Altay, L.; Schick, T.; Daha, M.R.; Fauser, S.; Hoyng, C.B.; et al. Genome-Wide Association Study Reveals Variants in CFH and CFHR4 Associated with Systemic Complement Activation: Implications in Age-Related Macular Degeneration. *Ophthalmology* **2018**, *125*, 1064–1074. [CrossRef]

36. Gaya da Costa, M.; Poppelaars, F.; van Kooten, C.; Mollnes, T.E.; Tedesco, F.; Wurzner, R.; Trouw, L.A.; Truedsson, L.; Daha, M.R.; Roos, A.; et al. Age and Sex-Associated Changes of Complement Activity and Complement Levels in a Healthy Caucasian Population. *Front. Immunol.* **2018**, *9*, 2664. [CrossRef]

37. Moreno-Navarrete, J.M.; Martinez-Barricarte, R.; Catalan, V.; Sabater, M.; Gomez-Ambrosi, J.; Ortega, F.J.; Ricart, W.; Bluher, M.; Fruhbeck, G.; Rodriguez de Cordoba, S.; et al. Complement factor H is expressed in adipose tissue in association with insulin resistance. *Diabetes* **2010**, *59*, 200–209. [CrossRef]

38. Higgins, V.; Omidi, A.; Tahmasebi, H.; Asgari, S.; Gordanifar, K.; Nieuwesteeg, M.; Adeli, K. Marked Influence of Adiposity on Laboratory Biomarkers in a Healthy Cohort of Children and Adolescents. *J. Clin. Endocrinol. Metab.* **2020**, *105*, e1781–e1797. [CrossRef]

39. Kersten, E.; Paun, C.C.; Schellevis, R.L.; Hoyng, C.B.; Delcourt, C.; Lengyel, I.; Peto, T.; Ueffing, M.; Klaver, C.C.W.; Dammeier, S.; et al. Systemic and ocular fluid compounds as potential biomarkers in age-related macular degeneration. *Surv. Ophthalmol.* **2018**, *63*, 9–39. [CrossRef]

40. Fanelli, G.; Gonzalez-Cordero, A.; Gardner, P.J.; Peng, Q.; Fernando, M.; Kloc, M.; Farrar, C.A.; Naeem, A.; Garred, P.; Ali, R.R.; et al. Human stem cell-derived retinal epithelial cells activate complement via collectin 11 in response to stress. *Sci. Rep.* **2017**, *7*, 14625. [CrossRef]

41. Grassmann, F.; Cantsilieris, S.; Schulz-Kuhnt, A.S.; White, S.J.; Richardson, A.J.; Hewitt, A.W.; Vote, B.J.; Schmied, D.; Guymer, R.H.; Weber, B.H.; et al. Multiallelic copy number variation in the complement component 4A (C4A) gene is associated with late-stage age-related macular degeneration (AMD). *J. Neuroinflamm.* **2016**, *13*, 81. [CrossRef] [PubMed]

42. Biasella, F.; PLoSsl, K.; Karl, C.; Weber, B.H.F.; Friedrich, U. Altered Protein Function Caused by AMD-associated Variant rs704 Links Vitronectin to Disease Pathology. *Investig. Ophthalmol. Vis. Sci.* **2020**, *61*, 2. [CrossRef] [PubMed]

43. Orozco, L.D.; Chen, H.H.; Cox, C.; Katschke, K.J., Jr.; Arceo, R.; Espiritu, C.; Caplazi, P.; Nghiem, S.S.; Chen, Y.J.; Modrusan, Z.; et al. Integration of eQTL and a Single-Cell Atlas in the Human Eye Identifies Causal Genes for Age-Related Macular Degeneration. *Cell Rep.* **2020**, *30*, 1246–1259. [CrossRef] [PubMed]

44. De Jong, S.; Volokhina, E.B.; de Breuk, A.; Nilsson, S.C.; de Jong, E.K.; van der Kar, N.; Bakker, B.; Hoyng, C.B.; van den Heuvel, L.P.; Blom, A.M.; et al. Effect of rare coding variants in the CFI gene on Factor I expression levels. *Hum. Mol. Genet.* **2020**, *29*, 2313–2324. [CrossRef]

45. Willems, E.; Lores-Motta, L.; Zanichelli, A.; Suffritti, C.; van der Flier, M.; van der Molen, R.G.; Langereis, J.D.; van Drongelen, J.; van den Heuvel, L.P.; Volokhina, E.; et al. Quantitative multiplex profiling of the complement system to diagnose complement-mediated diseases. *Clin. Transl. Immunol.* **2020**, *9*, e1225. [CrossRef]

46. Colijn, J.M.; den Hollander, A.I.; Demirkan, A.; Cougnard-Gregoire, A.; Verzijden, T.; Kersten, E.; Meester-Smoor, M.A.; Merle, B.M.J.; Papageorgiou, G.; Ahmad, S.; et al. Increased High-Density Lipoprotein Levels Associated with Age-Related Macular Degeneration: Evidence from the EYE-RISK and European Eye Epidemiology Consortia. *Ophthalmology* **2019**, *126*, 393–406. [CrossRef]

47. De Jong, S.; Gagliardi, G.; Garanto, A.; de Breuk, A.; Lechanteur, Y.T.E.; Katti, S.; van den Heuvel, L.P.; Volokhina, E.B.; den Hollander, A.I. Implications of genetic variation in the complement system in age-related macular degeneration. *Prog. Retin. Eye Res.* **2021**, *84*, 100952. [CrossRef]

48. Kumar-Singh, R. The role of complement membrane attack complex in dry and wet AMD - From hypothesis to clinical trials. *Exp. Eye Res.* **2019**, *184*, 266–277. [CrossRef]

49. Fauser, S.; Smailhodzic, D.; Caramoy, A.; van de Ven, J.P.; Kirchhof, B.; Hoyng, C.B.; Klevering, B.J.; Liakopoulos, S.; den Hollander, A.I. Evaluation of serum lipid concentrations and genetic variants at high-density lipoprotein metabolism loci and TIMP3 in age-related macular degeneration. *Investig. Ophthalmol. Vis. Sci.* **2011**, *52*, 5525–5528. [CrossRef]

50. MacLean, B.; Tomazela, D.M.; Shulman, N.; Chambers, M.; Finney, G.L.; Frewen, B.; Kern, R.; Tabb, D.L.; Liebler, D.C.; MacCoss, M.J. Skyline: An open source document editor for creating and analyzing targeted proteomics experiments. *Bioinformatics* **2010**, *26*, 966–968. [CrossRef]

Journal of
Personalized
Medicine

Article

Specific Autoantibodies in Neovascular Age-Related Macular Degeneration: Evaluation of Morphological and Functional Progression over Five Years

Michelle Prasuhn [1,2,]*, Caroline Hillers [1,2], Felix Rommel [1,2], Gabriela Riemekasten [3], Harald Heidecke [4], Khaled Nassar [1], Mahdy Ranjbar [1,2], Salvatore Grisanti [1] and Aysegül Tura [1]

[1] Department of Ophthalmology, University Hospital Schleswig-Holstein, University of Lübeck, 23562 Lübeck, Germany; caroline.hillers@student.uni-luebeck.de (C.H.); felix.rommel@uksh.de (F.R.); khaled.nassar@uksh.de (K.N.); eye.research101@gmail.com (M.R.); Salvatore.Grisanti@uksh.de (S.G.); Ayseguel.Tura@uksh.de (A.T.)

[2] Laboratory for Angiogenesis & Ocular Cell Transplantation, 23562 Lübeck, Germany

[3] Clinic of Rheumatology and Clinical Immunology, University Hospital Schleswig-Holstein, University of Lübeck, 23562 Lübeck, Germany; Gabriela.Riemekasten@uksh.de

[4] CellTrend GmbH, 14943 Luckenwalde, Germany; heidecke@celltrend.de

* Correspondence: michelle.prasuhn@uksh.de

Citation: Prasuhn, M.; Hillers, C.; Rommel, F.; Riemekasten, G.; Heidecke, H.; Nassar, K.; Ranjbar, M.; Grisanti, S.; Tura, A. Specific Autoantibodies in Neovascular Age-Related Macular Degeneration: Evaluation of Morphological and Functional Progression over Five Years. *J. Pers. Med.* **2021**, *11*, 1207. https://doi.org/10.3390/jpm11111207

Academic Editors: Peter D. Westenskow and Andreas Ebneter

Received: 22 September 2021
Accepted: 14 November 2021
Published: 16 November 2021

Abstract: (1) Background: Altered levels of autoantibodies (aab) and their networks have been identified as biomarkers for various diseases. Neovascular age-related macular degeneration (nAMD) is a leading cause for central vision loss worldwide with highly variable inter- and intraindividual disease courses. Certain aab networks could help in daily routine to identify patients with a high disease activity who need to be visited and treated more regularly. (2) Methods: We analyzed levels of aab against Angiotensin II receptor type 1 (AT1-receptor), Protease-activated receptors (PAR1), vascular endothelial growth factor (VEGF) -A, VEGF-B, and VEGF-receptor 2 in sera of 164 nAMD patients. In a follow-up period of five years, we evaluated changes in functional and morphological characteristics. Using correlation analyses, multiple regression models, and receiver operator characteristics, we assessed whether the five aab have a clinical significance as biomarkers that correspond to the clinical properties. (3) Results: Neither the analyzed aab individually nor taken together as a network showed statistically significant results that would allow us to draw conclusions on the clinical five-year course in nAMD patients. (4) Conclusions: The five aab that we analyzed do not correspond to the clinical five-year course of nAMD patients. However, larger, prospective studies should reevaluate different and more aab to gain deeper insights.

Keywords: autoantibodies; age-related macular degeneration; biomarkers; AT1-receptor; PAR1; VEGF-A; VEGF-B; VEGF-receptor 2

1. Introduction

Specific autoantibodies (aab) are known to be associated with autoimmune diseases. However, several studies showed elevated levels of aab also in healthy donors who never develop inflammatory disorders [1,2]. Their exact clinical roles are still unclear but are a fascinating approach for new diagnostic and therapeutic options.

Antibody (ab) levels can be elevated or reduced in patients with inflammatory disease and healthy donors, which strengthens the idea of physiological levels and a balanced generation of aab in human physiology and pathophysiology. Riemekasten et al. introduced the term "antibodiom" to understand networks of ab levels and their interactions. Antibody levels as well as their correlations can serve as biomarkers for diseases [3]. This study is a first approach to characterize antibodies in patients with neovascular age-related macular degeneration (nAMD), in which the pathogenesis is at least partially mediated by immunological factors, including a possible autoimmune response [4].

Age-related macular degeneration (AMD) is a major cause of central vision loss worldwide and becomes even more common due to the demographic change of developed countries. Thus, along with severe individual impairment, it also causes a high economic burden to society. In nAMD, patients regularly receive intravitreal injections (IVIs) with anti-vascular endothelial growth factor therapy (anti-VEGF). Even though adverse events are rare, endophthalmitis and severe vision loss or even blindness can occur. Additionally, the regular visit rate is a high burden for elderly patients not only financially and socially but also psychologically. Besides the discomfort during the injection, especially anxiety before the treatment and fear of losing eyesight as a complication have a high impact on the patients [5]. A wide variability concerning interindividual disease progression causes several patients to be visited too frequently or not frequently enough and therefore to be over- or undertreated.

Consequently, there is a high need for markers that allow us to identify patients who need to be followed up more regularly or who can gain more individual freedom by expanding clinical visit intervals. In this study, we investigated whether certain aab are indicative of the disease current and whether they can help us differentiate between those patients.

Therefore, for this study, we analyzed aab against Angiotensin II receptor type 1 (AT1-receptor), Protease-activated receptors (PAR1), VEGF-A, VEGF-B, and VEGF-receptor 2. These factors are known to have effects on angiogenesis, which is a main pathomechanism in nAMD. AT1-receptor is a G protein-coupled receptor (GPCR) that is widely expressed in the human body. It is activated by the binding of Angiotensin II and regulates important processes, such as blood flow, sodium retention, and aldosterone secretion [6]. AT1-receptor could also promote tumor angiogenesis by inducing the VEGF expression [7]. Furthermore, AT1-receptor ab has been shown to have detrimental effects in several diseases and cause acute or chronic rejection and graft loss [8]. PAR1 is also part of the GPCR superfamily and participates in vascular development mediating the angiogenetic activity of thrombin and promoting the VEGF expression [9,10]. Vascular endothelial growth factor in turn plays a key role in nAMD pathogenesis and is the target of regular intravitreal injections. VEGF-A shows prominent activity with vascular endothelial cells, primarily through its interactions with the VEGFR1 and -R2 receptors. The latter appears to mediate almost all of the known cellular responses to VEGF. VEGF-B seems to play a role only in the maintenance of newly formed blood vessels during pathological conditions [11].

2. Materials and Methods

Study design and participants: Patients with active nAMD in one or both eyes were identified, and written informed consent was obtained. We excluded patients with any chronic systemic inflammatory disease. Patients with diabetic retinopathy, glaucoma, inflammatory ocular diseases, and other disorders of the vitreoretinal interface were also not considered. Blood was taken at one of the regular visits at our ophthalmological clinic. The patients were followed up monthly as part of their routine examinations following the pro re nata treatment regimen for anti-VEGF-IVIs. This study was conducted in accordance with the 1964 Declaration of Helsinki, with all participants providing written informed consent. Approval by the ethics committee of the University of Lübeck, Germany (vote reference number: 16–199) was given.

Outcome measures: The primary endpoint was the number of intravitreal injections administered over the course of one and five years. For this purpose, we reviewed the medical records of patients whose ab levels were analyzed between 2011 and 2014. We differentiated between good and poor responders; the cut-off was 6 IVIs per year or 30 IVIs over five years. Investigated secondary outcome parameters were changes in best corrected visual acuity (BCVA) and optical coherence tomography (OCT) morphology. Correct automated measurements were reviewed by at least two experienced raters and manually corrected if needed. Central retinal thickness (CRT) was automatically calculated by the OCT device using the central circle of the Early Treatment Diabetic Retinopathy

Study (ETDRS) grid (Heidelberg Eye Explorer, Version 1.9.10; Heidelberg Engineering, Heidelberg, Germany). Two raters reviewed OCT images to evaluate whether intraretinal or subretinal fluid (IRF; SRF), fibrosis, or macular bleeding was present at baseline and in the five-year period.

Blood preparation: Blood was drawn into serum tubes and centrifuged at $2000 \times g$ for ten minutes at room temperature. The methods to measure the aab have been previously described in detail [12]. Briefly, individual serum aab were assessed using commercially available solid-phase sandwich ELISA Kits according to the manufacturer's instructions (all CellTrend GmbH, Luckenwalde, Germany). The aab concentrations were calculated as arbitrary units (U) by extrapolation from a standard curve of five standards ranging from 2.5 to 40 U/mL. The ELISAs were validated according to the Food and Drug Administration's Guidance for Industry: Bioanalytical Method Validation. We analyzed antibodies against AT1-receptor, Protease-activated PAR1, VEGF-A, VEGF-B, and VEGF-receptor 2.

Statistical analysis: In patients in which both eyes had active nAMD, the study eye was assigned by chance. Snellen VA was converted to logarithm of the minimum angle of resolution (logMAR) for statistical analysis. Data were analyzed using IBM SPSS (Version 24.0) and GraphPad Prism (Version 9.0). Testing for normality was done via Shapiro–Wilk test for correlation analyses and via QQ-Plots for the multiple linear regression models. Correlation analyses were carried out with Pearson's tests and corrected for multiple testing by computing adjusted p-values (false-discovery rate). We carried out multiple linear and multiple logistic regressions as implemented on SPSS. Antibody levels were reflected by creating a single variable using a principal component analysis. In addition, we performed a Principal Component Analysis (PCA) to reduce the dimensionality of our independent variables. We included five independent variables (the ab mentioned above) in a correlation matrix and subsequent factor analysis (based on eigenvalues larger than one; necessary assumptions tested by KMO and Bartlett tests; Rotation: Varimax; Kaiser normalization). Our PCA resulted in two significant factor scores (based on regression analyses). Here, we used the factor score that reflected the variance observed in our data to the highest possible degree (43.3%, VEGF_R2 ab excluded). The new variable was termed "ab score". ROC plots were plotted via SPSS, and the area under the curve (AUC) was automatically calculated. For all tests, values of $p < 0.05$ were considered statistically significant.

3. Results

A total of 164 eyes of 164 patients with nAMD was included. Unilateral nAMD was present in 90 patients, and the remaining 74 eyes of patients with bilateral disease were assigned by chance for further analysis. A complete five-year follow-up was achieved in 59 patients. Demographic and clinical data are summarized in Table 1. Mean age of our cohort was 78.32 years, and more women than men were included. Our patients had a mean BCVA of 0.34 logMar and CRT was 346.01 µm.

Table 1. Epidemiological and clinical baseline data of included patients. BCVA, best-corrected visual acuity; CRT, central retinal thickness; f, female; m, male; OD, right eye; OS, left eye; OU, both eyes; SD, standard deviation.

	$n = 164$
Age (years), mean $\pm$ SD	78.32 $\pm$ 8.17
gender (m/f)	63 (38.4%)/101 (61.6%)
Laterality (OD/OS/OU)	41 (25.0%), 49 (29.9%), 74(45.1%)
Study eye (OD/OS)	84 (51.2%)/80 (49.8%)
Baseline BCVA (logMar)	0.34 $\pm$ 0.31
Baseline CRT (µm)	346.01 $\pm$ 114.81

Antibody levels were analyzed from blood samples at baseline in nAMD patients that met the inclusion and exclusion criteria and are listed in Table 2. The injection rate, BCVA, and CRT over the five years are displayed in Figure 1. Mean BCVA decreased after five years from 0.34 ± 0.31 to 0.69 ± 0.5. The mean annual injection rate remained relatively steady from year one (5.15 ± 2.91) to year five (4.48 ± 3.14). The CRT decreased from 346 ± 115 to 299 ± 103 μm. After adjusting for multiple testing, Pearson's correlation analyses revealed no statistically significant values concerning correlation of aab with any of the clinical values. A multiple linear regression analysis was carried out to investigate whether the ab could significantly predict clinical outcome measures after five years. The results of the regression are summarized in Table 3. A multiple logistic regression analysis was carried out to analyze whether the ab predict the dichotomous clinical outcomes over five years (Table 4). Both regression models showed that neither the single aab nor taken together have a predictive value for the clinical course in the five-year time frame. We carried out ROC analyses to illustrate the diagnostic ability of the ab and ab score. The AUC values are listed in Table 5 and show low values concerning sensitivity and specificity for the evaluated aab and the clinical parameters.

Table 2. Antibody levels in nAMD patients. ab, antibody; AT1, Angiotensin II receptor type 1; PAR1, Protease-activated receptors; VEGF, vascular endothelial growth factor.

	Antibody Level (Units/mL) $n = 164$	Standard Deviation
AT1-receptor ab	8.531	10.35
PAR1 ab	3.398	7.79
VEGF-A ab	9.262	13.96
VEGF-B ab	5.998	15.13
VEGF-receptor 2 ab	6.020	9.73

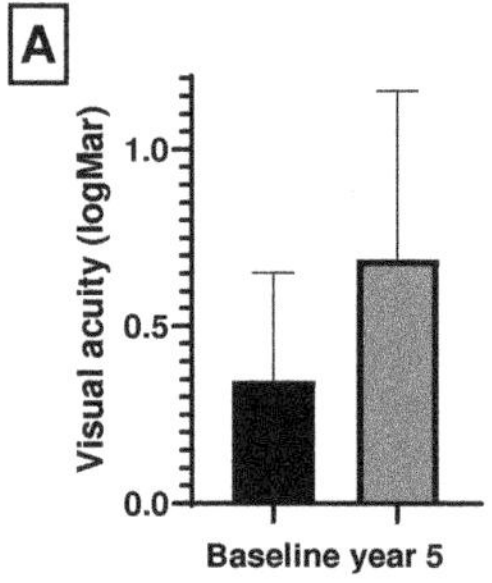
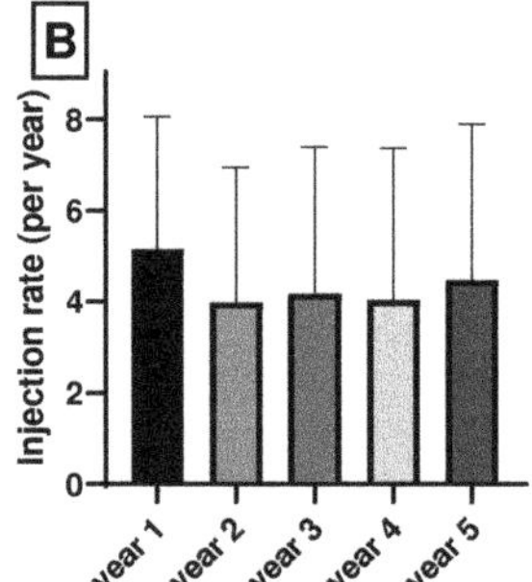
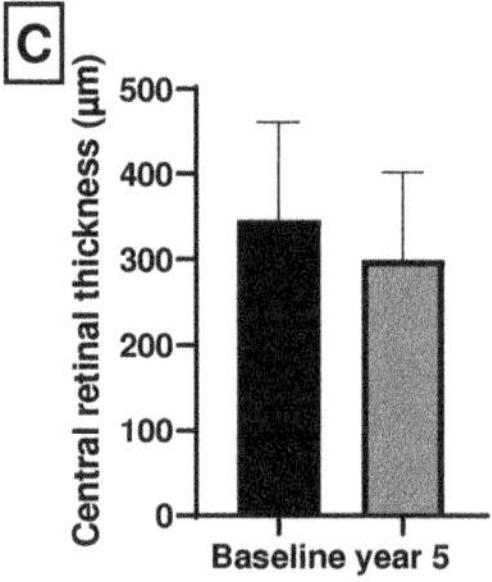

Figure 1. (**A**) Best corrected visual acuity (BCVA) in logMar at baseline and after five years. (**B**) Number of intravitreal injections per year. (**C**) Change in central retinal thickness.

Table 3. Multiple linear regression model. Clinical outcomes were used as dependent variables; the analyzed antibodies served as independent variables. BCVA, best-corrected visual acuity; CRT, central retinal thickness; IVI, intravitreal injections; SD, standard deviation.

	Mean $\pm$ SD	F	R^2	P
Number of IVIs year 1	5.15 $\pm$ 2.91	(5; 109) = 0.671	0.30	0.646
Number of IVIs year 2	3.98 $\pm$ 2.97	(5; 88) = 0.473	0.26	0.795
Number of IVIs year 3	4.16 $\pm$ 3.23	(5; 67) = 0.384	0.28	0.858
Number of IVIs year 4	4.03 $\pm$ 3.33	(5; 59) = 0.436	0.77	0.436
Number of IVIs year 5	4.48 $\pm$ 3.41	(5; 52) = 0.402	0.04	0.845
Number of IVIs total	22.42 $\pm$ 12.21	(5; 53) = 0.611	0.05	0.692
CRT change	−39.06 $\pm$ 128.686	(5; 44) = 0.686	0.72	0.636
BCVA change	0.345 $\pm$ 0.416	(5; 53) = 0.637	0.06	0.672

Table 4. Multiple logistic regression model. Clinical outcomes were used as dependent variables; the analyzed antibodies served as independent variables. BCVA, best-corrected visual acuity; CRT, central retinal thickness; IRF, intraretinal fluid; IVI, intravitreal injections; SD, standard deviation; SRF, subretinal fluid.

	Test Statistics	Nagelkerke's R^2	P
SRF development	χ^2 (5) = 1.409	0.043	0.923
IRF development	χ^2 (5) = 10.268	0.234	0.068
Fibrosis development	χ^2 (5) = 4.952	0.132	0.422
Macular bleeding	χ^2 (5) = 6.739	0.116	0.241
IVI response 1 year	χ^2 (5) = 3.902	0.045	0.564
IVI response 5 years	χ^2 (5) = 8.320	0.186	0.139

Table 5. Receiver operator characteristic. Values are given as area under the curve analysis. ab, antibody; AT1, Angiotensin II receptor 1; IRF, intraretinal fluid; IVI, intravitreal injection; PAR1, Protease-activated receptors; SRF, subretinal fluid; VEGF, vascular endothelial growth factor.

	AT1-Receptor ab	PAR1 ab	VEGF-A ab	VEGF-B ab	VEGF-Receptor 2 ab	Antibody Score
IVI response 1 year	0.443	0.522	0.416	0.418	0.435	0.430
IVI response 5 years	0.697	0.625	0.650	0.638	0.539	0.696
SRF development	0.482	0.433	0.460	0.506	0.518	0.550
IRF development	0.303	0.479	0.304	0.493	0.399	0.707
Fibrosis development	0.509	0.418	0.493	0.467	0.463	0.505
Macular bleeding	0.535	0.489	0.580	0.493	0.635	0.465

4. Discussion

The antibodiom can serve as a biomarker for autoimmune and non-autoimmune disease, and so far, only little is known about aab in patients with nAMD. This is the first study to examine specific aab concerning their role in disease progression in nAMD patients over a five-year follow-up period.

At baseline, we included a total of 164 patients with active nAMD. Over the course of five years, we were able to analyze complete datasets of 59 eyes of 59 patients. With a mean age of 78.32 years and more women than men affected, the characteristics of our study group correspond to other epidemiological data on nAMD [13]. Intravitreal injection frequency remained steady around four IVIs per year. As we included patients with different previous disease durations, and since nAMD activity varies extremely over the years, the IVI treatment frequency in our study group is hardly comparable to other studies.

Mean BCVA at baseline was 0.34 logMar and decreased over the five-year period, as pictured in Figure 1. As patients had active nAMD with macular edema at the beginning of the study, CRT was thickened to 346.01 µm at baseline and decreased to 299.48 µm over five years. A course in BCVA and CRT like this corresponds to other data on the nAMD disease current in patients with anti-VEGF treatment [13–15].

As previously described, altered aab levels can occur physiologically in healthy humans who do not develop inflammatory diseases. Nonetheless, elevated or lowered aab levels can directly impact different kinds of diseases [3]. There is increasing evidence for the presence of aab in association with nAMD although it is unclear whether these ab play an active role in the etiology of the disease or if they are generated in a response to retinal injury from the underlying disease processes [16]. However, more and more evidence accentuates that autoimmunity plays an important role in the pathogenesis of AMD [16–19].

Undoubtedly, the relevance to identify specific biomarkers like aab is of high interest to the research community, as many study groups started different attempt in various study designs. Morohoshi et al. identified a pathogenic role of specific autoantibody profiles in dry and wet AMD patients [19]. They observed that ab specific to several autoantigens were significantly elevated in sera of AMD patients compared to normal controls and that these antibodies caused alterations in endothelial cell function. However, the exact clinical role and their diagnostic value towards the patient's prognosis remains unclear. Nonetheless, this is one of the most important questions to address, as nAMD lacks specific and easily available prognostic markers. This would help us in clinical routine to identify patients who need to be followed up more regularly, as nAMD shows highly variable interindividual disease courses. The same study group around Morohoshi later focused on a microarray analysis comparing sera of AMD patients and a healthy control [20]. They detected an elevated ratio of IgG/IgM to phosphatidylserine in AMD patients that even corresponded to disease stages and also determined Vitronectin and Fibronectin as biomarker candidates. Joachim et al. and Adamus et al. also identified promising candidates in AMD patients like anti-enolase aab that even corresponded to different disease stages, or GFAP, Carboxyethylpyrrole, Cellular Retinaldehyde Binding Protein, and Retinol Binding Protein-3 [18,21]. Taking a closer look at clinical parameters and their relevance concerning the individual disease progress would be an interesting question to address in larger prospective studies. Umeda et al. conducted extensive studies on cynomolgus monkeys and identified annexin II and μ-crystallin to be elevated in early-onset macular degeneration compared to the healthy control [22].

All these studies are important steps towards larger prospective long-term studies. However, we chose five new promising candidates for aab in sera and especially focused on their diagnostic value for a long-term clinical course.

Our study examined serum levels of five different aab and reviewed medical records of the included nAMD cohort over five years. As neovascularization formation is a key characteristic of nAMD, we chose ab for our study that are known to be associated with alterations of the vascular system as described in the introduction.

We carried out a PCA to summarize these ab levels and received the ab score as described above. This model allows dimensionality reduction and enables further analyses, but it has to be noted that one ab is left out in this model, and not all ab contribute to the score to the same extent. In our multiple regression models, we were able to show that the five antibodies against AT1-receptor, PAR1, VEGF-A, VEGF-B, and VEGF-receptor 2 cannot predict clinical activity parameters for nAMD, like the number of injections, development of intra- or subretinal fluid, macular bleeding, and development of fibrosis.

We utilized ROC analyses to receive information on the diagnostic ability of the ab. Neither the single aab themselves nor the ab score were able to receive sufficient values. However, when looking at our main endpoint, the total IVI response over five years, we detected a tendency concerning AT1-receptor antibodies (AUC 0.679). Therefore, this serum marker would be particularly interesting to be investigated further in prospective studies. Pathophysiologically, the AT1-receptor, as part of the renin-angiotensin system (RAS), has been shown to play a role in angiogenesis [6]. Additionally, in mice models, ACE (angiotensin converting enzyme) inhibition with Icatibant led to significant suppression of CNV development seen in AT1-receptor knockouts [23,24]. Therefore, the RAS and its components are interesting targets for future studies, and examining further aab along with AT1-receptor ab might lead to new insights on nAMD pathophysiology.

It has to be noted that serum levels of the examined ab may be influenced by anti-VEGF intravitreal injections. Several study groups focused on systemic effects of IVI treatment and showed reduced systemic VEGF levels following an anti-VEGF injection [25–28]. Matsuyama et al. and Carneiro et al. both reported that the effect was mostly present after one day of a bevacizumab injection, but VEGF levels were still reduced even after one month [25,26]. However, in the study by Carneiro et al. and Zehetner et al., this effect was not noticeable after ranibizumab and pegaptanib injections [26,27].

Future studies should take these data into consideration and should also differentiate between the administered IVIs, as they result in varying systemic effects. In our study design, we did not differ between the different anti-VEGF agents that the patients had received in the five-year follow-up period, which may have an impact on the serum levels that we measured. However, blood was drawn on the day of an IVI before the therapy was administered. Therefore, the blood withdrawal took place at least four weeks after the last injection, which limits the systemic effect of previous anti-VEGF treatments. Additionally, it is not known to what extent the systemic VEGF levels influence the levels of our analyzed ab.

The limitations of our study include the retrospective nature, the lack of masking, and a potential selection bias, as only patients who attended monthly follow ups for five years were included. Especially regarding the latter factor, it can be argued that patients with a high disease activity resulting in more injections and worse visual outcome tend to continue follow-ups more closely. Moreover, patients with controlled disease, good vision, and no further need for injections become lost to follow-up at our treatment center over time and continue follow-up visits at their local ophthalmologist. Additionally, as the mean age of nAMD patients is relatively high, some patients died in the five-year time frame after the study started.

In addition to these limiting factors, for this study, we only chose five aab against targets that are of high relevance in angiogenetic processes. Nevertheless, angiogenesis and neovascularization are multifactorial and complex developments that can be influenced by various pathways. Therefore, the present study only gives limited insight into the complex world of aab in nAMD and opened the door to further, more widespread study approaches.

5. Conclusions

We identified a relatively large group of nAMD patients that we were able to follow up over a five-year time frame after five angiogenesis-specific aab were analyzed (ab against AT1-receptor, PAR1, VEGF-A, VEGF-B, and VEGF-Receptor 2). Neither the ab by themselves nor taken together had a predictive value for the nAMD current over five years. Identifying specific ab in nAMD can help us gain a deeper understanding of the disease, which is the basis for better therapeutic strategies. The antibodiom reflects important individual processes, which makes it interesting for precision medicine concerning diagnostics as well as therapeutic approaches. Future research should focus on wider antibody networks and their levels in nAMD to gain deeper insights into this highly relevant disease entity.

Author Contributions: Conceptualization, S.G., M.R., G.R. and A.T.; methodology, F.R., K.N., C.H., H.H. and M.P.; validation, M.R. and A.T.; formal analysis, C.H., M.P. and F.R.; investigation, K.N. and C.H.; resources, S.G., H.H. and M.R.; data curation, F.R. and M.P.; writing—original draft preparation, M.P.; writing—review and editing, A.T., M.R. and S.G.; visualization, M.P.; supervision, K.N. and M.P.; funding acquisition, M.P. All authors have read and agreed to the published version of the manuscript.

Funding: This research received no external funding.

Institutional Review Board Statement: The study was conducted according to the guidelines of the Declaration of Helsinki and approved by the ethics committee of the University of Lübeck, Germany (vote reference number: 16-199).

Informed Consent Statement: Informed consent was obtained from all subjects involved in the study.

Data Availability Statement: Raw data were generated at the department of ophthalmology, University clinic Schleswig Holstein, Lübeck.

Conflicts of Interest: The authors declare no conflict of interest.

References

1. Nagele, E.P.; Han, M.; Acharya, N.K.; DeMarshall, C.; Kosciuk, M.C.; Nagele, R.G. Natural IgG Autoantibodies Are Abundant and Ubiquitous in Human Sera, and Their Number Is Influenced by Age, Gender, and Disease. *PLoS ONE* **2013**, *8*, e60726. [CrossRef]
2. Plotz, P.H. The Autoantibody Repertoire: Searching for Order. *Nat. Rev. Immunol.* **2003**, *3*, 73–78. [CrossRef] [PubMed]
3. Riemekasten, G.; Petersen, F.; Heidecke, H. What Makes Antibodies Against G Protein-Coupled Receptors so Special? A Novel Concept to Understand Chronic Diseases. *Front. Immunol.* **2020**, *11*, 564526. [CrossRef]
4. Ardeljan, D.; Chan, C.-C. Aging Is Not a Disease: Distinguishing Age-Related Macular Degeneration from Aging. *Prog. Retin. Eye Res.* **2013**, *37*, 68–89. [CrossRef]
5. Baxter, J.M.; Fotheringham, A.J.; Foss, A.J.E. Determining Patient Preferences in the Management of Neovascular Age-Related Macular Degeneration: A Conjoint Analysis. *Eye Lond. Engl.* **2016**, *30*, 698–704. [CrossRef]
6. Philogene, M.C.; Johnson, T.; Vaught, A.J.; Zakaria, S.; Fedarko, N. Antibodies against Angiotensin II Type 1 and Endothelin A Receptors: Relevance and Pathogenicity. *Hum. Immunol.* **2019**, *80*, 561–567. [CrossRef]
7. Imai, N.; Hashimoto, T.; Kihara, M.; Yoshida, S.; Kawana, I.; Yazawa, T.; Kitamura, H.; Umemura, S. Roles for Host and Tumor Angiotensin II Type 1 Receptor in Tumor Growth and Tumor-Associated Angiogenesis. *Lab. Investig. J. Tech. Methods Pathol.* **2007**, *87*, 189–198. [CrossRef] [PubMed]
8. Banasik, M.; Boratyńska, M.; Kościelska-Kasprzak, K.; Kamińska, D.; Zmonarski, S.; Mazanowska, O.; Krajewska, M.; Bartoszek, D.; Żabińska, M.; Myszka, M.; et al. Non-HLA Antibodies: Angiotensin II Type 1 Receptor (Anti-AT1R) and Endothelin-1 Type A Receptor (Anti-ETAR) Are Associated With Renal Allograft Injury and Graft Loss. *Transplant. Proc.* **2014**, *46*, 2618–2621. [CrossRef] [PubMed]
9. Yin, Y.-J.; Salah, Z.; Maoz, M.; Even Ram, S.C.; Ochayon, S.; Neufeld, G.; Katzav, S.; Bar-Shavit, R. Oncogenic Transformation Induces Tumor Angiogenesis: A Role for PAR1 Activation. *FASEB J. Off. Publ. Fed. Am. Soc. Exp. Biol.* **2003**, *17*, 163–174. [CrossRef]
10. Duncan, M.B.; Kalluri, R. Parstatin, a Novel Protease-Activated Receptor 1-Derived Inhibitor of Angiogenesis. *Mol. Interv.* **2009**, *9*, 168–170. [CrossRef]
11. Apte, R.S.; Chen, D.S.; Ferrara, N. VEGF in Signaling and Disease: Beyond Discovery and Development. *Cell* **2019**, *176*, 1248–1264. [CrossRef]
12. Riemekasten, G.; Philippe, A.; Näther, M.; Slowinski, T.; Müller, D.N.; Heidecke, H.; Matucci-Cerinic, M.; Czirják, L.; Lukitsch, I.; Becker, M.; et al. Involvement of Functional Autoantibodies against Vascular Receptors in Systemic Sclerosis. *Ann. Rheum. Dis.* **2011**, *70*, 530–536. [CrossRef] [PubMed]
13. Lim, L.S.; Mitchell, P.; Seddon, J.M.; Holz, F.G.; Wong, T.Y. Age-Related Macular Degeneration. *Lancet Lond. Engl.* **2012**, *379*, 1728–1738. [CrossRef]
14. Comparison of Age-related Macular Degeneration Treatments Trials (CATT) Research Group; Maguire, M.G.; Martin, D.F.; Ying, G.-S.; Jaffe, G.J.; Daniel, E.; Grunwald, J.E.; Toth, C.A.; Ferris, F.L.; Fine, S.L. Five-Year Outcomes with Anti-Vascular Endothelial Growth Factor Treatment of Neovascular Age-Related Macular Degeneration: The Comparison of Age-Related Macular Degeneration Treatments Trials. *Ophthalmology* **2016**, *123*, 1751–1761. [CrossRef] [PubMed]
15. Wecker, T.; Ehlken, C.; Bühler, A.; Lange, C.; Agostini, H.; Böhringer, D.; Stahl, A. Five-Year Visual Acuity Outcomes and Injection Patterns in Patients with pro-Re-Nata Treatments for AMD, DME, RVO and Myopic CNV. *Br. J. Ophthalmol.* **2017**, *101*, 353–359. [CrossRef]
16. Adamus, G. Can Innate and Autoimmune Reactivity Forecast Early and Advance Stages of Age-Related Macular Degeneration? *Autoimmun. Rev.* **2017**, *16*, 231–236. [CrossRef]
17. Camelo, S. Potential Sources and Roles of Adaptive Immunity in Age-Related Macular Degeneration: Shall We Rename AMD into Autoimmune Macular Disease? *Autoimmune Dis.* **2014**, *2014*, 532487. [CrossRef]
18. Adamus, G.; Chew, E.Y.; Ferris, F.L.; Klein, M.L. Prevalence of Anti-Retinal Autoantibodies in Different Stages of Age-Related Macular Degeneration. *BMC Ophthalmol.* **2014**, *14*, 154. [CrossRef]
19. Morohoshi, K.; Goodwin, A.M.; Ohbayashi, M.; Ono, S.J. Autoimmunity in Retinal Degeneration: Autoimmune Retinopathy and Age-Related Macular Degeneration. *J. Autoimmun.* **2009**, *33*, 247–254. [CrossRef]
20. Morohoshi, K.; Patel, N.; Ohbayashi, M.; Chong, V.; Grossniklaus, H.E.; Bird, A.C.; Ono, S.J. Serum Autoantibody Biomarkers for Age-Related Macular Degeneration and Possible Regulators of Neovascularization. *Exp. Mol. Pathol.* **2012**, *92*, 64–73. [CrossRef]

21. Joachim, S.C.; Bruns, K.; Lackner, K.J.; Pfeiffer, N.; Grus, F.H. Analysis of IgG Antibody Patterns against Retinal Antigens and Antibodies to α-Crystallin, GFAP, and α-Enolase in Sera of Patients with "Wet" Age-Related Macular Degeneration. *Graefes Arch. Clin. Exp. Ophthalmol.* **2006**, *245*, 619. [CrossRef] [PubMed]

22. Umeda, S.; Suzuki, M.T.; Okamoto, H.; Ono, F.; Mizota, A.; Terao, K.; Yoshikawa, Y.; Tanaka, Y.; Iwata, T. Molecular Composition of Drusen and Possible Involvement of Anti-Retinal Autoimmunity in Two Different Forms of Macular Degeneration in Cynomolgus Monkey (Macaca Fascicularis). *FASEB J.* **2005**, *19*, 1683–1685. [CrossRef] [PubMed]

23. Nagai, N.; Oike, Y.; Izumi-Nagai, K.; Koto, T.; Satofuka, S.; Shinoda, H.; Noda, K.; Ozawa, Y.; Inoue, M.; Tsubota, K.; et al. Suppression of Choroidal Neovascularization by Inhibiting Angiotensin-Converting Enzyme: Minimal Role of Bradykinin. *Invest. Ophthalmol. Vis. Sci.* **2007**, *48*, 2321–2326. [CrossRef] [PubMed]

24. Nagai, N.; Oike, Y.; Izumi-Nagai, K.; Urano, T.; Kubota, Y.; Noda, K.; Ozawa, Y.; Inoue, M.; Tsubota, K.; Suda, T.; et al. Angiotensin II Type 1 Receptor-Mediated Inflammation Is Required for Choroidal Neovascularization. *Arterioscler. Thromb. Vasc. Biol.* **2006**, *26*, 2252–2259. [CrossRef]

25. Matsuyama, K.; Ogata, N.; Matsuoka, M.; Wada, M.; Takahashi, K.; Nishimura, T. Plasma Levels of Vascular Endothelial Growth Factor and Pigment Epithelium-Derived Factor before and after Intravitreal Injection of Bevacizumab. *Br. J. Ophthalmol.* **2010**, *94*, 1215–1218. [CrossRef]

26. Carneiro, A.M.; Costa, R.; Falcão, M.S.; Barthelmes, D.; Mendonça, L.S.; Fonseca, S.L.; Gonçalves, R.; Gonçalves, C.; Falcão-Reis, F.M.; Soares, R. Vascular Endothelial Growth Factor Plasma Levels before and after Treatment of Neovascular Age-Related Macular Degeneration with Bevacizumab or Ranibizumab. *Acta Ophthalmol. Copenh.* **2012**, *90*, e25–e30. [CrossRef] [PubMed]

27. Zehetner, C.; Kirchmair, R.; Huber, S.; Kralinger, M.T.; Kieselbach, G.F. Plasma Levels of Vascular Endothelial Growth Factor before and after Intravitreal Injection of Bevacizumab, Ranibizumab and Pegaptanib in Patients with Age-Related Macular Degeneration, and in Patients with Diabetic Macular Oedema. *Br. J. Ophthalmol.* **2013**, *97*, 454–459. [CrossRef]

28. Avery, R.L. What Is the Evidence for Systemic Effects of Intravitreal Anti-VEGF Agents, and Should We Be Concerned? *Br. J. Ophthalmol.* **2014**, *98*, i7–i10. [CrossRef]

Journal of
*Personalized
Medicine*

MDPI

Review

Quantitative Imaging Biomarkers in Age-Related Macular Degeneration and Diabetic Eye Disease: A Step Closer to Precision Medicine

Gagan Kalra [1,2], Sudeshna Sil Kar [2,3], Duriye Damla Sevgi [1,2], Anant Madabhushi [3,4], Sunil K. Srivastava [1,2] and Justis P. Ehlers [1,2,*]

1   Cole Eye Institute, Cleveland Clinic, Cleveland, OH 44195, USA; kalrag2@ccf.org (G.K.);
     ddamlasevgi@gmail.com (D.D.S.); srivass2@ccf.org (S.K.S.)
2   Tony and Leona Campane Center for Excellence in Image-Guided Surgery & Advanced, Cleveland Clinic,
     Cleveland, OH 44195, USA; SXS2440@case.edu
3   Department of Biomedical Engineering, Case Western Reserve University, Cleveland, OH 44106, USA;
     axm788@case.edu
4   Louis Stokes Cleveland Veterans Administration Medical Center, Cleveland, OH 44106, USA
*   Correspondence: ehlersj@ccf.org

Abstract: The management of retinal diseases relies heavily on digital imaging data, including optical coherence tomography (OCT) and fluorescein angiography (FA). Targeted feature extraction and the objective quantification of features provide important opportunities in biomarker discovery, disease burden assessment, and predicting treatment response. Additional important advantages include increased objectivity in interpretation, longitudinal tracking, and ability to incorporate computational models to create automated diagnostic and clinical decision support systems. Advances in computational technology, including deep learning and radiomics, open new doors for developing an imaging phenotype that may provide in-depth personalized disease characterization and enhance opportunities in precision medicine. In this review, we summarize current quantitative and radiomic imaging biomarkers described in the literature for age-related macular degeneration and diabetic eye disease using imaging modalities such as OCT, FA, and OCT angiography (OCTA). Various approaches used to identify and extract these biomarkers that utilize artificial intelligence and deep learning are also summarized in this review. These quantifiable biomarkers and automated approaches have unleashed new frontiers of personalized medicine where treatments are tailored, based on patient-specific longitudinally trackable biomarkers, and response monitoring can be achieved with a high degree of accuracy.

Keywords: retinal imaging; quantitative biomarkers; diabetic retinopathy; diabetic macular edema; age-related macular degeneration; precision medicine; anti-VEGF therapy

Citation: Kalra, G.; Kar, S.S.; Sevgi, D.D.; Madabhushi, A.; Srivastava, S.K.; Ehlers, J.P. Quantitative Imaging Biomarkers in Age-Related Macular Degeneration and Diabetic Eye Disease: A Step Closer to Precision Medicine. *J. Pers. Med.* **2021**, *11*, 1161. https://doi.org/10.3390/jpm11111161

Academic Editors: Peter D. Westenskow and Andreas Ebneter

Received: 29 September 2021
Accepted: 4 November 2021
Published: 8 November 2021

Publisher's Note: MDPI stays neutral with regard to jurisdictional claims in published maps and institutional affiliations.

1. Introduction

Ophthalmology and the field of retinal diseases relies heavily on information derived from ophthalmic imaging for diagnosis, treatment and disease activity monitoring. The development of different imaging modalities, including optical coherence tomography (OCT) and ultra-widefield fluorescein angiography (UWFA), have provided incredible visualization of retinal microstructures and abnormalities, which has helped to build new insights for the management of retinal diseases, including diabetic eye disease (diabetic retinopathy, DR; diabetic macular edema, DME) and age-related macular degeneration [1,2].

Optical coherence tomography (OCT) is a rapid, non-invasive diagnostic test that provides outstanding visualization of cross-sectional and 3D morphological characteristics in addition to high-definition anatomy. OCT has become the backbone for the diagnosis and management of retinal diseases, with more than 30 million OCT scans being performed annually [3–5]. Due to its widespread utilization for retinal disease, OCT has become a key source for the exploration of imaging biomarkers through computational and deep learning techniques. The assessment of targeted features, such as retinal compartment volumes or volumetric pathology characterization, has been shown to be associated with disease burden and has the potential to enhance personalized treatment decisions [6–8]. OCT angiography (OCTA) uses non-invasive OCT technology to obtain vascular structural information by assessing decorrelation in the OCT signal due to vascular flow.

There are some important limitations to consider related to current OCT technology. With a limited field of view in most widely available OCT devices, the primary imaging location is the macula, and peripheral changes may be missed, especially in early disease [9,10]. Further, artifacts due to inconsistent montage techniques, motion-blur, and projection shadows may impact interpretation [9,10].

UWFA is an emerging imaging technique which enables visualization of panretinal vascular abnormalities including leakage, microaneurysms, and nonperfusion [2]. UWFA is a critical tool in the panretinal evaluation of retinal vascular and inflammatory disorders. With up to a 200-degree field of view, this imaging modality is the gold standard for detecting peripheral disease, especially early on in the disease process [11–14]. However, the technique does require the intravenous injection of fluorescein dye, which poses potential systemic risks [11–14]. Additionally, peripheral shadowing, eyelash artifacts, and image quality control can limit the consistency of interpretations [11–13].

Optical coherence tomography angiography (OCTA) is a major leap forward in this regard as it is completely non-invasive and provides high-resolution 3D binarized vessel maps that are objective and easy to interpret. The depth-encoded nature of the OCTA vascular data provides a unique advantage for evaluating the location of vascular abnormalities. However, current technology is primarily limited to macular pathology and can be subject to significant quality challenges, such as motion artifacts [11,12,14]. Additionally, OCTA does not provide information on leakage.

Current imaging systems provide outstanding details of disease burdens and the impact of different retinal diseases. Traditionally, this information has been utilized in a qualitative manner and relies on an ophthalmologist's interpretation and expertise. This inherently introduces bias and subjectivity in the assessment of these images, and therefore may limit consistency and the opportunities for precision medicine. Additionally, all of these images encode incredible amounts of data related to the underlying imaging phenotype of a given disease. These features, such as the location and type of leakage in UWFA or the reflectivity features of cysts on OCT, may carry critical information regarding the underlying pathophysiology and driving cellular pathways of a given disease [15–19].

Recently, machine learning (ML) based algorithms has gained traction for use in several medical image processing operations such as organ segmentation [20], cancer detection [21] and numerous diseases including diabetic eye diseases [20,22,23]. Deep learning (DL) is a subfield of ML and uses multi-layered neural-network structures. Most typical ML models employ pre-defined or engineered features, while DL models can learn useful representations of data and features directly from the raw data itself [20,23]. Hence DL approaches are also referred to as unsupervised feature generation-based ML approaches. The opportunities for the application of DL for different ophthalmologic diseases is quite rich. DL models are not without their challenges. The opacity of DL models creates unique issues in transparency of understanding the underpinnings of classification and model performance. DL models consider segmentation or classification problem as a binary problem and does not evaluate the heterogeneity within the tissue. Optimization of the deep neural network hyperparameters is a significant challenge. The search space for the model parameters is generally very high. Also in a data scarce environment, DL models tend to perform only marginally better than random guessing [24].

Radiomics is an emerging field of medical image processing that refers to the computerized data extraction from medical images and aims to capture the subvisual image attributes that may not be identified by the human experts. It provides opportunity to physicians to interpret images better regarding individualized therapy, surveillance, diagnosis, and prognosis [25]. These advanced image analysis techniques have been described broadly in the domain of brain tumor [26], breast cancer [27], prostate cancer [28] and several other diseases. The role of radiomics features in predicting therapeutic response and prognosis in ophthalmic diseases is emerging as an exciting opportunity for enhanced personalized care [17,18].

The boom in this image analysis space over the past decade has made it possible to automate the quantification and interpretation of ophthalmic imaging biomarkers. These computational imaging elements can then be evaluated for their role as biomarkers for disease diagnosis, prognosis, treatment initiation, and therapeutic response. For this review, we describe these measured features that are found to have clinical applications for the management of disease as "quantitative imaging biomarkers", which may serve as objective tools for the future in the context of diabetic eye disease and age-related macular degeneration (Figure 1).

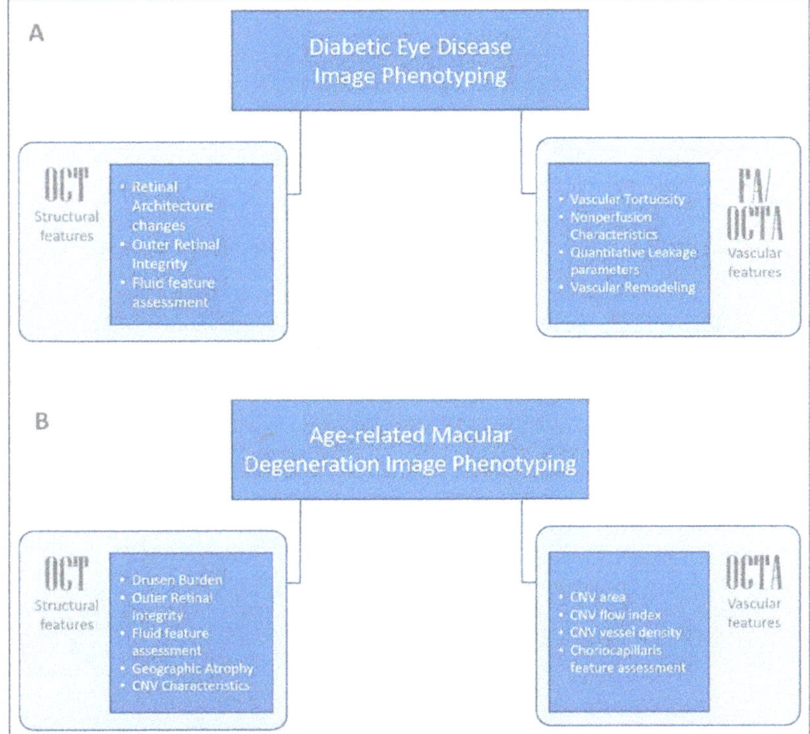

Figure 1. Schematic summarizing changes in various quantitative imaging biomarkers in (**A**) diabetic eye disease and (**B**) age-related macular degeneration. CNV: choroidal neovascularization.

Review Methodology. A literature search was performed using the key words "quantitative imaging", "diabetic retinopathy", "age related macular degeneration", "OCT", "OCTA", "fluorescein angiography", and "quantitative biomarkers" on databases, including PubMed Central and Google Scholar. Studies reporting quantitative imaging biomarkers using OCT, OCTA, and FA in diabetic eye disease and age-related macular degeneration. Studies that included only qualitative findings or that focused on pathologies other than diabetic eye disease and AMD were not included in this study.

2. Diabetic Eye Disease: Diabetic Retinopathy and Diabetic Macular Edema

2.1. Structural Biomarkers: Optical Coherence Tomography (OCT)

2.1.1. Characterizing Disease Burden and Functional Significance

In diabetic retinopathy, increased central subfield retinal thickness (CST) and decreased retinal nerve fiber layer thickness have been associated with increased severity of retinopathy (DR) [29–31]. Furthermore, disruption of retinal inner layers (i.e., DRIL) has been shown to be associated with worse visual acuity in DR patients [32,33]. The presence of DRIL has been shown to have very high specificity for macular nonperfusion in DR [34]. DRIL, as well as outer retinal disruption (e.g., ellipsoid zone and external limiting membrane loss), have been shown to be associated with visual acuity in both DR and diabetic macular edema (DME) (Figure 2) [33,35]. Morphological signs such as hyperreflective foci (HRF) have been described in diabetic retinopathy and diabetic macular edema as a sign of lipid extravasation and inflammatory cellular aggregates [36–38]. They often initially appear in the inner retina adjacent to the native microglia, only appearing in the outer retina in more advanced stages of the disease [38]. These HRF have been shown to be aggregated activated microglial cells with numbers significantly higher in diabetic eyes when compared to controls [39,40]. HRF count has been explored as a potential biomarker to assess inflammation in diabetic eye disease. Manual and automated approaches of the segmentation of these HRF have been tested [40–42]. A recent study monitored the HRF counts in diabetic retinopathy and diabetic macular edema in eyes that received anti-VEGF and steroid injections. This study reported a decrease in the number of HRF with either treatment, but a more marked decrease in the steroid group [42]. This biomarker provides an interesting avenue to monitor inflammatory activity in diabetic eye disease.

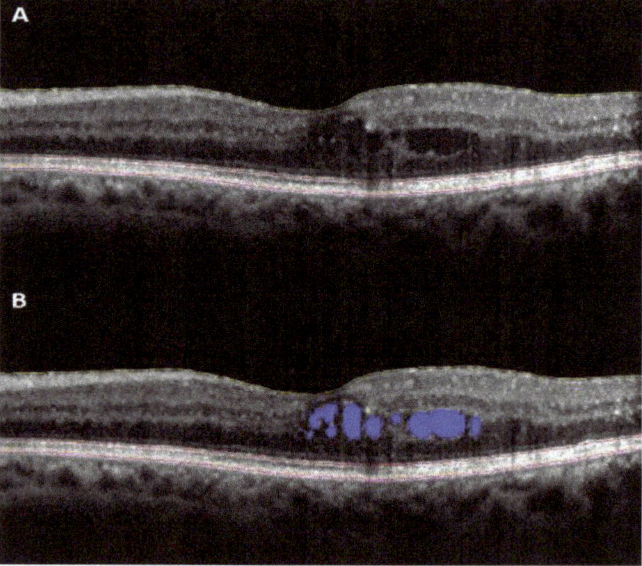

Figure 2. Spectral domain OCT scan of a patient with diabetic macular edema. (**A**) B-scan showing segmented retinal layers and (**B**) B-scan showing segmented retinal layers and intra-retinal fluid (shown in blue). OCT: optical coherence tomography.

Th extraction of quantitative fluid features and the assessment of retinal multi-layer segmentation has provided insights into disease prognosis and overall longitudinal disease dynamics. A recent study confirmed quantitative improvement in ellipsoid zone integrity subsequent to anti-VEGF therapy for DME [1]. This measurable improvement in ellipsoid zone integrity correlated significantly with visual function recovery. Novel higher-order imaging biomarkers, such as the retinal fluid index (RFI), are continuing to be discovered, which may help in the precise monitoring of treatment response [1,42]. Recent studies have shown that RFI volatility in the early follow-up period is correlated significantly with instability in long-term visual response to treatment [43].

2.1.2. Imaging Biomarkers and Disease Pathway Expression

Utilizing these techniques, various imaging biomarkers may be able to be linked to the underlying pathways involved in disease pathogenesis. In a recent study assessing quantitative imaging biomarkers and cytokine expression, the levels of multiple cytokines (e.g., vascular endothelial growth factor (VEGF), monocyte chemotactic protein-1 (MCP-1), and interleukin-6 (IL-6)) were linked with various imaging biomarkers, such as fluid parameters and outer retinal integrity [15]. The identification of these critical components of imaging phenotype and cytokine expression may help to identify eyes that may tolerate longer intervals in-between treatments or eyes that may benefit from emerging therapeutics with novel targets.

2.1.3. Predicting Future Treatment Need and Treatment Response Characteristics

Utilizing an attention-based convolutional neural network (CNN) model using pre-treatment OCT scans that preserved and highlighted the global structures in OCT images and enhanced local features from fluid/exudate-affected regions, Rasti et al. utilized retinal thickness information for the prediction of the response to intravitreal anti-VEGF treatment [44]. An additional DL algorithm developed by Prahs et al. distinguished retinal OCT B-scans that required an intravitreal injection from those that did not [45].

Beyond evaluating for treatment need, additional studies have assessed specific retinal compartment radiomics features that may predict therapeutic response. In a recent study [18], the relevance of radiomics features extracted from different spatial compartments of the retina on OCT scans to identify the patients with DME who tolerate an extension in the intervals between treatment with anti-VEGF treatment were evaluated. Texture-based radiomic features within the intraretinal fluid subcompartment were found to be most associated with a response to anti-VEGF therapy and most strongly associated in discriminating rebounders from the non-rebounders of anti-VEGF treatment following treatment interval extension.

2.2. Vascular Biomarkers: Ultra-Widefield Fluorescein Angiography (UWFA)

Ultra-widefield fluorescein angiography (UWFA) can capture 200° field of view (FOV) compared to conventional imaging with 30–60° FOV, enabling a more comprehensive disease evaluation. [46,47] Visualization of specific vascular features that enhance assessment of disease burden and optimize diagnostic accuracy make this modality an essential tool for the evaluation of posterior segment disorders. Areas of nonperfusion, vascular leakage, microaneurysm count, and neovascularization are among known clinically apparent biomarkers that assist diagnosis, choice of treatment and assessment of treatment response. Emerging image analysis methods provide the opportunity for manual and automated quantification of known angiographic features and discovery of novel and more complex features. The labor-intensive nature of manual feature assessment is a barrier to more widespread use. Recently, methods and systems have been developed to provide in-depth evaluation of leakage features, microaneurysm counts, ischemic burden and vascular characteristics (Figure 3) [48,49]. Machine learning systems have provided the ability to perform enhanced vascular segmentation, feature extraction, and more efficient methods for evaluating imaging characteristics [50–52].

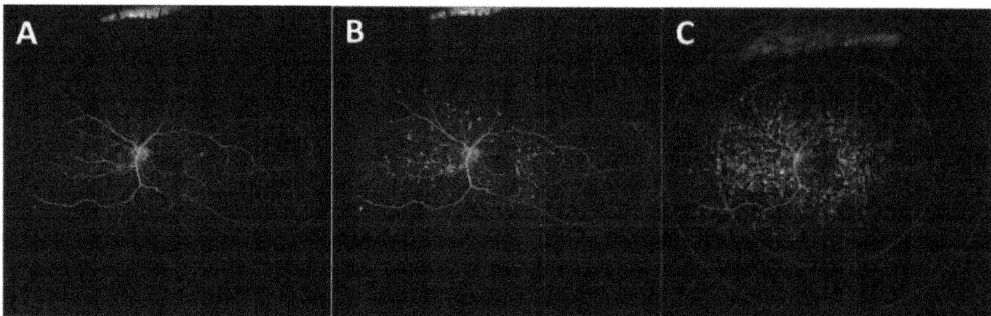

Figure 3. (**A**) Dewarped late-phase UWFA scan; (**B**) dewarped late-phase UWFA scan with overlay (yellow) indicating the area occupied by leakage; (**C**) dewarped early phase UWFA scan with overlay (green) indicating individual MAs in 3 mm (macular), 6 mm, and 9 mm zones (yellow circles). UWFA: ultra-widefield fluorescein angiography.

2.2.1. Biomarkers for Disease Severity and Disease Burden

Various biomarkers are investigated for severity grading, progression, and treatment response. Nonperfusion area, ischemic index, leakage, and microaneurysm counts have been shown to correlate strongly with the clinical severity of DR and treatment response [53]. Ehlers et al. demonstrated quantitative UWFA parameters, including panretinal MA count, ischemia, and leakage index, that were strongly associated with DR severity in 339 eyes [54]. Assessment of these disease burden metrics may help in predicting the risk of progression or DR-related complications. The panretinal leakage index has shown promise as a potential predictor of disease-related complications, such as vitreous hemorrhage and DME [53,55]. Quantification of these features allows for the longitudinal tracking of numerical changes that can be used to guide clinical decisions and assess response to treatment.

The spatial distribution of DR lesions on ultra-widefield photography including MA, cotton wool spots, intraretinal microvascular abnormalities, neovascularization, and fibrovascular proliferation was investigated in a large study with 1406 eyes demonstrating a predominantly central distribution in 63% of eyes [56]. Silva et al.'s study on nonperfusion distribution reported higher DR severity in eyes with predominantly peripheral lesions [57].

2.2.2. Evaluating and Predicting Treatment Response Characteristics

In addition to the assessment of disease burden, quantitative feature characterization can also be used to assess treatment response. In an automated UWFA approach, intravitreal anti-VEGF therapy demonstrated significant and stark improvements in leakage index and total microaneurysm counts in DR [55,58]. Wykoff et al. reported that the ischemic index increased by 34% in one year with quarterly aflibercept ($p = 0.009$) and 10% in monthly aflibercept ($p = 0.18$) treatment [59]. In a prospective clinical trial, the authors studied the change in the panretinal leakage index in eyes with DME with aflibercept therapy to quantify therapeutic response. The authors noted a dramatic reduction in the leakage index (from 3.5% at baseline to 0.4% at 12 months) with aflibercept therapy [58]. Utilizing quantitative UWFA in the RECOVERY study, which evaluated eyes with severe PDR, quantitative UWFA demonstrated a dramatic reduction of 68% to 79% in leakage index reduction at 1 year, with similar outcomes in both monthly and quarterly dosing [55]. In a randomized controlled trial comparing leakage-index-guided treatment and Diabetic Retinopathy Severity Scale (DRSS)-level-guided treatment with intravitreal aflibercept for DR, the authors found that deteriorations in the leakage index preceded those in the DRSS level, thereby providing a potentially higher sensitivity marker for the need for retreatment [60].

2.2.3. Imaging Biomarkers and Disease Pathway Expression

Another recent study assessed the correlation between UWFA imaging phenotype and cytokine expression in eyes with DME from the IMAGINE study [16]. The authors noted that an increased panretinal leakage index correlated strongly with VEGF, angiopoietin-like 4, and interleukin-6 levels, while the panretinal ischemic index was positively correlated with the tissue inhibitor of metalloproteinases 1 (TIMP-1) and VEGF [16]. Further research is needed to understand the implications of these phenotype–cytokine expression correlations in assessing response to treatment.

2.2.4. Radiomics Angiographic Biomarkers for DR Severity

In addition to clinically apparent biomarkers, Fan et al. demonstrated the branching complexity of peripheral vessels and the distribution of nonperfusion areas correlated with DR severity [61]. Fractal dimension (FD) depicts the complexity of vascular geometry, such that higher values indicate dense, intricate, space-filling branching patterns [62]. Peripheral retinal vessels of diabetic eyes have been demonstrated to have lower complexity in their branching pattern (fractal dimension) compared to healthy controls. FD was shown to be negatively associated with the nonperfusion area [63]. A significantly lower FD is noted in the retinal vasculature in DR, especially in the far peripheral fields when compared to normal eyes. Additionally, a decrease in panretinal FD was shown to be associated with an increase in the global nonperfusion area [64]. In addition to FD, the skewness of retinal vasculature density distribution has also been associated with DR severity [65].

2.2.5. Angiographic Biomarkers for DME Presence

Quantitative UWFA has also been explored in DME pathogenesis. The leakage index and MA count in the posterior pole have been associated with the presence and severity of DME [53]. The nonperfusion distribution pattern in DR was observed in DME, being more extensive in mid-periphery ischemia compared to the posterior pole and far periphery. Fang et al. classified ischemic areas and investigated nonperfusion with and without leakage in DME eyes [66]. Nonperfusion areas with leakage were found more extensively in the posterior retina compared to nonperfusion without leakage, which is predominantly in the mid-periphery [66]. A nonperfusion area with leakage positively correlated whereas nonperfusion without leakage negatively correlated with DME severity.

2.2.6. Evaluating and Predicting Therapeutic Response: From Quantitative UWFA to Radiomics

Quantitative UWFA biomarkers have been explored as assessment tools for therapeutic response in eyes with DME treated with aflibercept in the PERMEATE study [58]. Aflibercept injections resulted in a 78% decrease in the leakage index of eyes with DME. Similar to the outcome in eyes with DR, the nonperfusion area is increased despite anti-VEGF therapy [58].

Beyond characterizing the longitudinal quantitative UWFA feature alterations in response to therapy, radiomics features have been utilized to predict treatment response and durability. Prasanna et al. developed novel radiomic CIBs that characterized different morphological properties of leakage nodes and vascular tortuosity on UWFA, which were linked to the durability of anti-VEGF treatment [17]. The distribution of leakage nodes in eyes that did not tolerate treatment extension was found to be more disordered than eyes that tolerated an extension in the intervals between treatment. Vessel tortuosity was increased and more complex in eyes that experienced clinical worsening following treatment extension. In a supportive assessment of radiomics features for predicting treatment response characteristics, Moosavi et al. identified that the proximity of leakage foci to the vessels has a higher variance in eyes who have more durable treatment response, whereas increased local vascular tortuosity was linked to reduction in tolerance of treatment extension [67].

2.3. Vascular Biomarkers: OCTA

OCTA uses non-invasive OCT technology to obtain vascular structural information by assessing decorrelation in the OCT signal due to vascular flow. As a result of the depth resolution of OCTA, different chorioretinal vascular plexuses, such as the nerve fiber layer plexus (NFLP), ganglion cell layer plexus (GCLP), intermediate capillary plexus (ICP) and deep capillary plexus (DCP), have been studied using this technology. NFLP and GCLP form the superficial vascular complex while ICP and DCP form the deep vascular complex [68].

Biomarkers for Disease Severity and Burden: From Quantitative Features to Radiomics

An increased foveal avascular zone (FAZ) size is noted in patients with DR compared to normal [69–71]. Recent OCTA studies have provided evidence for a correlation between FAZ size and visual acuity, such that an increase in FAZ size is associated with decreased visual acuity [72–74]. In addition to FAZ area, the shape of the FAZ has been shown to change in various DR grades [75].

Vessel density, as calculated from OCTA, has been shown to be inversely correlated with DR grade in multiple trials [70,76,77]. In a study characterizing the association between visual acuity and vessel density in DR, vessel density was reduced in eyes with decreased visual acuity [78].

Vessel diameter index (VDI) is a representation of vessel diameter obtained by calculating a ratio of the total area of the scan occupied by blood vessels and the total skeletonized length of blood vessels in the scan. In a recent study, the VDI obtained using OCTA has been shown to positively correlate with the severity of DR and blood glucose levels [79–81].

Similar to UWFA, retinal vessel tortuosity in OCTA is another important metric that holds high potential for the evaluation of DR. Vascular tortuosity on OCTA positively correlates with the severity of DR in superficial and deep retinal vascular plexuses in moderate to severe DR [75]. In a recent study, vessel tortuosity demonstrated a positive correlation with DR severity in NPDR, but decreased significantly in PDR [75]. Recently, three-dimensional volume-rendering biomarkers such as vessel sphericity and cylindricity were used to assess blood vessel shape, demonstrating potential differences between normal eyes and eyes with DR [82]. Geometric features, such as vessel branching angle and vessel-width-based features have also been noted to be significantly different between normal eyes and eyes with DR [83].

3. Age-Related Macular Degeneration (AMD): Neovascular and Non-Neovascular AMD

3.1. Structural Biomarkers: Optical Coherence Tomography (OCT)

3.1.1. Features for Predicting Progression in AMD

Non-neovascular (i.e., dry) age-related macular degeneration has been extensively evaluated for numerous imaging biomarkers such as intraretinal hyper-reflective foci (HRF), complex drusenoid lesions (DL, i.e., heterogeneous reflectivity), subretinal drusenoid deposits (SDDs), and drusen burden. SD-OCT has been used to qualitatively describe these biomarkers and has confirmed that each of these features confers a greater risk of disease progress [84,85]. In a recent study, quantitative EZ integrity measures, EZ mapping, and sub-RPE compartment quantification were shown to be important predictors of progression to geographic atrophy in nonexudative AMD patients [86]. Specifically, the reduced EZ integrity and increased sub-RPE compartment thickness was identified in eyes that progressed to subfoveal geographic atrophy. These quantitative biomarkers were more strongly associated with progression than qualitative features, such as HRF and SDD. Utilizing a ML classifier, a high-performance system was developed for predicting progression to subfoveal GA based on multiple quantitative outer retinal features [87,88].

Automated drusen volume quantification has been enabled by multi-layer segmentation platforms that provide isolation of the sub-RPE compartment. One study demonstrated that an increase in the drusen volume was associated with a significant increase in the risk of developing geographic atrophy or conversion to neovascular AMD [89]. ML-enhanced systems for advanced segmentation and feature extraction are creating new opportunities for automated disease characterization and longitudinal monitoring of therapeutic response in AMD. Multiple studies have demonstrated volumetric fluid characterization, compartment-specific OCT feature evaluation (such as ellipsoid zone integrity), and volumetric quantification of subretinal fibrosis as well as subretinal hyperreflective material [6,90,91]. In a recent study utilizing deep learning for the extraction of quantitative features in AMD patients, the authors noted that an increase in drusen volume, SRF, IRF, serous pigment epithelium detachment, HRF and subretinal hyperreflective material was associated with worse visual acuity [7].

3.1.2. Deep Learning and Radiomics Biomarkers in AMD

DL-based analysis systems have been explored to detect the presence of disease. Multiple other studies have shown the effectiveness of DL models in classifying normal versus AMD eyes from OCT images [92,93]. Automated SD-OCT image analysis using DL techniques are currently widely used for predicting disease progression in AMD. Predicting conversion from early or intermediate non-neovascular AMD to neovascular AMD using quantitative imaging features (e.g., drusen shape, drusen volume) in SD-OCT images has been previously explored [94,95]. Banerjee et al. proposed a hybrid sequential model integrating hand-crafted size-based and shape-based radiomics features (related to the relationship of image intensity between voxels), demographic and visual acuity data, and DL with a recursive neural network (RNN) model in the same platform to predict the probability of future neovascular conversion [22].

3.2. Vascular Biomarkers: OCTA

In neovascular AMD, CNV is a major cause of vision loss due to photoreceptor damage that results from exudation processes [96,97]. Although FA has traditionally been the gold standard to characterize and identify CNV lesions, OCT has now become the benchmark evaluation for the presence of CNV and exudation. OCTA is also emerging as a promising technology for the high-level visualization of neovascular membranes in neovascular AMD and for evaluating the choriocapillaris in non-neovascular AMD [98–100].

3.2.1. Quantitative Biomarkers of CNV Features

In one study aimed at characterizing CNV using quantitative biomarkers on OCTA, the CNV area and flow index using outer retinal choriocapillaris OCTA slabs for assessment of CNV characterization [101]. The study identified a higher flow in larger CNVs and those that were type II [101]. In a recent study, the quantification of CNV and other vascular characteristics was evaluated to assess treatment response to anti-VEGF therapy in neovascular AMD patients [102]. Eyes requiring more frequent dosing of anti-VEGF agents had lower CNV vessel density compared to groups with longer duration intervals between doses [102]. Further, the CNV area was noted to be higher in eyes with fovea involvement and core vessel presence. Absence of these findings may therefore be suggestive of inactive CNV.

3.2.2. Choriocapillaris Biomarkers in Non-Neovascular AMD

In non-neovascular AMD, OCTA has been explored to study many aspects of the disease process such as drusen, reticular pseudodrusen, and geographic atrophy, in addition to exploring its utility for the monitoring of disease progression [100]. Choriocapillaris flow depletion in eyes with drusen has been shown on OCTA [103,104]. Reduced flow may result in relative hypoxia of outer retinal layers and disease progression. In a recent study, quantitative assessment of choriocapillaris flow deficits demonstrated reduced flow in eyes with drusen with hyporeflective cores compared with eyes with drusen without hyporeflective cores, suggesting that the presence of hyporeflective cores may indeed indicate a more advanced disease process in intermediate AMD [105]. OCTA has been used to characterize geographic atrophy (GA) as well, particularly choriocapillaris flow deficits [106]. Focal perfusion loss (FPL) on OCTA has been used to evaluate choriocapillaris flow features in AMD, which has been identified to be higher in AMD eyes compared to controls [107].

4. Conclusions

Quantitative imaging biomarkers derived from multiple imaging modalities may provide a critical platform for the future in providing objective and trackable metrics that enable precision medicine in ophthalmic care through the comprehensive characterization of the "imaging phenotype", Figure 4.

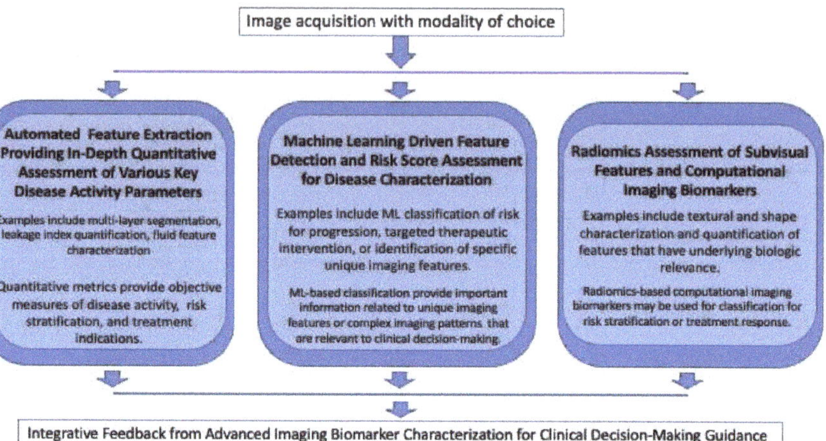

Figure 4. Developing the "Imaging Phenotype". A potential multi-factorial approach for integrative imaging biomarker characterization utilizing multiple advanced feature interrogation and extraction methods.

OCT imaging biomarkers provide valuable structural information of retinal layers, such as retinal compartment thickness, layer integrity maps, fluid volume, and the fluid index. UWFA and OCTA imaging biomarkers provide key information regarding the retinal and choroidal vasculature, such as measures of vessel density, ischemic area, flow voids, leakage area, leakage index, ischemic index, and the CNV area. Radiomics is an emerging field in ophthalmology and is having an increasingly high impact on personalized medicine. As the field matures in the future, a combination of different novel DL networks and advanced radiomic methods may be of high value for a more complete decision support system (Figure 4). The implementation of deep learning, advanced feature interrogation methods, and radiomics characterization provides an exciting opportunity for enhanced understanding of and new insights into retinal disease. The field of computational imaging biomarker discovery and exploration in AMD and diabetic eye disease is emerging as a major opportunity for personalized care and precision medicine.

Author Contributions: Conceptualization, J.P.E. and A.M.; writing—original draft preparation, G.K.; writing—review and editing, D.D.S., S.S.K., A.M., S.K.S. and J.P.E.; supervision, J.P.E. All authors have read and agreed to the published version of the manuscript.

Funding: NIH/NEI K23-EY022947-01A1. S.K.S. has research support from Regeneron, Allergan, and Gilead; is a consultant for Bausch and Lomb, Novartis, and Regeneron. A.M. has research funding from Astrazeneca, Bristol Myers-Squibb, and Philips. They have equity in Inspirata Inc., Elucid Bioimaging. They are also a consultant for Aiforia, Caris, Roche, and Cernostics. J.P.E. has research support from the following: Aerpio, Alcon, Thrombogenics/Oxurion, Regeneron, Genentech, Novartis, Allergan; is a consultant for the following: Aerpio, Adverum, Alcon, Allegro, Allergan, Genentech/Roche, Stealth, Novartis, Thrombogenics/Oxurion, Leica, Zeiss, Regeneron, and Santen; holds a patent with Leica. All other authors do not have any financial disclosures.

Institutional Review Board Statement: Not applicable.

Informed Consent Statement: Not applicable.

Conflicts of Interest: The authors declare no conflict of interest.

References

1. Ehlers, J.P.; Uchida, A.; Hu, M.; Figueiredo, N.; Kaiser, P.K.; Heier, J.S.; Brown, D.M.; Boyer, D.S.; Do, D.V.; Gibson, A.; et al. Higher-Order Assessment of OCT in Diabetic Macular Edema from the VISTA Study: Ellipsoid Zone Dynamics and the Retinal Fluid Index. *Ophthalmol. Retin.* **2019**, *3*, 1056–1066. [CrossRef] [PubMed]
2. Querques, L.; Parravano, M.; Sacconi, R.; Rabiolo, A.; Bandello, F.; Querques, G. Ischemic index changes in diabetic retinopathy after intravitreal dexamethasone implant using ultra-widefield fluorescein angiography: A pilot study. *Acta Diabetol.* **2017**, *54*, 769–773. [CrossRef] [PubMed]
3. Bourne, R.R.; Flaxman, S.R.; Braithwaite, T.; Cicinelli, M.V.; Das, A.; Jonas, J.B.; Keeffe, J.; Kempen, J.H.; Leasher, J.; Limburg, H.; et al. Magnitude, temporal trends, and projections of the global prevalence of blindness and distance and near vision impairment: A systematic review and meta-analysis. *Lancet Glob. Health* **2017**, *5*, e888–e897. [CrossRef]
4. Kurmann, T.; Yu, S.; Márquez-Neila, P.; Ebneter, A.; Zinkernagel, M.S.; Munk, M.R.; Wolf, S.; Sznitman, R. Expert-level Automated Biomarker Identification in Optical Coherence Tomography Scans. *Sci. Rep.* **2019**, *9*, 13605. [CrossRef] [PubMed]
5. Schmidt-Erfurth, U.; Klimscha, S.; Waldstein, S.M.; Bogunović, H. A view of the current and future role of optical coherence tomography in the management of age-related macular degeneration. *Eye* **2016**, *31*, 26–44. [CrossRef] [PubMed]
6. Ehlers, J.P.; Clark, J.; Uchida, A.; Figueiredo, N.; Babiuch, A.; Talcott, K.E.; Lunasco, L.; Le, T.K.; Meng, X.; Hu, M.; et al. Longitudinal Higher-Order OCT Assessment of Quantitative Fluid Dynamics and the Total Retinal Fluid Index in Neovascular AMD. *Transl. Vis. Sci. Technol.* **2021**, *10*, 29. [CrossRef] [PubMed]
7. Moraes, G.; Fu, D.J.; Wilson, M.; Khalid, H.; Wagner, S.K.; Korot, E.; Ferraz, D.; Faes, L.; Kelly, C.J.; Spitz, T.; et al. Quantitative Analysis of OCT for Neovascular Age-Related Macular Degeneration Using Deep Learning. *Ophthalmology* **2021**, *128*, 693–705. [CrossRef]
8. Ehlers, J.P.; Khan, M.; Petkovsek, D.; Stiegel, L.; Kaiser, P.; Singh, R.P.; Reese, J.L.; Srivastava, S.K. Outcomes of Intraoperative OCT–Assisted Epiretinal Membrane Surgery from the PIONEER Study. *Ophthalmol. Retin.* **2018**, *2*, 263–267. [CrossRef]
9. Reznicek, L.; Kolb, J.P.; Klein, T.; Mohler, K.J.; Wieser, W.; Huber, R.; Kernt, M.; Märtz, J.; Neubauer, A.S. Wide-Field Megahertz OCT Imaging of Patients with Diabetic Retinopathy. *J. Diabetes Res.* **2015**, *2015*, 1–5. [CrossRef]
10. De Pretto, L.R.; Moult, E.M.; Alibhai, A.Y.; Carrasco-Zevallos, O.M.; Chen, S.; Lee, B.K.; Witkin, A.J.; Baumal, C.R.; Reichel, E.; de Freitas, A.Z.; et al. Controlling for artifacts in widefield optical coherence tomography angiography measurements of non-perfusion area. *Sci. Rep.* **2019**, *9*, 1–15. [CrossRef] [PubMed]
11. de Carlo, T.E.; Bonini Filho, M.A.; Baumal, C.R.; Reichel, E.; Rogers, A.; Witkin, A.J.; Duker, J.S.; Waheed, N.K. Evaluation of preretinal neovascularization in proliferative diabetic retinopathy using optical coherence tomography angiography. *Ophthalmic Surg. Lasers Imaging Retin.* **2016**, *47*, 115–119. [CrossRef] [PubMed]
12. Sawada, O.; Ichiyama, Y.; Obata, S.; Ito, Y.; Kakinoki, M.; Sawada, T.; Saishin, Y.; Ohji, M. Comparison between wide-angle OCT angiography and ultra-wide field fluorescein angiography for detecting non-perfusion areas and retinal neovascularization in eyes with diabetic retinopathy. *Graefe's Arch. Clin. Exp. Ophthalmol.* **2018**, *256*, 1275–1280. [CrossRef] [PubMed]
13. Kaines, A.; Tsui, I.; Sarraf, D.; Schwartz, S. The Use of Ultra Wide Field Fluorescein Angiography in Evaluation and Management of Uveitis. *Semin. Ophthalmol.* **2009**, *24*, 19–24. [CrossRef]
14. Couturier, A.; Rey, P.-A.; Erginay, A.; Lavia, C.; Bonnin, S.; Dupas, B.; Gaudric, A.; Tadayoni, R. Widefield OCT-Angiography and Fluorescein Angiography Assessments of Nonperfusion in Diabetic Retinopathy and Edema Treated with Anti–Vascular Endothelial Growth Factor. *Ophthalmology* **2019**, *126*, 1685–1694. [CrossRef]
15. Abraham, J.R.; Wykoff, C.C.; Arepalli, S.; Lunasco, L.; Hannah, J.Y.; Hu, M.; Reese, J.; Krovastava, S.K.; Brown, D.M.; Ehlers, J.P. Aqueous cytokine expression and higher order OCT biomarkers: Assessment of the Anatomic-Biologic bridge in the IMAGINE DME study. *Am. J. Ophthalmol.* **2021**, *222*, 328–339. [CrossRef]
16. Abraham, J.R.; Wykoff, C.C.; Arepalli, S.; Lunasco, L.; Hannah, J.Y.; Martin, A.; Mugnaini, C.; Hu, M.; Reese, J.; Strivastava, S.K.; et al. Exploring the angiographic-biologic phenotype in the IMAGINE study: Quantitative UWFA and cytokine expression. *Br. J. Ophthalmol.* **2021**. Available online: https://pubmed.ncbi.nlm.nih.gov/34099465/ (accessed on 31 July 2021).
17. Prasanna, P.; Bobba, V.; Figueiredo, N.; Sevgi, D.D.; Lu, C.; Braman, N.; Alilou, M.; Sharma, S.; Srivastava, S.K.; Madabhushi, A.; et al. Radiomics-based assessment of ultra-widefield leakage patterns and vessel network architecture in the PERMEATE study: Insights into treatment durability. *Br. J. Ophthalmol.* **2020**, *105*, 1155. [CrossRef] [PubMed]
18. Kar, S.S.; Sevgi, D.D.; Dong, V.; Srivastava, S.K.; Madabhushi, A.; Ehlers, J.P. Multi-Compartment Spatially-Derived Radiomics From Optical Coherence Tomography Predict Anti-VEGF Treatment Durability in Macular Edema Secondary to Retinal Vascular Disease: Preliminary Findings. *IEEE J. Transl. Eng. Health Med.* **2021**, *9*, 1–13. [CrossRef]
19. Sil, K.S.; Sevji, D.D.; Dong, V.; Srivastava, S.K.; Madabhushi, A.; Ehlers, J.P. Multi-Compartment OCT-derived Radiomics Features to predict Anti-VEGF Treatment Durability for Diabetic Macular Edema. *Investig. Ophthalmol. Vis. Sci.* **2021**, *62*, 3.
20. Akkus, Z.; Galimzianova, A.; Hoogi, A.; Rubin, D.L.; Erickson, B.J. Deep Learning for Brain MRI Segmentation: State of the Art and Future Directions. *J. Digit. Imaging* **2017**, *30*, 449–459. [CrossRef]

21. Sumathipala, Y.; Lay, N.S.; Turkbey, B. Prostate cancer detection from multi-institution multiparametric MRIs using deep convolutional neural networks. *J. Med. Imaging* **2018**, *5*, 044507. [CrossRef] [PubMed]

22. Banerjee, I.; De Sisternes, L.; Hallak, J.A.; Leng, T.; Osborne, A.; Rosenfeld, P.J.; Gregori, G.; Durbin, M.; Rubin, D. Prediction of age-related macular degeneration disease using a sequential deep learning approach on longitudinal SD-OCT imaging bi-omarkers. *Sci. Rep.* **2020**, *10*, 1–16. [CrossRef] [PubMed]

23. Lundervold, A.; Lundervold, A. An overview of deep learning in medical imaging focusing on MRI. *Z. Med. Phys.* **2019**, *29*, 102–127. [CrossRef] [PubMed]

24. Dong, V.; Sevgi, D.D.; Kar, S.S.; Srivastava, S.K.; Ehlers, J.P.; Madabhushi, A. Evaluating the utility of deep learning using ultra-widefield fluorescein angiography for predicting need for anti-VEGF therapy in diabetic eye disease. *Investig. Ophthalmol. Visual Sci.* **2021**, *62*, 2114.

25. Rizzo, S.; Botta, F.; Raimondi, S.; Origgi, D.; Fanciullo, C.; Morganti, A.G.; Bellomi, M. Radiomics: The facts and the challenges of image analysis. *Eur. Radiol. Exp.* **2018**, *2*, 36. [CrossRef]

26. Wu, G.; Chen, Y.; Wang, Y.; Yu, J.; Lv, X.; Ju, X.; Shi, Z.; Chen, L.; Chen, Z. Sparse Representation-Based Radiomics for the Diagnosis of Brain Tumors. *IEEE Trans. Med. Imaging* **2018**, *37*, 893–905. [CrossRef] [PubMed]

27. Parekh, V.S.; Jacobs, M.A. Integrated radiomic framework for breast cancer and tumor biology using advanced machine learning and multiparametric MRI. *NPJ Breast Cancer* **2017**, *3*, 1–9. [CrossRef]

28. Penzias, G.; Singanamalli, A.; Elliott, R.; Gollamudi, J.; Shih, N.; Feldman, M.; Stricker, P.; Delprado, W.; Tiwari, S.; Böhm, M.; et al. Identifying the morphologic basis for radiomic features in distinguishing different Gleason grades of prostate cancer on MRI: Preliminary findings. *PLoS ONE* **2018**, *13*, e0200730. [CrossRef]

29. Vujosevic, S.; Midena, E. Retinal Layers Changes in Human Preclinical and Early Clinical Diabetic Retinopathy Support Early Retinal Neuronal and Müller Cells Alterations. *J. Diabetes Res.* **2013**, *2013*, 1–8. [CrossRef]

30. Shi, R.; Guo, Z.; Wang, F.; Lin, R.; Zhao, L. Alterations in retinal nerve fiber layer thickness in early stages of diabetic reti-nopathy and potential risk factors. *Curr. Eye Res.* **2018**, *43*, 244–253. [CrossRef]

31. Deák, G.G.; Schmidt-Erfurth, U.; Jampol, L.M. Correlation of Central Retinal Thickness and Visual Acuity in Diabetic Macular Edema. *JAMA Ophthalmol.* **2018**, *136*, 1215–1216. [CrossRef]

32. Joltikov, K.; Sesi, C.A.; De Castro, V.M.; Davila, J.R.; Anand, R.; Khan, S.M.; Farbman, N.; Jackson, G.R.; Johnson, C.A.; Gardner, T.W. Disorganization of Retinal Inner Layers (DRIL) and Neuroretinal Dysfunction in Early Diabetic Retinopathy. *Investig. Opthalmol. Vis. Sci.* **2018**, *59*, 5481–5486. [CrossRef] [PubMed]

33. Sun, J.K.; Lin, M.M.; Lammer, J.; Prager, S.; Sarangi, R.; Silva, P.S.; Aiello, L.P. Disorganization of the Retinal Inner Layers as a Predictor of Visual Acuity in Eyes With Center-Involved Diabetic Macular Edema. *JAMA Ophthalmol.* **2014**, *132*, 1309–1316. [CrossRef]

34. Nicholson, L.; Ramu, J.; Triantafyllopoulou, I.; Patrao, N.V.; Comyn, O.; Hykin, P.; Sivaprasad, S. Diagnostic accuracy of disorganization of the retinal inner layers in detecting macular capillary non-perfusion in diabetic retinopathy. *Clin. Exp. Ophthalmol.* **2015**, *43*, 735–741. [CrossRef]

35. Eliwa, T.F.; Hussein, M.A.; Zaki, M.A.; Raslan, O.A. Outer retinal layer thickness as good visual predictor in patients with diabetic macular edema. *Retina* **2018**, *38*, 805–811. [CrossRef] [PubMed]

36. Bolz, M.; Schmidt-Erfurth, U.; Deak, G.; Mylonas, G.; Kriechbaum, K.; Scholda, C. Optical Coherence Tomographic Hyperreflective Foci: A Morphologic Sign of Lipid Extravasation in Diabetic Macular Edema. *Ophthalmology* **2009**, *116*, 914–920. [CrossRef] [PubMed]

37. Vujosevic, S.; Bini, S.; Midena, G.; Berton, M.; Pilotto, E.; Midena, E. Hyperreflective Intraretinal Spots in Diabetics without and with Nonproliferative Diabetic Retinopathy: AnIn VivoStudy Using Spectral Domain OCT. *J. Diabetes Res.* **2013**, *2013*, 1–5. [CrossRef]

38. Lee, H.; Jang, H.; A Choi, Y.; Kim, H.C.; Chung, H. Association Between Soluble CD14 in the Aqueous Humor and Hyperreflective Foci on Optical Coherence Tomography in Patients With Diabetic Macular Edema. *Investig. Opthalmol. Vis. Sci.* **2018**, *59*, 715–721. [CrossRef]

39. De Benedetto, U.; Sacconi, R.; Pierro, L.; Lattanzio, R.; Bandello, F. Optical coherence tomographic hyperreflective foci in early stages of diabetic retinopathy. *Retina* **2015**, *35*, 449–453. [CrossRef] [PubMed]

40. Okuwobi, I.P.; Ji, Z.; Fan, W.; Yuan, S.; Bekalo, L.; Chen, Q. Automated Quantification of Hyperreflective Foci in SD-OCT With Diabetic Retinopathy. *IEEE J. Biomed. Health Inform.* **2019**, *24*, 1125–1136. [CrossRef] [PubMed]

41. Rübsam, A.; Wernecke, L.; Rau, S.; Pohlmann, D.; Müller, B.; Zeitz, O.; Joussen, A.M. Behavior of SD-OCT Detectable Hyperreflec-tive Foci in Diabetic Macular Edema Patients after Therapy with Anti-VEGF Agents and Dexamethasone Implants. *J. Diabetes Res.* **2021**, *2021*, 8820216. [CrossRef] [PubMed]

42. Roberts, P.K.; Vogl, W.D.; Gerendas, B.S.; Glassman, A.R.; Bogunovic, H.; Jampol, L.M.; Schmidt-Erfurth, U.M. Quantification of fluid resolution and visual acuity gain in patients with diabetic macular edema using deep learning: A post hoc analysis of a randomized clinical trial. *JAMA Ophthalmol.* **2020**, *138*, 945–953. [CrossRef] [PubMed]

43. Ehlers, J.P.; Uchida, A.; Sevgi, D.D.; Hu, M.; Reed, K.; Berliner, A.; Vitti, R.; Chu, K.; Srivastava, S.K. Retinal Fluid Volatility Associated with Interval Tolerance and Visual Outcomes in Diabetic Macular Edema in the VISTA Phase III Trial. *Am. J. Ophthalmol.* **2021**, *224*, 217–227. [CrossRef]

44. Rasti, R.; Allingham, M.J.; Mettu, P.S.; Kavusi, S.; Govind, K.; Cousins, S.W.; Farsiu, S. Deep learning-based single-shot pre-diction of differential effects of anti-VEGF treatment in patients with diabetic macular edema. *Biomed. Opt. Express* **2020**, *11*, 1139–1152. [CrossRef] [PubMed]
45. Prahs, P.; Radeck, V.; Mayer, C.; Cvetkov, Y.; Cvetkova, N.; Helbig, H.; Märker, D. OCT-based deep learning algorithm for the evaluation of treatment indication with anti-vascular endothelial growth factor medications. *Graefe's Arch. Clin. Exp. Ophthalmol.* **2018**, *256*, 91–98. [CrossRef]
46. Manivannan, A.; Plskova, J.; Farrow, A.; Mckay, S.; Sharp, P.F.; Forrester, J.V. Ultra-Wide-Field Fluorescein Angiography of the Ocular Fundus. *Am. J. Ophthalmol.* **2005**, *140*, 525–527. [CrossRef] [PubMed]
47. Falavarjani, K.G.; Wang, K.; Khadamy, J.; Sadda, S.R. Ultra-wide-field imaging in diabetic retinopathy; an overview. *J. Curr. Ophthalmol.* **2016**, *28*, 57–60. [CrossRef] [PubMed]
48. Rabbani, H.; Allingham, M.J.; Mettu, P.S.; Cousins, S.W.; Farsiu, S. Fully Automatic Segmentation of Fluorescein Leakage in Subjects With Diabetic Macular Edema. *Investig. Opthalmol. Vis. Sci.* **2015**, *56*, 1482–1492. [CrossRef]
49. Ehlers, J.P.; Wang, K.; Vasanji, A.; Hu, M.; Srivastava, S.K. Automated quantitative characterisation of retinal vascular leakage and microaneurysms in ultra-widefield fluorescein angiography. *Br. J. Ophthalmol.* **2017**, *101*, 696–699. [CrossRef] [PubMed]
50. O'Connell, M.; Sevgi, D.D.; Srivastava, S.K.; Whitney, J.; Hach, J.M.; Atwood, R.; Springer, Q.; Williams, J.; Vasanji, A.; Reese, J.; et al. Longitudinal precision of vasculature parameter assessment on ultra-widefield fluorescein angiography using a deep-learning model for vascular segmentation in eyes without vascular pathology. *Investig. Ophthalmol. Vis. Sci.* **2020**, *61*, 2010.
51. Sevgi, D.D.; Hach, J.; Srivastava, S.K.; Wykoff, C.; O'connell, M.; Whitney, J.; Reese, J.; Ehlers, J.P. Automated quality optimized phase selection in ultra-widefield angiography using retinal vessel segmentation with deep neural networks. *Investig. Ophthalmol. Vis. Sci.* **2020**, *61*, PB00125.
52. Sevgi, D.D.; Scott, A.W.; Martin, A.; Mugnaini, C.; Patel, S.; Linz, M.O.; Nti, A.; Reese, J.; Ehlers, J.P. Longitudinal assessment of quantitative ultra-widefield ischaemic and vascular parameters in sickle cell retinopathy. *Br. J. Ophthalmol.* **2020**. [CrossRef] [PubMed]
53. Jiang, A.C.; Srivastava, S.K.; Hu, M.; Figueiredo, N.; Babiuch, A.; Boss, J.D.; Reese, J.L.; Ehlers, J.P. Quantitative Ultra-Widefield Angiographic Features and Associations with Diabetic Macular Edema. *Ophthalmol. Retin.* **2020**, *4*, 49–56. [CrossRef] [PubMed]
54. Ehlers, J.P.; Jiang, A.C.; Boss, J.D.; Hu, M.; Figueiredo, N.; Babiuch, A.; Talcott, K.; Sharma, S.; Hach, J.; Le, T.K.; et al. Quantitative Ultra-Widefield Angiography and Diabetic Retinopathy Severity. *Ophthalmology* **2019**, *126*, 1527–1532. [CrossRef]
55. Babiuch, A.S.; Wykoff, C.C.; Srivastava, S.K.; Talcott, K.; Zhou, B.; Hach, J.; Hu, M.; Reese, J.L.; Ehlers, J.P. Retinal Leakage Index Dynamics On Ultra-Widefield Fluorescein Angiography In Eyes Treated With Intravitreal Aflibercept For Proliferative Diabetic Retinopathy In The Recovery Study. *Retina* **2020**, *40*, 2175–2183. [CrossRef] [PubMed]
56. Verma, A.; Indian Retina Research Associates (IRRA); Alagorie, A.R.; Ramasamy, K.; van Hemert, J.; Yadav, N.; Pappuru, R.R.; Tufail, A.; Nittala, M.G.; Sadda, S.R.; et al. Distribution of peripheral lesions identified by mydriatic ultra-wide field fundus imaging in diabetic retinopathy. *Graefe's Arch. Clin. Exp. Ophthalmol.* **2020**, *258*, 725–733. [CrossRef]
57. Silva, P.S.; Cruz, A.J.D.; Ledesma, M.G.; van Hemert, J.; Radwan, A.; Cavallerano, J.; Aiello, L.M.; Sun, J.K. Diabetic Retinopathy Severity and Peripheral Lesions Are Associated with Nonperfusion on Ultrawide Field Angiography. *Ophthalmology* **2015**, *122*, 2465–2472. [CrossRef]
58. Figueiredo, N.; Srivastava, S.K.; Singh, R.P.; Babiuch, A.; Sharma, S.; Rachitskaya, A.; Talcott, K.; Reese, J.; Hu, M.; Ehlers, J.P. Longitudinal Panretinal Leakage and Ischemic Indices in Retinal Vascular Disease after Aflibercept Therapy. *Ophthalmol. Retin.* **2020**, *4*, 154–163. [CrossRef]
59. Wykoff, C.C.; Nittala, M.G.; Zhou, B.; Fan, W.; Velaga, S.B.; Lampen, S.I.; Rusakevich, A.; Ehlers, J.P.; Babiuch, A.; Brown, D.M.; et al. Intravitreal Aflibercept for Retinal Nonperfusion in Proliferative Diabetic Retinopathy. *Ophthalmol. Retin.* **2019**, *3*, 1076–1086. [CrossRef]
60. Yu, H.J.; Ehlers, J.P.; Sevgi, D.D.; Hach, J.; O'Connell, M.; Reese, J.L.; Srivastava, S.K.; Wykoff, C.C. Real-Time Photographic- and Fluorescein Angiographic-Guided Management of Diabetic Retinopathy: Randomized PRIME Trial Outcomes. *Am. J. Ophthalmol.* **2021**, *226*, 126–136. [CrossRef] [PubMed]
61. Fan, W.; Nittala, M.G.; Velaga, S.B.; Hirano, T.; Wykoff, C.C.; Ip, M.; Lampen, S.I.; van Hemert, J.; Fleming, A.; Verhoek, M.; et al. Distribution of Nonperfusion and Neovascularization on Ultrawide-Field Fluorescein Angiography in Proliferative Diabetic Retinopathy (RECOVERY Study): Report 1. *Am. J. Ophthalmol.* **2019**, *206*, 154–160. [CrossRef] [PubMed]
62. Mainster, M.A. The fractal properties of retinal vessels: Embryological and clinical implications. *Eye* **1990**, *4*, 235–241. [CrossRef] [PubMed]
63. Fan, W.; Uji, A.; Wang, K.; Falavarjani, K.G.; Wykoff, C.C.; Brown, D.M.; Van Hemert, J.; Sagong, M.; Sadda, S.R.; Ip, M. Severity Of Diabetic Macular Edema Correlates With Retinal Vascular Bed Area On Ultra-Wide Field Fluorescein Angiography: DAVE Study. *Retina* **2020**, *40*, 1029–1037. [CrossRef]
64. Fan, W.; Nittala, M.G.; Fleming, A.; Robertson, G.; Uji, A.; Wykoff, C.C.; Brown, D.M.; van Hemert, J.; Ip, M.; Wang, K.; et al. Relationship Between Retinal Fractal Dimension and Nonperfusion in Diabetic Retinopathy on Ultrawide-Field Fluorescein Angiography. *Am. J. Ophthalmol.* **2020**, *209*, 99–106. [CrossRef] [PubMed]
65. Sevgi, D.D.; Srivastava, S.K.; Whitney, J.; O'Connell, M.; Kar, S.S.; Hu, M.; Reese, J.; Madabhushi, A.; Ehlers, J.P. Characterization of Ultra-Widefield Angiographic Vascular Features in Diabetic Retinopathy with Automated Severity Classification. *Ophthalmol. Sci.* **2021**, *1*, 100049. [CrossRef]

66. Fang, M.; Fan, W.; Shi, Y.; Ip, M.S.; Wykoff, C.C.; Wang, K.; Falavarjani, K.G.; Brown, D.M.; van Hemert, J.; Sadda, S.R. Classification of Regions of Nonperfusion on Ultra-widefield Fluorescein Angiography in Patients with Diabetic Macular Edema. *Am. J. Ophthalmol.* **2019**, *206*, 74–81. [CrossRef] [PubMed]
67. Moosavi, A.; Figueiredo, N.; Prasanna, P.; Srivastava, S.K.; Sharma, S.; Madabhushi, A.; Ehlers, J.P. Imaging Features of Vessels and Leakage Patterns Predict Extended Interval Aflibercept Dosing Using Ultra-Widefield Angiography in Retinal Vascular Disease: Findings From the PERMEATE Study. *IEEE Trans. Biomed. Eng.* **2021**, *68*, 1777–1786. [CrossRef]
68. Hormel, T.T.; Jia, Y.; Jian, Y.; Hwang, T.S.; Bailey, S.T.; Pennesi, M.E.; Wilson, D.J.; Morrison, J.C.; Huang, D. Plexus-specific retinal vascular anatomy and pathologies as seen by projection-resolved optical coherence tomographic angiography. *Prog. Retin. Eye Res.* **2021**, *80*, 100878. [CrossRef]
69. Shahlaee, A.; Pefkianaki, M.; Hsu, J.; Ho, A.C. Measurement of Foveal Avascular Zone Dimensions and its Reliability in Healthy Eyes Using Optical Coherence Tomography Angiography. *Am. J. Ophthalmol.* **2016**, *161*, 50–55.e1. [CrossRef] [PubMed]
70. Barraso, M.; Alé-Chilet, A.; Hernández, T.; Oliva, C.; Vinagre, I.; Ortega, E.; Figueras-Roca, M.; Sala-Puigdollers, A.; Esquinas, C.; Esmatjes, E.; et al. Optical Coherence Tomography Angiography in Type 1 Diabetes Mellitus. Report 1: Diabetic Retinopathy. *Transl. Vis. Sci. Technol.* **2020**, *9*, 34. [CrossRef] [PubMed]
71. Salz, D.A.; De Carlo, T.E.; Adhi, M.; Moult, E.M.; Choi, W.; Baumal, C.R.; Witkin, A.J.; Duker, J.S.; Fujimoto, J.G.; Waheed, N.K. Select Features of Diabetic Retinopathy on Swept-Source Optical Coherence Tomographic Angiography Compared with Fluorescein Angiography and Normal Eyes. *JAMA Ophthalmol.* **2016**, *134*, 644–650. [CrossRef]
72. Freiberg, F.J.; Pfau, M.; Wons, J.; Wirth, M.A.; Becker, M.D.; Michels, S. Optical coherence tomography angiography of the foveal avascular zone in diabetic retinopathy. *Graefe's Arch. Clin. Exp. Ophthalmol.* **2016**, *254*, 1051–1058. [CrossRef]
73. Balaratnasingam, C.; Inoue, M.; Ahn, S.; McCann, J.; Dhrami-Gavazi, E.; Yannuzzi, L.A.; Freund, K.B. Visual Acuity Is Correlated with the Area of the Foveal Avascular Zone in Diabetic Retinopathy and Retinal Vein Occlusion. *Ophthalmology* **2016**, *123*, 2352–2367. [CrossRef] [PubMed]
74. Samara, W.A.; Shahlaee, A.; Adam, M.; Khan, M.A.; Chiang, A.; Maguire, J.I.; Hsu, J.; Ho, A.C. Quantification of Diabetic Macular Ischemia Using Optical Coherence Tomography Angiography and Its Relationship with Visual Acuity. *Ophthalmology* **2017**, *124*, 235–244. [CrossRef]
75. Lee, H.; Lee, M.; Chung, H.; Kim, H.C. Quantification Of Retinal Vessel Tortuosity In Diabetic Retinopathy Using Optical Coherence Tomography Angiography. *Retina* **2018**, *38*, 976–985. [CrossRef] [PubMed]
76. Zarranz-Ventura, J.; Barraso, M.; Alé-Chilet, A.; Hernandez, T.; Oliva, C.; Gascón, J.; Sala-Puigdollers, A.; Figueras-Roca, M.; Vinagre, I.; Ortega, E.; et al. Evaluation of microvascular changes in the perifoveal vascular network using optical coherence tomography angiography (OCTA) in type I diabetes mellitus: A large scale prospective trial. *BMC Med. Imaging* **2019**, *19*, 1–6. [CrossRef] [PubMed]
77. Chu, Z.; Lin, J.; Gao, C.; Xin, C.; Zhang, Q.; Chen, C.-L.; Roisman, L.; Gregori, G.; Rosenfeld, P.J.; Wang, R. Quantitative assessment of the retinal microvasculature using optical coherence tomography angiography. *J. Biomed. Opt.* **2016**, *21*, 066008. [CrossRef]
78. Dupas, B.; Minvielle, W.; Bonnin, S.; Couturier, A.; Erginay, A.; Massin, P.; Gaudric, A.; Tadayoni, R. Association Between Vessel Density and Visual Acuity in Patients with Diabetic Retinopathy and Poorly Controlled Type 1 Diabetes. *JAMA Ophthalmol.* **2018**, *136*, 721–728. [CrossRef]
79. Nguyen, T.T.; Wang, J.J.; Sharrett, A.R.; Islam, F.A.; Klein, R.; Klein, B.E.; Cotch, M.F.; Wong, T.Y. Relationship of Retinal Vascular Caliber with Diabetes and Retinopathy: The Multi-Ethnic Study of Atherosclerosis (MESA). *Diabetes Care* **2007**, *31*, 544–549. [CrossRef]
80. Tsai, A.S.; Wong, T.Y.; Lavanya, R.; Zhang, R.; Hamzah, H.; Tai, E.S.; Cheung, C. Differential association of retinal arteriolar and venular caliber with diabetes and retinopathy. *Diabetes Res. Clin. Pr.* **2011**, *94*, 291–298. [CrossRef] [PubMed]
81. Tang, F.Y.; Ng, D.S.; Lam, A.; Luk, F.; Wong, R.; Chan, C.; Mohamed, S.; Fong, A.; Lok, J.; Tso, T.; et al. Determinants of Quantitative Optical Coherence Tomography Angiography Metrics in Patients with Diabetes. *Sci. Rep.* **2017**, *7*, 1–10. [CrossRef] [PubMed]
82. Maloca, P.M.; IOB Study Group; Spaide, R.F.; De Carvalho, E.R.; Studer, H.P.; Hasler, P.W.; Scholl, H.P.N.; Heeren, T.; Schottenhamml, J.; Balaskas, K.; et al. Novel biomarker of sphericity and cylindricity indices in volume-rendering optical coherence tomography angiography in normal and diabetic eyes: A preliminary study. *Graefe's Arch. Clin. Exp. Ophthalmol.* **2020**, *258*, 711–723. [CrossRef]
83. Le, D.; Alam, M.; Miao, B.A.; Lim, J.I.; Yao, X. Fully automated geometric feature analysis in optical coherence tomography angiography for objective classification of diabetic retinopathy. *Biomed. Opt. Express* **2019**, *10*, 2493–2503. [CrossRef] [PubMed]
84. Nassisi, M.; Lei, J.; Abdelfattah, N.; Karamat, A.; Balasubramanian, S.; Fan, W.; Uji, A.; Marion, K.M.; Baker, K.; Huang, X.; et al. OCT Risk Factors for Development of Late Age-Related Macular Degeneration in the Fellow Eyes of Patients Enrolled in the HARBOR Study. *Ophthalmology* **2019**, *126*, 1667–1674. [CrossRef] [PubMed]
85. Toth, C.A.; Tai, V.; Chiu, S.J.; Winter, K.; Sevilla, M.B.; Daniel, E.; Grunwald, J.E.; Jaffe, G.J.; Martin, D.F.; Ying, G.-S.; et al. Linking OCT, Angiographic, and Photographic Lesion Components in Neovascular Age-Related Macular Degeneration. *Ophthalmol. Retin.* **2017**, *2*, 481–493. [CrossRef] [PubMed]
86. Lunasco, L.; Abraham, J.R.; Sarici, K.; Sevgi, D.D.; Hanumanthu, A.; Cetin, H.; Hu, M.; Srivastava, S.; Reese, J.; Ehlers, J.P. Comparative Assessment of Long-Term Longitudinal Multi-Layer Retinal Dynamics in Non-neovascular Age-Related Macular

Degeneration in Eyes Progressing to Subfoveal Geographic Atrophy and Eyes without Progression. *Investig. Ophthalmol. Vis. Sci.* **2021**, *62*, 2548.

87. Hanumanthu, A.; Sarici, K.; Abraham, J.R.; Whitney, J.; Lunasco, L.; Sevgi, D.D.; Cetin, H.; Srivastava, S.K.; Reese, J.; Ehlers, J.P. Utilizing Higher-Order Quantitative SD-OCT Biomarkers in a Machine Learning Prediction Model for the Development of Subfoveal Geographic Atrophy in Age-Related Macular Degeneration. *Investig. Ophthalmol. Vis. Sci.* **2021**, *62*, 98.

88. Lunasco, L.; Abraham, J.R.; Sarici, K.; Sevgi, D.D.; Hanumanthu, A.; Cetin, H.; Hu, M.; Srivastava, S.K.; Reese, J.; Ehlers, J.P. Risk Classification for Progression to Subfoveal Geographic Atrophy in Dry Age-Related Macular Degeneration Using Machine Learning-Enabled Outer Retinal Feature Extraction. *OSLI Retin.* **2021**, in press.

89. Abdelfattah, N.; Zhang, H.; Boyer, D.S.; Rosenfeld, P.J.; Feuer, W.J.; Gregori, G.; Sadda, S.R. Drusen Volume as a Predictor of Disease Progression in Patients with Late Age-Related Macular Degeneration in the Fellow Eye. *Investig. Opthalmol. Vis. Sci.* **2016**, *57*, 1839–1846. [CrossRef]

90. Ehlers, J.P.; Zahid, R.; Kaiser, P.K.; Heier, J.S.; Brown, D.M.; Meng, X.; Reese, J.; Le, T.K.; Lunasco, L.; Hu, M.; et al. Longitudinal Assessment of Ellipsoid Zone Integrity, Subretinal Hyperreflective Material, and Subretinal Pigment Epithelium Disease in Neovascular Age-Related Macular Degeneration. *Ophthalmol. Retin.* **2021**. [CrossRef]

91. Waldstein, S.; Philip, A.-M.; Leitner, R.; Simader, C.; Langs, G.; Gerendas, B.S.; Schmidt-Erfurth, U. Correlation of 3-Dimensionally Quantified Intraretinal and Subretinal Fluid with Visual Acuity in Neovascular Age-Related Macular Degeneration. *JAMA Ophthalmol.* **2016**, *134*, 182–190. [CrossRef] [PubMed]

92. De Fauw, J.; Ledsam, J.R.; Romera-Paredes, B.; Nikolov, S.; Tomasev, N.; Blackwell, S.; Askham, H.; Glorot, X.; O'Donoghue, B.; Visentin, D.; et al. Clinically applicable deep learning for diagnosis and referral in retinal disease. *Nat. Med.* **2018**, *24*, 1342–1350. [CrossRef]

93. Lee, C.S.; Baughman, D.M.; Lee, A.Y. Deep Learning Is Effective for Classifying Normal versus Age-Related Macular Degeneration OCT Images. *Ophthalmol. Retin.* **2017**, *1*, 322–327. [CrossRef] [PubMed]

94. De Sisternes, L.; Simon, N.; Tibshirani, R.; Leng, T.; Rubin, D.L. Quantitative SD OCT Imaging Biomarkers as Indicators of Age-Related Macular Degeneration Progression. *Investig. Opthalmology Vis. Sci.* **2014**, *55*, 7093–7103. [CrossRef] [PubMed]

95. Schmidt-Erfurth, U.; Waldstein, S.M.; Klimscha, S.; Sadeghipour, A.; Hu, X.; Gerendas, B.S.; Osborne, A.; Bogunović, H. Prediction of Individual Disease Conversion in Early AMD Using Artificial Intelligence. *Investig. Ophthalmol. Vis. Sci.* **2018**, *59*, 3199–3208. [CrossRef] [PubMed]

96. Wong, W.L.; Su, X.; Li, X.; Cheung, C.M.G.; Klein, R.; Cheng, C.Y.; Wong, T.Y. Global prevalence of age-related macular degeneration and disease burden projection for 2020 and 2040: A systematic review and meta-analysis. *Lancet Glob. Health* **2014**, *2*, e106–e116. [CrossRef]

97. Freund, K.B.; Yannuzzi, L.A.; Sorenson, J.A. Age-related Macular Degeneration and Choroidal Neovascularization. *Am. J. Ophthalmol.* **1993**, *115*, 786–791. [CrossRef]

98. Jia, Y.; Bailey, S.T.; Hwang, T.S.; McClintic, S.M.; Gao, S.S.; Pennesi, M.E.; Flaxel, C.J.; Lauer, A.K.; Wilson, D.J.; Hornegger, J.; et al. Quantitative optical coherence tomography angiography of vascular abnormalities in the living human eye. *Proc. Natl. Acad. Sci. USA* **2015**, *112*, E2395–E2402. [CrossRef]

99. Spaide, R.F.; Klancnik, J.M.; Cooney, M.J. Retinal Vascular Layers Imaged by Fluorescein Angiography and Optical Coherence Tomography Angiography. *JAMA Ophthalmol.* **2015**, *133*, 45–50. [CrossRef]

100. Cicinelli, M.V.; Rabiolo, A.; Sacconi, R.; Carnevali, A.; Querques, L.; Bandello, F.; Querques, G. Optical coherence tomography angiography in dry age-related macular degeneration. *Surv. Ophthalmol.* **2018**, *63*, 236–244. [CrossRef]

101. Jia, Y.; Bailey, S.T.; Wilson, D.J.; Tan, O.; Klein, M.L.; Flaxel, C.J.; Potsaid, B.; Liu, J.J.; Lu, C.D.; Kraus, M.F.; et al. Quantitative Optical Coherence Tomography Angiography of Choroidal Neovascularization in Age-Related Macular Degeneration. *Ophthalmology* **2014**, *121*, 1435–1444. [CrossRef]

102. Uchida, A.; Hu, M.; Babiuch, A.; Srivastava, S.K.; Singh, R.P.; Kaiser, P.K.; Talcott, K.; Rachitskaya, A.; Ehlers, J.P. Optical coherence tomography angiography characteristics of choroidal neovascularization requiring varied dosing frequencies in treat-and-extend management: An analysis of the AVATAR study. *PLoS ONE* **2019**, *14*, e0218889. [CrossRef]

103. Chatziralli, I.; Theodossiadis, G.; Panagiotidis, D.; Pousoulidi, P.; Theodossiadis, P. Choriocapillaris Vascular Density Changes in Patients with Drusen: Cross-Sectional Study Based on Optical Coherence Tomography Angiography Findings. *Ophthalmol. Ther.* **2018**, *7*, 101–107. [CrossRef] [PubMed]

104. Lane, M.; Moult, E.M.; Novais, E.A.; Louzada, R.N.; Cole, E.D.; Lee, B.; Husvogt, L.; Keane, P.A.; Denniston, A.K.; Witkin, A.J.; et al. Visualizing the Choriocapillaris Under Drusen: Comparing 1050-nm Swept-Source Versus 840-nm Spectral-Domain Optical Coherence Tomography Angiography. *Investig. Opthalmol. Vis. Sci.* **2016**, *57*, OCT585–OCT590. [CrossRef] [PubMed]

105. Byon, I.; Ji, Y.; Alagorie, A.R.; Tiosano, L.; Sadda, S.R. Topographic Assessment Of Choriocapillaris Flow Deficits In The Intermediate Age-Related Macular Degeneration Eyes With Hyporeflective Cores Inside Drusen. *Retina* **2021**, *41*, 393–401. [CrossRef] [PubMed]

106. Choi, W.; Moult, E.M.; Waheed, N.K.; Adhi, M.; Lee, B.; Lu, C.D.; de Carlo, T.E.; Jayaraman, V.; Rosenfeld, P.J.; Duker, J.S.; et al. Ultrahigh-Speed, Swept-Source Optical Coherence Tomography Angiography in Nonexudative Age-Related Macular Degeneration with Geographic Atrophy. *Ophthalmology* **2015**, *122*, 2532–2544. [CrossRef] [PubMed]

107. Camino, A.; Guo, Y.; You, Q.S.; Wang, J.; Huang, D.; Bailey, S.T.; Jia, Y. Detecting and measuring areas of choriocapillaris low perfusion in intermediate, non-neovascular age-related macular degeneration. *Neurophotonics* **2019**, *6*, 041108. [CrossRef]

Journal of
Personalized
Medicine

Article

The 2-Year Leakage Index and Quantitative Microaneurysm Results of the RECOVERY Study: Quantitative Ultra-Widefield Findings in Proliferative Diabetic Retinopathy Treated with Intravitreal Aflibercept

Amy S. Babiuch [1,2], Charles C. Wykoff [3,4], Sari Yordi [2], Hannah Yu [3,4], Sunil K. Srivastava [1,2], Ming Hu [2,5], Thuy K. Le [2], Leina Lunasco [2], Jamie Reese [1,2], Muneeswar G. Nittala [6], SriniVas R. Sadda [6] and Justis P. Ehlers [1,2,*]

1 Cole Eye Institute, Cleveland Clinic, Cleveland, OH 44106, USA; babiuca@ccf.org (A.S.B.); SRIVASS2@ccf.org (S.K.S.); REESEJ3@ccf.org (J.R.)
2 The Tony and Leona Campane Center for Excellence for Image-Guided Surgery and Advanced Imaging Research, Cole Eye Institute, Cleveland Clinic, Cleveland, OH 44106, USA; yordis2@ccf.org (S.Y.); hum@ccf.org (M.H.); thuyle1561@gmail.com (T.K.L.); LUNASCL@ccf.org (L.L.)
3 Retina Consultants of Texas, Kingwood, TX 77339, USA; ccwmd@retinaconsultantstexas.com (C.C.W.); hannah.yu@houstonretina.com (H.Y.)
4 Blanton Eye Institute, Houston Methodist Hospital, Houston, TX 77030, USA
5 Department of Quantitative Health Sciences, Lerner Research Institute, Cleveland Clinic Foundation, Cleveland, OH 44195, USA
6 Doheny Eye Institute, Los Angeles, CA 90033, USA; mnittala@doheny.org (M.G.N.); ssadda@doheny.org (S.R.S.)
* Correspondence: ehlersj@ccf.org; Tel.: +1-(216)-636-0183

Citation: Babiuch, A.S.; Wykoff, C.C.; Yordi, S.; Yu, H.; Srivastava, S.K.; Hu, M.; Le, T.K.; Lunasco, L.; Reese, J.; Nittala, M.G.; et al. The 2-Year Leakage Index and Quantitative Microaneurysm Results of the RECOVERY Study: Quantitative Ultra-Widefield Findings in Proliferative Diabetic Retinopathy Treated with Intravitreal Aflibercept. *J. Pers. Med.* **2021**, *11*, 1126. https://doi.org/10.3390/jpm11111126

Academic Editors: Peter D. Westenskow and Andreas Ebneter

Received: 25 September 2021
Accepted: 27 October 2021
Published: 1 November 2021

Publisher's Note: MDPI stays neutral with regard to jurisdictional claims in published maps and institutional affiliations.

Abstract: Eyes with proliferative diabetic retinopathy (PDR) have been shown to improve in the leakage index and microaneurysm (MA) count after intravitreal aflibercept (IAI) treatment. The authors investigated these changes via automatic segmentation on ultra-widefield fluorescein angiography (UWFA). Forty subjects with PDR were randomized to receive either 2 mg IAI every 4 weeks (Arm 1) or every 12 weeks (Arm 2) through Year 1. After Year 1, Arm 1 switched to quarterly IAI and Arm 2 to monthly IAI through Year 2. By Year 2, the Arm 1 leakage index decreased by 43% from Baseline ($p = 0.03$) but increased by 59% from Year 1 ($p = 0.04$). Arm 2 decreased by 61% from Baseline ($p = 0.008$) and by 31% from Year 1 ($p = 0.12$). Both cohorts exhibited a significant decline in MAs from Baseline to Year 2 (871 to 410; $p < 0.001$; 776 to 207; $p < 0.001$, respectively). Subjects with an improved leakage and MA count showed a more significant improvement in the Diabetic Retinopathy Severity Scale (DRSS) score. Moreover, central subfield thickness (CST) was positively associated with changes in the leakage index. In conclusion, the leakage index and MA counts significantly improved from Baseline following IAI treatment, and monthly injections provided a more rapid and sustained reduction in these parameters compared with quarterly injections.

Keywords: anti-vascular endothelial growth factor; diabetic macular edema; diabetic retinopathy; leakage index; microaneurysms; intravitreal aflibercept; neovascularization; optical coherence tomography; ultra-widefield fluorescein angiography

1. Introduction

Currently, over 34.2 million people in the United States have Type I or Type II diabetes mellitus, and 88 million US adults have prediabetes [1]. Diabetic retinopathy (DR), the most common form of diabetic-related eye disease, is the leading cause of visual impairment and blindness in working-age Americans, and is expected nearly to double from 2010 to 2050 (7.7 million to 14.6 million) [2]. Approximately one-third of patients with DR are estimated to have vision-threatening complications such as proliferative diabetic retinopathy (PDR) or diabetic macular edema (DME) [3].

Vision-threatening complications in DR arise from prolonged hyperglycemia causing a cascade of biochemical pathways. These pathways lead to oxidative stress in the retinal vasculature and subsequently cause microvascular dysfunction. Ultimately, the development of retinal ischemia in DR triggers the release of the Vascular Endothelial Growth Factor (VEGF), a signaling protein that potentiates this cycle, leading to progressive ischemia and resultant neovascularization, the hallmark of proliferative diabetic retinopathy (PDR). The use of fluorescein angiography (FA) can demonstrate microaneurysms (MAs), vascular leakage, retinal ischemia, and neovascularization (NV) [3]. Retinal vision threatening complications of PDR can include traction retinal detachment, vitreous hemorrhage, macular edema, and ischemic changes [4–6]. If neovascularization occurs in the anterior chamber, neovascular glaucoma may develop with potential complications such as hyphema, optic nerve disease, and ensuing vision loss [7].

Classically, treatment for PDR has been panretinal photocoagulation (PRP). Its efficacy in halting disease progression was demonstrated in multiple studies over the last 50 years [8–11]. Therapies specifically targeting the neovascularization process, in particular the VEGF blockade, recently showed the potential to preserve and possibly reverse underlying retinal damage and vision loss that can occur in the setting of PDR. The Diabetic Retinopathy Clinical Research Network (DRCR.net) Protocol S study demonstrated that intravitreal anti-VEGF therapy was non-inferior to PRP and was potentially associated with fewer complications [12]. Unlike PRP, which slows disease progression, anti-VEGF treatment can not only slow disease progression but can also improve the Diabetic Retinopathy Severity Scale (DRSS) score, as defined by the Early Treatment Diabetic Retinopathy Study (ETDRS). Several longitudinal studies demonstrated at least a 2-step improvement in DRSS with anti-VEGF therapy compared to sham or PRP treatment over 2, 3, and 5 years [13–16], with some showing improved visual acuity, slower progression, and decreased macular edema when compared to PRP [14,15].

The ETDRS research group previously defined DRSS using 7-frame fundus photographs to characterize and monitor changes in the retina over time [15,17,18]. More recently, ultra-widefield photography and ultra-widefield fluorescein angiography (UWFA) have allowed for a wider field of view of up to 200°, making it possible to monitor panretinal changes with a single image. In order to standardize and more effectively track these changes, automated programs were employed to measure angiographic parameters such as microaneurysm count, leakage, and ischemia [19–25]. Although the effects of anti-VEGF on PDR features are recognized, there is limited information regarding the effect on quantitative angiographic features on UWFA. The purpose of this study is to assess longitudinally the MA count and leakage occurring on UWFA images over a 2-year period in patients with PDR being treated with a fixed-interval intravitreal aflibercept.

2. Materials and Methods

RECOVERY (Intravitreal Aflibercept for Retinal Non-Perfusion in Proliferative Diabetic Retinopathy) is a prospective, randomized, multicenter, and open-label clinical trial (NCT02863354; IND131056; https://clinicaltrials.gov/ct2/show/NCT02863354, accessed on 8 September 2021) as previously described [26–28]. Institutional Review Board (IRB)/Ethics Committee approval was obtained (Sterling IRB); the tenets of the Declaration of Helsinki were followed, and the study is in accord with the Health Insurance Portability and Accountability Act of 1996. All subjects were enrolled at the Retina Consultants of Texas (Houston, Katy, and Woodlands, TX, USA).

2.1. Inclusion and Exclusion Criteria

Inclusion in the study required PDR, an EDTRS best corrected visual acuity of ≥19, and substantial nonperfusion (≥20 disk areas). Subjects with previous anti-VEGF treatment, a history of PRP or vitreoretinal surgery, a clinically relevant DME, or $\geq$320 μm central retinal thickness in the study eye were excluded.

2.2. Study Design

At enrollment, subjects were consented for participation in the trial and then randomized 1:1. Arm 1 (n = 20) received 2 mg (0.05 mL) intravitreal aflibercept injection (IAI) every month (q4weeks; q28 $\pm$ 7 days). Arm 2 (n = 20) received 2 mg IAI quarterly (q12weeks; q3 months). After Year 1, treatment regimens were crossed over, and the Arm 1 began receiving quarterly injections whereas the Arm 2 began receiving monthly injections until the study end at Year 2. Each visit consisted of BCVA testing with the ETDRS chart, a slit lamp examination, an indirect ophthalmoscope examination, and spectral-domain optical coherence tomography (SD-OCT) scanning.

If a patient undergoing quarterly treatment met prespecified criteria at any visit, they were to be treated every 4 weeks with IAI. These criteria were (1) increased neovascularization, (2) BCVA decrease by $\geq$5 letters due to progressive DME or PDR, (3) worsening of DME causing vision loss, (4) total area of retinal ischemia increasing by 10%. If a patient was then determined to be either stable or improved by these criteria, they were continued on their pre-specified treatment. All subjects could also receive rescue treatment with PRP if PDR progressed despite IAI treatment.

UWFA images were obtained after dilation at Baseline, Year 1, and Year 2 via the Optos 200Tx (Optos plc, Dunfermline, UK). Images were taken during early (~0–60 s after fluorescein dye injection), middle (~60–180 s), and late (~300 s) phases of the angiogram. The images were then transformed to stereographic projection images using proprietary software available from the manufacturer based on ray tracing each pixel with a combined optical model of the imaging device, as previously described [15].

2.3. Outcome Measures

Early and late images were analyzed by an automated assessment platform to provide quantitative image feature extraction, including the MA count and leakage index [21,24,25]. Two masked readers reviewed the program's automated segmentation and manually corrected any errors. Three pre-specified macula-centered zones were also created to enable the calculation of zonal changes. Zone 1 defines a posterior zone with a 3-disc-diameter boundary including the fovea. Zone 2 defines a midperiphery zone with a 6-disc-diameter boundary centered at the fovea. Zone 3 defines a far periphery zone with a 9-disc-diameter boundary centered at the fovea.

The MA count analysis was performed in the mid-arteriovenous phase of the fluorescein angiogram. MAs were defined as small dots which were significantly hyperfluorescent compared to the surrounding choroidal fluorescence in early-mid-phase fluorescein angiography images [21,28]. For leakage analysis, an early and late phase UWFA image was selected. Leakage was defined as an area of increased hyperfluorescence in the late phase compared to the early phase. The panretinal leakage index was calculated as the area of leakage divided by the total analyzable retinal area. Values were multiplied by 100 to express as a percentage [27,28]. DRSS scores were derived in accordance with the Early Treatment Diabetic Retinopathy Study (ETDRS).

2.4. Statistical Analysis

Statistical analysis was performed with R software version 3.4.3 (www.r-project.org; last accessed on 8 September 2021) and the Microsoft Excel statistical function to compare mean differences in each group between Baseline and Year 1, Baseline and Year 2, and between Year 1 and Year 2. Differences in groups were also determined per the region of interest, including the total analyzable retina, Zone 1, Zone 2, and Zone 3. RECOVERY data were also analyzed by visit: Baseline, Year 1, and Year 2.

3. Results

3.1. Clinical Characteristics

Forty subjects were randomly allocated 1:1 to Arm 1 ($n = 20$) and Arm 2 ($n = 20$). All characteristics were similar between Arm 1 and Arm 2 at Baseline. In Arm 1, the mean age was 47.7 ± 12.1 years, with 9 (45%) male and 11 (55%) female subjects. Four (20%) subjects had Type I Diabetes Mellitus (DM) while 16 (80%) were diagnosed with Type II DM, with an average Baseline glycated hemoglobin A1C (HbA1c) of 9.7 ± 2.2%. Baseline BCVA was 20/32 (ETDRS 77.8 ± 6.6), and the central subfield thickness (CST) was 279.7 ± 38.8 μm. In Arm 2, the mean age was 48.3 ± 12.0 years. Furthermore, 12 (60%) were male and 8 (40%) were female; 5 (25%) subjects had Type I DM while 15 (75%) had Type II DM. Mean Baseline HbA1c was 9.2 ± 2.7%. At Baseline, mean BCVA was 20/25 (ETDRS 79.0 ± 8.2), and the mean CST was 276.4 ± 22.7 μm.

All subjects received an average of 14.4 ± 3.8 total injections up to Year 2, with no overall significant difference between the cohorts ($p = 0.16$). Arm 1 received a mean 10.95 ± 1.8 injections between Baseline and Year 1, and 4.25 ± 1.5 injections between Year 1 and Year 2. Arm 2 received an average of 3.9 ± 0.5 injections by Year 1, and 9.6 ± 4.4 injections between Year 1 and Year 2. All Baseline characteristics are listed in Table 1.

Table 1. Baseline Demographics and Ocular Characteristics.

	Arm 1 ($n = 20$)	Arm 2 ($n = 20$)	Total ($n = 40$)	*p*-Value, Arm 1 vs. Arm 2
Age	47.7 ± 12.1	48.3 ± 12.0	48.0 ± 12.1	$p = 0.88$ [†]
Eye (%)				
Right	10 (50)	10 (50)	20 (50)	$p = 1.00$ [*]
Left	10 (50)	10 (50)	20 (50)	
Sex (%)				
Male	9 (45)	12 (60)	21 (52.5)	$p = 0.53$ [*]
Female	11 (55)	8 (40)	19 (47.5)	
Diabetes type (%)				
Type 1	4 (20)	5 (25)	9 (23)	$p = 1.00$ [*]
Type 2	16 (80)	15 (75)	31 (77)	
Years with Diabetes	16.4 ± 8.9	15.8 ± 9.4	16.1 ± 9.0	$p = 0.82$ [†]
BMI, kg/m^2	33.1 ± 6.9	31.8 ± 6.4	32.4 ± 6.6	$p = 0.54$ [†]
HbA1C, %	9.7 ± 2.2	9.2 ± 2.7	9.4 ± 2.5	$p = 0.56$ [†]
Lens status (%)				
Phakic	19 (95)	18 (90)	37 (92.5)	$p = 1.00$ [*]
Pseudophakic	1 (5)	2 (10)	3 (7.5)	
CST, μm	279.7 ± 38.8	276.4 ± 22.7	278 ± 31.8	$p = 0.75$ [†]
ETDRS, letters	77.8 ± 6.6	79 ± 8.2	78.4 ± 7.5	$p = 0.64$ [†]
Injections, n	15.2 ± 2.8	13.5 ± 4.5	14.4 ± 3.8	$p = 0.16$ [†]

[*] Chi-square test; [†] Two-sample *t*-test; BMI = Body Mass Index (kg/m^2); HbA1c = Glycated hemoglobin A1c (%); CST = Central Subfield Thickness (μm); ETDRS = Early Treatment Diabetic Retinopathy Score.

3.2. Panretinal Leakage Index

At Baseline, the mean panretinal leakage index in Arm 1 and Arm 2 was 6.4 ± 5.4% and 4.4 ± 2.5% ($p = 0.15$), respectively. At Year 1, the total leakage index of Arm 1 improved significantly to 1.5 ± 1.45% ($p < 0.001$). Following crossover to q12week IAI dosing, the Arm 1 leakage index increased to 3.7 ± 3.5% at Year 2, which is a 43% decrease from Baseline ($p = 0.03$), but a 59% increase following the transition to quarterly dosing ($p = 0.03$). In Arm 2, panretinal leakage significantly improved to 2.5 ± 2.7% ($p = 0.04$) at Year 1. Following treatment crossover to q12week IAI, further improvement was demonstrated as 1.7 ± 3.4% (31% decrease; $p = 0.12$) through Year 2, but this was not a significant change from Year 1. However, the improvement from Baseline to Year 2 remained significant, demonstrating a 61% decrease from Baseline ($p = 0.008$). A representative case is shown in

Figure 1, and changes in the mean leakage index in both arms are presented graphically in Figure 2.

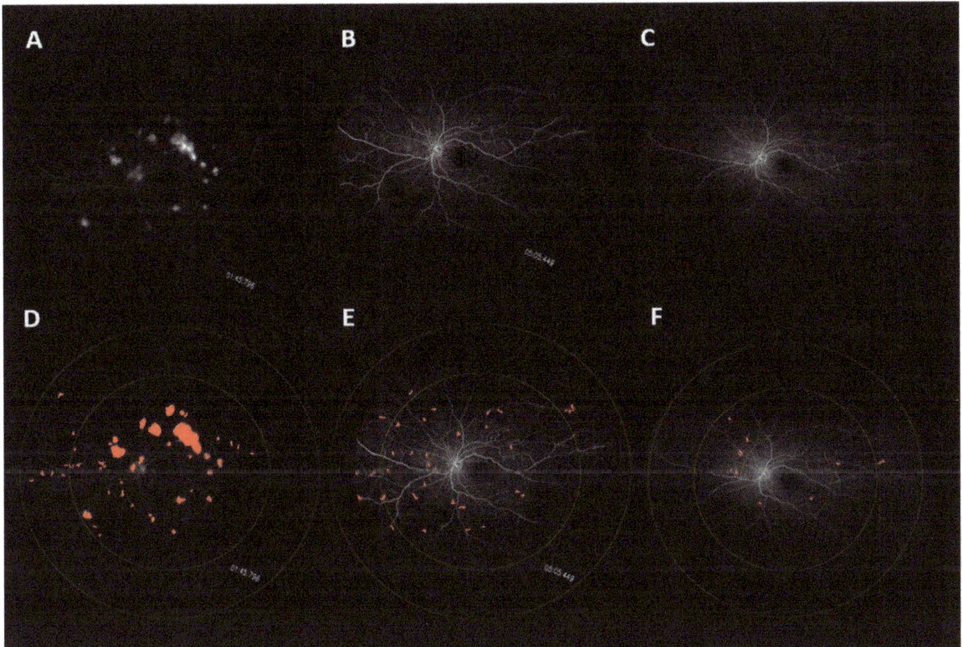

Figure 1. UWFA fundus images of a representative case showing changes in leakage over time. Fundus images are shown at (**A**) Baseline, (**B**) Year 1, (**C**) Year 3. Leakage mask over-lays are shown at (**D**) Baseline, (**E**) Year 1, and (**F**) Year 2. (**D–F**) also show the macula-centered concentric rings: Zone 1 defines a posterior zone with a 3-disc-diameter boundary including the fovea. Zone 2 defines a mid-periphery zone with a 6-disc-diameter boundary centered at the fovea. Zone 3 defines a far periphery zone with a 9-disc-diameter boundary centered at the fovea.

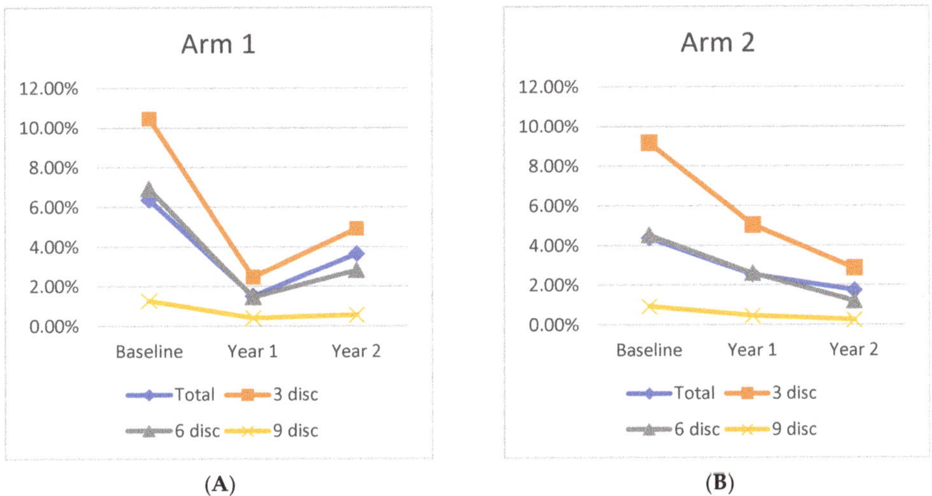

Figure 2. Change in leakage index over time for (**A**) Arm 1 and (**B**) Arm 2 in each region.

Overall, 26 subjects, 46% (12/26) from Arm 1 and 54% (14/26) from Arm 2, showed an improvement in the panretinal leakage index from Baseline to Year 2, while 6 subjects, 83% (5/6) from Arm 1 and 17% (1/6) from Arm 2, worsened in the leakage index. Those with the most improved leakage index had a significantly higher leakage index at Baseline (6.08 ± 4.93% vs. 3.60 ± 1.92%; p = 0.02) and also demonstrated a greater decrease in CST from Baseline to Year 1 (−29.17 ± 28.2 vs. −24.3 ± 16.0; p = 0.009) and to Year 2 (−31.83 ± 32.6 vs. −28.3 ± 16.1; p = 0.03). Notably, all subjects in Arm 1 (19/19) showed an improvement in the panretinal leakage index from Baseline to Year 1. However, in Arm 2, only 13/18 (72%) subjects showed an improvement in the leakage index before treatment crossover, but these subjects showed an overall improvement in the panretinal leakage index from Baseline to Year 2 (−4.10 ± 1.64%). The other five subjects in Arm 2 who had a worsening panretinal leakage index before crossover (5/18; 28%) demonstrated an overall increase in the panretinal leakage index from Baseline to Year 2 (1.42 ± 3.50%; p = 0.046).

3.3. Zonal Leakage Index

At Baseline, mean leakage index values in Zone 1, Zone 2, and Zone 3 were similar between Arm 1 and Arm 2. By Year 1, both cohorts exhibited a significant leakage index improvement in Zone 1, as Arm 1 decreased from 10.5 ± 6.3% at Baseline to 2.5 ± 2.3% (p < 0.01) at Year 1, while Arm 2 decreased from 9.2 ± 4.2% at Baseline to 5.1 ± 5.9% (p = 0.02) at Year 1. Moreover, in Arm 1, there was significant leakage index improvement in Zone 2 (6.9% to 1.5%; p = 0.03) and Zone 3 (1.3% to 0.4%; p = 0.05). Arm 2 had a significant improvement in Zone 2 (4.5% to 2.6%; p = 0.045) from Baseline to Year 1, and Zone 3 had a nonsignificant decrease in the leakage index (0.9% to 0.5%; p = 0.17). Differences in the leakage index per Zone are illustrated in Figure 3.

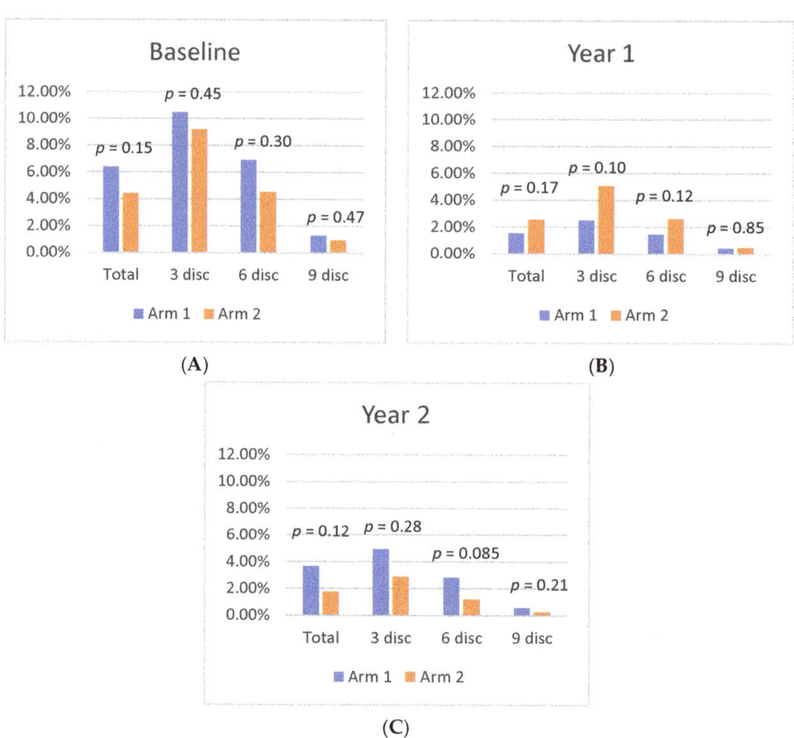

Figure 3. Differences in leakage index per region between Arm 1 and Arm 2 at (**A**) Baseline, (**B**) Year 1, and (**C**) Year 2.

Following treatment crossover, the Zone 1 leakage index increased significantly in Arm 1, but continued to decrease in Arm 2. At Year 1, the leakage index of Arm 1 in Zone 1 was 2.5% and increased to 4.9% ($p = 0.01$) by Year 2, representing a 53% decrease from Baseline ($p = 0.006$). In Arm 2, the leakage index of Zone 1 decreased from 5.1% to 2.9% ($p = 0.05$) by Year 2, which represents a 68% decrease from Baseline ($p = 0.001$). This pattern was also observed in Zone 2 and Zone 3 in both cohorts from Year 1 to Year 2, although the only significant change was seen in Zone 2 of Arm 2, which decreased to 1.2% at Year 2, a 54% decrease from Year 1 ($p = 0.01$), and a 73% decrease from Baseline ($p < 0.001$).

3.4. Microaneurysms

From Baseline to Year 1, the panretinal MA count decreased significantly and within each zone. A representative case is shown in Figure 4. In Arm 1, the mean panretinal MA count decreased by 48% (870.7 to 455.4; $p < 0.001$), 63% in Zone 1 (57.5 to 21.2; $p < 0.001$), 62% in Zone 2 (148.8 to 60.6; $p < 0.001$), and 52% in Zone 3 (380 to 181; $p < 0.001$). In Arm 2, the panretinal MA count decreased by 50% (775.7 to 388.6; $p < 0.001$), 38% in Zone 1 (39.6 to 25.1; $p = 0.009$), 47% in Zone 2 (105.8 to 56.4; $p < 0.001$), and 54% in Zone 3 (326.3 to 151.8; $p < 0.001$). However, it is important to note that there was a significant difference in the mean MA count between Arm 1 and Arm 2 at Baseline in Zone 1 (57.5 vs. 39.6 respectively; $p = 0.03$) and in Zone 2 (148.8 vs. 105.8 respectively; $p = 0.04$). The changes in the mean MA count over time can be seen in Figure 5.

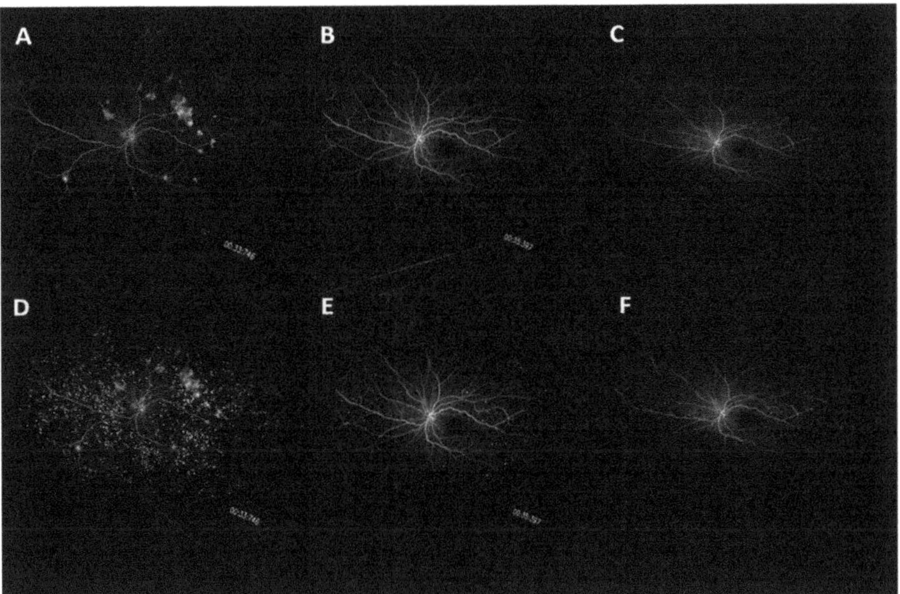

Figure 4. UWFA fundus images of a representative case showing changes in microaneurysm count over time. Fundus images are shown at (**A**) Baseline, (**B**) Year 1, (**C**) Year 3. Microaneurysm mask overlays are shown at (**D**) Baseline, (**E**) Year 1, and (**F**) Year 2.

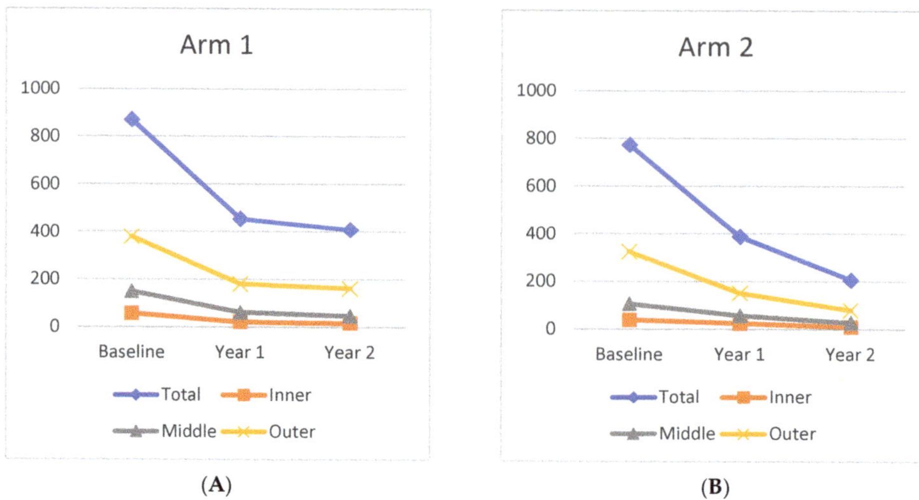

(**A**)   (**B**)

Figure 5. Change in microaneurysm count over time for (**A**) Arm 1 and (**B**) Arm 2 in each region.

The mean MA count continued to decrease after treatment crossover in both cohorts. When compared to Baseline, the MA count showed significant panretinal improvement and in all zones by Year 2. However, from Year 1 to Year 2, only Arm 2 had significant decreases in the mean MA count in all regions (panretinal, $p < 0.001$; Zone 1, $p = 0.002$; Zone 2, $p < 0.001$; Zone 3, $p = 0.002$). When compared to Baseline, Arm 1 showed a significant panretinal decrease of 53% (870.7 to 409.9; $p < 0.001$), 70% in Zone 1 (57.5 to 17.5; $p < 0.001$), 68% in Zone 2 (148.8 to 46.6; $p < 0.001$), and 57% in Zone 3 (379.6 to 162.9; $p < 0.001$) by Year 2. In Arm 2, the mean panretinal MA burden significantly decreased by 73% (775.7 to 388.6; $p < 0.001$), by 75% in Zone 1 (39.6 to 9.7; $p < 0.001$), by 75% in Zone 2 (105.8 to 26.6; $p < 0.001$), and by 75% in Zone 3 (326.3 to 80.5; $p < 0.001$) from Baseline to Year 2. These region differences are illustrated in Figure 6.

Between Year 1 and 2, eight subjects had both an increase in the MA count and panretinal leakage index, 7/8 (87.5%) of which were from Arm 1 while 1/8 (12.5%) were from Arm 2. The subjects who exhibited a concurrent increased MA count and panretinal leakage index after treatment crossover showed a worsening in DRSS by 1.63 ± 1.68 steps in that time period, whereas those that did not had an improvement of 0.48 ± 1.21 steps ($p = 0.009$). Moreover, between Baseline and Year 2, those with a concurrent worsening MA and panretinal leakage index after crossover had a less significant improvement in DRSS than those that did not (+1.25 ± 1.67 vs. +3.35 ± 1.83; $p = 0.01$).

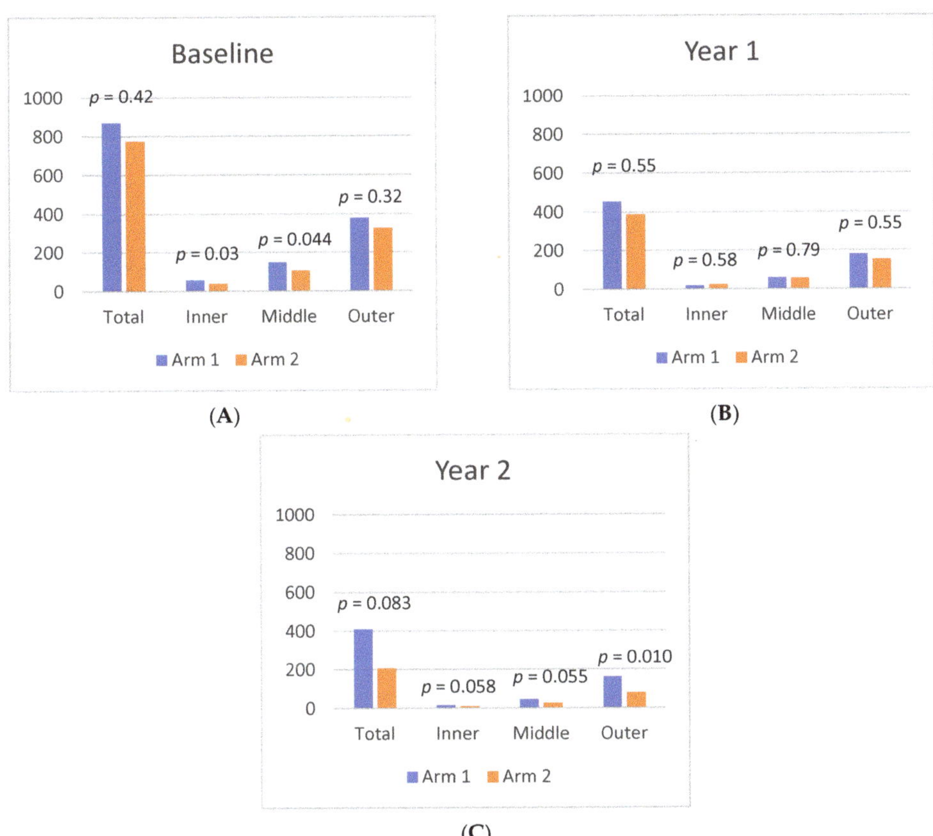

Figure 6. Differences in microaneurysm count per region between Arm 1 and Arm 2 at (**A**) Baseline, (**B**) Year 1, (**C**) and Year 2.

3.5. Functional and Anatomic Outcomes

At Baseline, the mean BCVA of Arm 1 was 20/32 (ETDRS 77.8 $\pm$ 6.6). Visual acuity significantly improved in Arm 1 from 20/32 to 20/25 (ETDRS 82.8 $\pm$ 7.8; $p = 0.01$) by Year 1, and remained at 20/25 (ETDRS 83.2 $\pm$ 9.7) by Year 2. Arm 2 remained at 20/25 between Baseline (ETDRS 79.0 $\pm$ 8.2) and Year 1 (ETDRS 80.3 $\pm$ 15.6; $p = 0.69$), but improved to 20/20 (ETDRS 87.6 $\pm$ 5.8) by Year 2, which is a significant improvement from Baseline ($p < 0.001$). Overall, both cohorts exhibited a significant improvement in visual acuity from Baseline to Year 2 ($p < 0.001$).

The mean central subfield thickness (CST) was similar between Arm 1 and Arm 2 at Baseline—279.7 $\pm$ 38.8 μm and 276.4 $\pm$ 22.7 μm ($p = 0.75$), respectively. Both groups improved significantly from Baseline to Year 1, decreasing by 11.5% (279.7 to 247.5 μm; $p < 0.001$) in Arm 1 and by 7.6% (276.4 to 255.3 μm; $p = 0.006$) in Arm 2. Arm 1 and Arm 2 also improved significantly from Baseline by Year 2 (246.3 $\pm$ 30.9 μm; $p < 0.001$, 249.6 $\pm$ 29.2 μm; $p = 0.01$, respectively). The mean change in CST was moderately positively correlated to the change in the leakage index in Arm 1 in all time periods, but was only moderately positively correlated in Arm 2 after switching to monthly dosing at Year 1 (see Figure 7).

	Years	Correlations with change in CST	
		Change in LI	Change in MA
Total	0 to 1	0.34	0.23
	1 to 2	0.43	0.05
	0 to 2	0.22	0.11
Cohort 1	0 to 1	0.47	0.5
	1 to 2	0.52	−0.14
	0 to 2	0.48	0.42
Cohort 2	0 to 1	−0.04	−0.13
	1 to 2	0.46	0.16
	0 to 2	−0.13	−0.32

Negative correlation — Positive correlation

Figure 7. Correlation heat map (red = negative correlation; green = positive correlation). Change in CST correlated to change in leakage index and MA count in the respective time period. Numbers correspond to the Pearson correlation coefficient (r).

DRSS improvement (i.e., decrease in DRSS value) was also apparent in both cohorts. From Baseline to Year 1, 100% (17/17) of subjects in Arm 1 had at least a 1-step improvement, and 70.6% (12/17) had at least a 2-step improvement. In Arm 2, 83.3% (14/16) had at least a 1-step improvement, with 75% (12/16) having improved by at least 2 steps. Between Year 1 and Year 2, 23.5% (4/17) of Arm 1 and 62.5% (10/16) of Arm 2 had at least a 1-step improvement. However, 47.1% (8/17) of subjects in Arm 1 exhibited a worsening of DRSS (i.e., increase in score value) by at least one step during that period compared to 6.25% (1/16) of subjects in Arm 2. Overall, between Baseline and Year 2, Arm 1 had 23.5% (4/17) of subjects maintain the same DRSS, 76.5% (13/17) improved by at least one step, and 58.8% (10/17) improved by at least 2 steps. In that period, 6.3% (1/16) of Arm 2 had stable DRSS, 92.7% (15/16) had at least a 1-step improvement, and 81.3% (13/16) had at least a 2-step improvement.

Subjects with at least a 1-step improvement in DRSS showed a greater reduction in the leakage index over time, as subjects that had an improved DRSS between Baseline and Year 1 showed a reduction of 2.70 ± 4.98% in the leakage index, while those without improvement had a mean increase of 1.38 ± 2.34% ($p = 0.035$) in that period. Similarly, between Year 1 and Year 2, subjects with at least a 1-step DRSS improvement had a mean reduction of the panretinal leakage index of 0.68 ± 1.62%, whereas subjects without improvement in DRSS had a mean increase of 1.86 ± 3.51% ($p = 0.013$) during that period. Although a greater reduction in the MA count over time was also observed in patients with improved DRSS, the change was not significantly different from the group with stable/worsening DRSS from Baseline to Year 1 (−395 vs. −316, respectively; $p = 0.55$), Year 1 to Year 2 (−206 vs. −100; $p = 0.21$), and Baseline to Year 2 (−568 vs. −429; $p = 0.52$).

Nine subjects had worsening DRSS between Year 1 and 2, 8/9 (88.9%) of whom were from Arm 1. When compared to those that did not have worsening DRSS during that period ($n = 24$), there was a significant difference in the mean change in the panretinal leakage index (+3.87 ± 3.75%, −0.46 ± 1.66%, respectively; $p = 0.008$) and MA count (+4 ± 246, −205 ± 217, respectively; $p = 0.04$) between Year 1 and 2. When comparing the 8 subjects from Arm 1 with worsening DRSS to the other subjects that had stable/improved DRSS ($n = 9$) after treatment crossover, there was a significant difference in the Baseline

MA count, as those in Arm 1 with worsening DRSS had a Baseline mean MA count of 695 ± 411 whereas those that did not worsen had 1098 ± 382 ($p = 0.05$).

In Arm 1, among the 9 subjects that had DRSS improvement from Baseline to Year 1, five (55.6%) had stable DRSS from Year 1 to Year 2 whereas 4 (44.4%) continued to improve. The panretinal leakage index at Baseline in the stable group was $4.32 \pm 3.12\%$ and was $9.25 \pm 1.97\%$ ($p = 0.02$) in the group that showed improvement, and CST measurements were 265.8 ± 36.57 μm and 310.25 ± 18.25 μm ($p = 0.05$), respectively. The change in the panretinal leakage index after crossover was also significantly different in these groups, as the stable group's leakage index decreased by $1.64 \pm 3.97\%$ whereas the improved group's leakage index decreased by $8.17 \pm 1.63\%$ ($p = 0.02$). The change in CST was also greater in the improved group, decreasing by a mean of 65.0 ± 22.73 μm by Year 1 and 71.0 ± 25.14 μm by Year 2, while the stable group decreased by 21.8 ± 16.67 μm ($p = 0.02$) from Baseline to Year 1, and by 25.8 ± 11.39 μm ($p = 0.03$) by Year 2 in the stable group.

3.6. Optimal Responders

Optimal responders were defined as subjects who show an improvement in the panretinal leakage index and MA count between Baseline and Year 2. In that group, 13/25 (52%) were male, 12/25 (48%) were current/previous smokers, and 14/25 (56%) had hypertension at Baseline. Overall, both Arm 1 and Arm 2 had a similar number of optimal responders (12/25, 48% vs. 13/25, 52%, respectively), with the majority having a DRSS of a moderate PDR (DRSS − 65). The distributions of DRSS at Baseline were as follows: 6 high-risk PDR (three subjects in Arm 1, three subjects in Arm 2), 16 moderate PDR (eight subjects in Arm 1, eight subjects in Arm 2), and 3 mild PDR (one subject in Arm 1, two subjects in Arm 2).

Optimal responders demonstrated a number of significant differences when compared to other subjects. Optimal responders had a significantly higher leakage index at Baseline compared to those with a worsening leakage index and/or MA count ($6.32 \pm 4.88\%$ vs. $3.60 \pm 1.92\%$; $p = 0.05$). Between Baseline and Year 1, the decrease in the leakage index was also greater in the optimal responders ($-4.63 \pm 5.3\%$ vs. $-0.12 \pm 3.28\%$; $p = 0.03$). From Year 1 to Year 2, the optimal responders had a slight increase in the average panretinal leakage, but it was significantly smaller than the other subjects (0.14% vs. 3.62%, respectively, $p = 0.018$).

With regards to DRSS, optimal responders improved significantly from Year 1 to Year 2 by a mean 0.42 ± 1.2 steps, while other subjects demonstrated a worsening of 2.17 ± 1.6 steps ($p = 0.01$). Between Baseline and Year 2, optimal responders also had overall a significantly greater improvement in DRSS, improving by a mean 3.17 ± 2.0 steps compared to other subjects, which only improved by 1.0 ± 0.82 ($p < 0.01$). All non-optimal responders either demonstrated a worsened DRSS after treatment crossover (4/6; 66.7%) or remained with a stable DRSS (2/6; 33.3%).

3.7. Complications

An analysis of vision-threatening complications in both cohorts revealed that no subjects developed center-involving DME at any point during the study. In addition, no subject required Panretinal Photocoagulation (PRP) over the 2-year time period; however, vitreous hemorrhage (VH) did occur. When excluding the 3 subjects with VH at Baseline (one from Arm 1, two from Arm 2), VH occurred in 7 out of 37 subjects (18.9%), with 5/7 (71.4%) occurring in Arm 1. Subjects who developed VH after Baseline contained the top 20% of the highest values for the Baseline leakage index, with one subject having a Baseline leakage index of 26.08%, the highest of all subjects. These subjects also had a smaller step improvement in DRSS between Baseline and Year 2 compared to those without VH (1 ± 0.89, 3.21 ± 1.93, respectively; $p < 0.01$), as well as a smaller decrease in CST (-20.3 ± 8.69 μm vs. -48.7 ± 58.4 μm; $p = 0.03$). The occurrence of VH did not show a significant difference in the Baseline leakage index or microaneurysm count. No significant difference in age,

BMI, Baseline HbA1C, or duration of diabetes was noted between subjects with VH and those without VH.

In the optimal responders, two (2/25, 8%) had VH at Baseline while three developed VH after Baseline (3/25, 12%). All three optimal responders that developed VH after Baseline were from Arm 1, while the two subjects with Baseline VH were from Arm 2. Notably, the optimal responders that developed VH after Baseline were reported to have VH on examination in an average of 10 out of 18.7 (53.5%) visits: 7.3 mean visits between Baseline and Year 1 and 2.7 mean visits between Year 1 and Year 2. Two subjects developed VH after Baseline that were not optimal responders (both in Arm 1), and they had VH reported in a mean 3.5 out of 20 (17.5%) visits; all of these were reported between Year 1 and Year 2.

4. Discussion

The semi-automated analysis of ultra-widefield angiography images has enabled an in-depth assessment of angiographic features. Changes in retinal leakage and microaneurysm burden were recently quantitatively analyzed to allow a more detailed understanding of the effects of anti-VEGF treatment on the retina [5,29–31]. Recent studies on patients with DR treated by anti-VEGF showed a significant decrease in retinal leakage and microaneurysm count on UWFA [32–36]. The 1-year RECOVERY results also highlighted the effects of dosing frequency on the leakage index and microaneurysm count dynamics over 52 weeks [27,28].

In this 2-year trial, subjects undergoing IAI therapy for DME exhibited a significant decrease in the leakage index and microaneurysm count from Baseline up to Year 2. However, each cohort had significant differences in the rate of improvement, which is further highlighted after treatment crossover at Year 1. Between Baseline and Year 1, Arm 1 (monthly IAI from Baseline to Year 1) showed significant improvement in panretinal and zonal leakage, while the Arm 2 (q12 week IAI from Baseline to Year 1) cohort had significant improvement in all regions except in Zone 3 (9-disc-diameter). More importantly, after treatment crossover, Arm 1, which switched to q12 week IAI injections, saw a worsening in the panretinal leakage index, increasing significantly from Year 1 to Year 2. In contrast, Arm 2, which switched to q4 week IAI, continued to show improvements in the leakage index in all zones, although those changes were not statistically significant. Microaneurysm burden differences were also apparent between cohorts when comparing changes before and after treatment crossover. From Baseline to Year 1, both cohorts showed significant decreases in the MA count in all zones. However, after crossover, only Arm 2 showed a significant improvement in the mean panretinal MA count and in all zones.

In addition to leakage and MA burden, changes in DRSS scores have been a parameter of significant interest in subjects undergoing anti-VEGF treatment. Specifically, the clinical importance of DRSS changes have been examined in relation to visual acuity outcomes [13,15,37]. A study in 2017 by Ip and colleagues showed that subjects with stable or improved DRSS levels had a greater improvement in BCVA letter scores, whereas those with worsening DRSS had a smaller improvement. Moreover, significant changes in the BCVA letter score (>/=15 letters) were more common in patients with at least a 2-step improvement in DRSS [37]. Another study in 2018 by Dhoot and colleagues demonstrated only Baseline DRSS having a strong association with a 2-step DRSS improvement by Year 2, while age, duration of diabetes, HbA1C, BMI, and BCVA had no correlation [38].

In RECOVERY, most subjects (85%) exhibited an improvement of at least 1 step in DRSS between Baseline and the end of the study at Year 2. Similar to the leakage index and MA count, differences were noted when comparing cohorts before and after treatment crossover. Between Baseline and Year 1, all subjects (100%) in Arm 1 and the majority of Arm 2 (83%) had improved DRSS (decrease in score value). However, after crossover, a substantially higher proportion of Arm 2 subjects continued to improve in DRSS by at least 1 step compared to Arm 1. Moreover, from Baseline to Year 2, a higher ratio of subjects in Arm 2 exhibited a 2-step improvement than subjects in Arm 1. Overall, by the end of the study, a larger proportion of subjects in Arm 2 had an improved DRSS than in Arm 1. Notably, the worsening of DRSS (increase in score value) was only seen after crossover, and the majority of cases were from Arm 1. Overall, the subjects with worsening DRSS after crossover showed a significant worsening in the MA count in that period compared to the significant improvement in the MA count seen in those with stable/improved DRSS. Subjects with worsening DRSS after crossover also showed a significantly larger worsening of the panretinal leakage index during that period. Moreover, there was no significant difference in the age, BMI, Baseline HbA1C, and duration of diabetes noted between subjects in those that improved by 1 step or by 2 steps compared to those who did not improve by Year 2.

Changes in DRSS in relation to the leakage index, MA count, and CST were also investigated. The subjects with an improvement in the DRSS score by at least 1 step showed a greater reduction in the leakage index compared to those without improvement. In Arm 1, subjects with an increased DRSS before treatment crossover and who continued to improve after crossover had a significantly larger leakage index and CST at Baseline compared to those with a stable DRSS after crossover. Moreover, those who continued to improve also had a significantly larger decrease in the leakage index and CST between Baseline and Year 1, and a greater decrease in mean CST was shown from Baseline to Year 2. The change in BCVA was also different between those two groups in Arm 1, as those who continued to improve in DRSS showed a significantly greater increase in ETDRS letters by Year 1. In contrast, subjects in Arm 2 who improved in DRSS before crossover and continued to improve had no significant differences in any parameter compared to subjects who improved before crossover and did not improve after crossover.

The results of this study highlight the potential worsening in the leakage index, MA counts, and DRSS associated with a decreasing IAI frequency from monthly to q12 week dosing. Both the PRIME [39] and PANORAMA [40,41] trials suggested that once an improvement in DRSS is achieved, it can be sustained with q12–16 week intravitreal anti-VEGF, but this was not the case in our study. In contrast, our study showed that a gradual induction of q12 week dosing followed by a more frequent q4 week dosing resulted in better functional and anatomic outcomes overall, especially in terms of leakage and DRSS. An important distinction from both the PRIME study and PANORAMA study is the greater severity of PDR in the current study. This difference in findings supports the need for the frequent monitoring of DR and the potential value of concurrent UWFA imaging. It also suggests that applying a treatment regimen across all patients with PDR may be less efficacious than image-guided personalized therapeutic decision-making.

Additionally, this study and recent publications demonstrate a clear link between the leakage index and DRSS [27,39]. Specifically, the progression of DR and hence the worsening of DRSS may be preceded by increases in the leakage index on UWFA, as evidenced by the PRIME Trial (a prospective, randomized, phase 2 trial that enrolled 40 eyes to assess the safety and efficacy of the real-time leakage index vs. DRSS scoring at monthly visits to guide PRN retreatment in both PDR and NPDR patients) [39]. However, repeated regular UWFAs to assess the leakage index in DR is not without drawbacks, and there is a clear appeal to using DRSS for regular DR monitoring because of its non-invasive nature, efficiency, and ease of image capture.

Traditionally, DRSS has been determined by using a central reading center (CRC). The PANORAMA study demonstrated the necessity for trained readers or CRC for the determination of DRSS, when the disparity between the investigator and CRC reader proved the investigator wrong in 50% of cases [40]. Further, the need for precise DRSS and the ability to monitor changes in DRSS is becoming of greater importance, not just in terms of DR progression, but for morbidity and mortality in DM. A recent study in Diabetes Care by Fahrmann demonstrated for the first time the strong correlation between DCCT-ETDRS levels ≥ 3 (Level 35; mild NPDR) and cardiovascular complications in young type 1 diabetics [42]. ACCORDION, a 9-year observation follow-up study to ACCORD, emphasizes the benefits of the personalization of DR treatment and the potential harm in treating all patients with a one-size-fits-all approach [43]. The authors concluded that patients should be stratified according to their microvascular complications, as their results revealed that DR patients who have undergone vitrectomy or retinal laser photocoagulation derived benefits from intensive glucose control, while those who had not were at an increased cardiovascular risk. While this study used a binary approach to DR, the opportunity for further research in the relationship between DRSS, cardiovascular disease, and glucose control is an unmet need. Multiple other studies demonstrated the relationship between DR and all-cause mortality, further highlighting the increasing importance of DRSS determination in clinical practice [44–47].

Specific limitations of this study include the relatively small number of subjects, presence of artifacts such as eyelashes and lenticular opacities, and errors that may occur analyzing different or low-quality images. Strengths of the study include its prospective nature and the standardization of image assessment for each parameter through the use of an image reading center and previously validated techniques in image analysis and quantification.

5. Conclusions

The quantitative assessments of UWF imaging in this study and others demonstrate the utility and importance of these algorithms in identifying DRSS and DR progression, but also for the prognostication of glucose control and overall morbidity and mortality in patients with DM [27–29,37].

Author Contributions: Conceptualization, C.C.W., S.R.S. and J.P.E.; methodology, C.C.W., S.R.S. and J.P.E.; formal analysis, A.S.B., C.C.W., S.Y., H.Y., M.H., T.K.L., L.L., M.G.N., S.R.S. and J.P.E.; writing—original draft preparation, A.S.B., S.Y. and J.P.E.; writing—review and editing, A.S.B., C.C.W., S.Y., H.Y., S.K.S., M.H., T.K.L., L.L., J.R., M.G.N., S.R.S. and J.P.E.; supervision, C.C.W., S.R.S. and J.P.E.; project administration, J.R., J.P.E. and C.C.W.; funding acquisition, C.C.W. and J.P.E. All authors have read and agreed to the published version of the manuscript.

Funding: This research was funded by NIH/NEI K23-EY022947-01A1 (JPE), Research to Prevent Blindness (Cole Eye Institutional Grant), and Regeneron (CCW, JPE, and RECOVERY Study Group).

Institutional Review Board Statement: Prospective Institutional Review Board (IRB)/Ethics Committee approval was obtained (IRB ID: 5598-CCWykoff), and the tenets of the Declaration of Helsinki were followed, in accordance with the Health Insurance Portability and Accountability Act of 1996.

Informed Consent Statement: Informed consent was obtained from all subjects involved in the study.

Data Availability Statement: All data relevant to the study are included in the article. Additional data may be available by request if for some reason that is required.

Conflicts of Interest: The authors declare no conflict of interest.

Financial Disclosures: Amy Babiuch, M.D. receives research support from Genentech, Regeneron and is a consultant for Genentech. Charles C. Wykoff, M.D., Ph.D. receives research support from Adverum, Allergan, Apellis, Clearside, EyePoint, Genentech/Roch, Neurotech, Novartis, Opthea, Regeneron, Regenxbio, Samsung, Santen, and is a consultant for Alimera Sciences, Allegro, Allergan, Alynylam, Apellis, Bayer, Clearside, D.O.R.C., EyePoint, Genentech/Roche, Kodiak, Notal Vision, Novartis, ONL Therapeutics, PolyPhotonix, RecensMedical, Regeneron, Regenxbio, Santen. Sari Yordi, M.D. is a Betty J. Powers Retina Research Fellow. Sunil Srivastava, M.D. receives research support from Regeneron, Allergan, Gilead, and is a consultant for Bausch and Lomb, Santen, and has a patent with Leica. Justis P. Ehlers, M.D. receives research support from Aerpio, Alcon, Thrombogenics, Regeneron, Genentech, Novartis, and is a consultant for Aerpio, Alcon, Allegro, Allergan, Genentech/Roche, Novartis, Thrombogenics, Leica, Zeiss, Regeneron, Santen, and has a patent with Leica.

RECOVERY Study Group: Charles C. Wykoff M.D., Ph.D.; Muneeswar G. Nittala M.D.; Brenda Zhou, B.S.; Shaun I.R. Lampen, B.S.; Michael Ip, M.D.; SriniVas Sadda, M.D.; Justis P. Ehlers. M.D.; Sunil K. Srivastava, M.D.; Jamie L. Reese, BSN; Amy Babiuch, M.D.; Katherine Talcott, M.D.; Natalia Figueiredo, M.D.; Jenna Hach, B.A.; Alexander Rusakevich, B.A.;

William C. Ou, B.S.; Richard H. Fish; Matthew S. Benz, M.D.; Eric Chen, M.D.; Rosa Y. Kim, M.D.; James C. Major Jr., M.D., Ph.D.; Ronan E. O'Malley, M.D.; David M. Brown, M.D.; Ankoor R. Shah, M.D.; Amy C. Schefler, M.D.; Tien P. Wong, M.D.; Christopher R. Henry, M.D.

References

1. Center for Disease Control. *National Diabetes Statistics Report*; Center for Disease Control: Atlanta, GA, USA, 2020.
2. National Eye Institute. *Diabetic Retinopathy Data and Statistics*; National Eye Institute: Bethesda, MD, USA, 2020. Available online: https://www.nei.nih.gov/learn-about-eye-health/outreach-campaigns-and-resources/eye-health-data-and-statistics/diabetic-retinopathy-data-and-statistics (accessed on 28 October 2021).
3. Lee, R.; Wong, T.Y.; Sabanayagam, C. Epidemiology of diabetic retinopathy, diabetic macular edema and related vision loss. *Eye Vis.* **2015**, *2*. [CrossRef]
4. Tolentino, M.J.; Miller, J.W.; Gragoudas, E.S.; Jakobiec, F.A.; Flynn, E.; Chatzstefanou, K.; Ferrara, N.; Adamis, A.P. Intravitreous Injections of Vascular Endothelial Growth Factor Produce Retinal Ischemia and Microangiopathy in an Adult Primate. *Ophthalmology* **1996**, *103*, 1820–1828. [CrossRef]
5. Campochiaro, P.A.; Wykoff, C.C.; Shapiro, H.; Rubio, R.G.; Ehrlich, J.S. Neutralization of Vascular Endothelial Growth Factor Slows Progression of Retinal Nonperfusion in Patients with Diabetic Macular Edema. *Ophthalmology* **2014**, *121*, 1783–1789. [CrossRef]
6. Aiello, L.P.; Avery, R.L.; Arrigg, P.G.; Keyt, B.A.; Jampel, H.D.; Shah, S.T.; Pasquale, L.R.; Thieme, H.; Iwamoto, M.A.; Park, J.E.; et al. Vascular Endothelial Growth Factor in Ocular Fluid of Patients with Diabetic Retinopathy and Other Retinal Disorders. *N. Engl. J. Med.* **1994**, *331*, 1480–1487. [CrossRef] [PubMed]
7. Fong, D.S.; Ferris, F.L.; Davis, M.D.; Chew, E.Y. Causes of severe visual loss in the early treatment diabetic retinopathy study: ETDRS report no. 24. *Am. J. Ophthalmol.* **1999**, *127*, 137–141. [CrossRef]
8. The Diabetic Retinopathy Study Research Group. Preliminary Report on Effects of Photocoagulation Therapy. *Am. J. Ophthalmol.* **1976**, *81*, 383–396. [CrossRef]
9. Early Treatment Diabetic Retinopathy Study Research Group. Early Photocoagulation for Diabetic Retinopathy: ETDRS Report Number 9. *Ophthalmology* **1991**, *98*, 766–785. [CrossRef]
10. Ferris, F. Early photocoagulation in patients with either type I or type II diabetes. *Trans. Am. Ophthalmol. Soc.* **1996**, *94*, 505–537. [CrossRef]
11. Riaskoff, S. Photocoagulation Treatment of Proliferative Diabetic Retinopathy. *Ophthalmology* **1981**, *88*, 583–600. [CrossRef]
12. Gross, J.G.; Glassman, A.R.; Jampol, L.M.; Inusah, S.; Aiello, L.P.; Antoszyk, A.N.; Baker, C.W.; Berger, B.B.; Bressler, N.M.; Browning, D.; et al. Panretinal Photocoagulation vs Intravitreous Ranibizumab for Proliferative Diabetic Retinopathy. *JAMA* **2015**, *314*, 2137–2146. [CrossRef]
13. Ip, M.S.; Domalpally, A.; Hopkins, J.J.; Wong, P.; Ehrlich, J.S. Long-term Effects of Ranibizumab on Diabetic Retinopathy Severity and Progression. *Arch. Ophthalmol.* **2012**, *130*, 1145–1152. [CrossRef]
14. Nguyen, Q.D.; Brown, D.M.; Marcus, D.M.; Boyer, D.S.; Patel, S.; Feiner, L.; Gibson, A.; Sy, J.; Rundle, A.C.; Hopkins, J.J.; et al. Ranibizumab for Diabetic Macular Edema. *Ophthalmology* **2012**, *119*, 789–801. [CrossRef]
15. Brown, D.M.; Schmidt-Erfurth, U.; Do, D.V.; Holz, F.G.; Boyer, D.S.; Midena, E.; Heier, J.S.; Terasaki, H.; Kaiser, P.; Marcus, D.M.; et al. Intravitreal Aflibercept for Diabetic Macular Edema. *Ophthalmology* **2015**, *122*, 2044–2052. [CrossRef]
16. Wykoff, C.C.; Eichenbaum, D.A.; Roth, D.B.; Hill, L.; Fung, A.E.; Haskova, Z. Ranibizumab Induces Regression of Diabetic Retinopathy in Most Patients at High Risk of Progression to Proliferative Diabetic Retinopathy. *Ophthalmol. Retin.* **2018**, *2*, 997–1009. [CrossRef]
17. Ip, M.S.; Domalpally, A.; Sun, J.; Ehrlich, J.S. Long-term Effects of Therapy with Ranibizumab on Diabetic Retinopathy Severity and Baseline Risk Factors for Worsening Retinopathy. *Ophthalmology* **2015**, *122*, 367–374. [CrossRef] [PubMed]
18. Early Treatment Diabetic Retinopathy Study Research Group. Grading Diabetic Retinopathy from Stereoscopic Color Fundus Photographs—An Extension of the Modified Airlie House Classification: ETDRS Report Number 10. *Ophthalmology* **1991**, *98*, 786–806. [CrossRef]
19. Ehlers, J.P.; Jiang, A.C.; Boss, J.D.; Hu, M.; Figueiredo, N.; Babiuch, A.; Talcott, K.; Sharma, S.; Hach, J.; Le, T.K.; et al. Quantitative Ultra-Widefield Angiography and Diabetic Retinopathy Severity: An Assessment of Panretinal Leakage Index, Ischemic Index and Microaneurysm Count. *Ophthalmology* **2019**, *126*, 1527–1532. [CrossRef] [PubMed]
20. Jiang, A.; Srivastava, S.; Figueiredo, N.; Babiuch, A.; Hu, M.; Reese, J.; Ehlers, J.P. Repeatability of automated leakage quantification and microaneurysm identification utilising an analysis platform for ultra-widefield fluorescein angiography. *Br. J. Ophthalmol.* **2019**, *104*, 500–503. [CrossRef] [PubMed]
21. Ehlers, J.P.; Wang, K.; Vasanji, A.; Hu, M.; Srivastava, S.K. Automated quantitative characterisation of retinal vascular leakage and microaneurysms in ultra-widefield fluorescein angiography. *Br. J. Ophthalmol.* **2017**, *101*, 696–699. [CrossRef]
22. Rabbani, H.; Allingham, M.J.; Mettu, P.S.; Cousins, S.W.; Farsiu, S. Fully Automatic Segmentation of Fluorescein Leakage in Subjects with Diabetic Macular Edema. *Investig. Opthalmology Vis. Sci.* **2015**, *56*, 1482–1492. [CrossRef]
23. Zhao, Y.; MacCormick, I.J.C.; Parry, D.G.; Leach, S.; Beare, N.; Harding, S.P.; Zheng, Y. Automated Detection of Leakage in Fluorescein Angiography Images with Application to Malarial Retinopathy. *Sci. Rep.* **2015**, *5*, 10425. [CrossRef] [PubMed]

24. Tanchon, C.; Srivastava, S.K.; Ehlers, J.P. Automated Quantitative Analysis of Leakage and Ischemia for Ultra-widefield Angiography in Retinal Vascular Disease. *Investig. Ophthalmol. Vis. Sci.* **2015**, *56*, 3067.
25. Tan, C.S.; Chew, M.C.; van Hemert, J.; Singer, M.; Bell, D.; Sadda, S.R. Measuring the precise area of peripheral retinal nonperfusion using ultra-widefield imaging and its correlation with the ischaemic index. *Br. J. Ophthalmol.* **2015**, *100*, 235–239. [CrossRef] [PubMed]
26. Wykoff, C.C.; Nittala, M.G.; Zhou, B.; Fan, W.; Velaga, S.B.; Lampen, S.I.; Rusakevich, A.; Ehlers, J.P.; Babiuch, A.; Brown, D.M.; et al. Intravitreal Aflibercept for Retinal Nonperfusion in Proliferative Diabetic Retinopathy: Outcomes from the Randomized RECOVERY Trial. *Ophthalmol. Retin.* **2019**, *3*, 1076–1086. [CrossRef]
27. Babiuch, A.S.; Wykoff, C.C.; Srivastava, S.K.; Talcott, K.; Zhou, B.; Hach, J.; Hu, M.; Reese, J.L.; Ehlers, J.P. Retinal leakage index dynamics on ultra-widefield fluorescein angiography in eyes treated with intravitreal aflibercept for proliferative diabetic retinopathy in the recovery study. *Retina* **2020**, *40*, 2175–2183. [CrossRef]
28. Babiuch, A.; Wykoff, C.C.; Hach, J.; Srivastava, S.; Talcott, K.E.; Yu, H.J.; Nittala, M.; Sadda, S.; Ip, M.S.; Le, T.; et al. Longitudinal panretinal microaneurysm dynamics on ultra-widefield fluorescein angiography in eyes treated with intravitreal aflibercept for proliferative diabetic retinopathy in the recovery study. *Br. J. Ophthalmol.* **2020**. [CrossRef]
29. Fan, W.; Nittala, M.G.; Velaga, S.B.; Hirano, T.; Wykoff, C.C.; Ip, M.; Lampen, S.I.; van Hemert, J.; Fleming, A.; Verhoek, M.; et al. Distribution of Nonperfusion and Neovascularization on Ultrawide-Field Fluorescein Angiography in Proliferative Diabetic Retinopathy (RECOVERY Study): Report 1. *Am. J. Ophthalmol.* **2019**, *206*, 154–160. [CrossRef]
30. Lange, J.; Hadziahmetovic, M.; Zhang, J.; Li, W. Region-specific ischemia, neovascularization and macular oedema in treatment-naïve proliferative diabetic retinopathy. *Clin. Exp. Ophthalmol.* **2018**, *46*, 757–766. [CrossRef]
31. Xue, K.; Yang, E.; Chong, N.V. Classification of diabetic macular oedema using ultra-widefield angiography and implications for response to anti-VEGF therapy. *Br. J. Ophthalmol.* **2016**, *101*, 559–563. [CrossRef]
32. Allingham, M.J.; Mukherjee, D.; Lally, E.B.; Rabbani, H.; Mettu, P.S.; Cousins, S.W.; Farsiu, S. A Quantitative Approach to Predict Differential Effects of Anti-VEGF Treatment on Diffuse and Focal Leakage in Patients with Diabetic Macular Edema: A Pilot Study. *Transl. Vis. Sci. Technol.* **2017**, *6*, 7. [CrossRef] [PubMed]
33. Chandra, S.; Sheth, J.; Anantharaman, G.; Gopalakrishnan, M. Ranibizumab-induced retinal reperfusion and regression of neovascularization in diabetic retinopathy: An angiographic illustration. *Am. J. Ophthalmol. Case Rep.* **2018**, *9*, 41–44. [CrossRef]
34. Gupta, M.P.; Kiss, S.; Chan, R.V.P. Reversal of Retinal Vascular Leakage and Arrest of Progressive Retinal Nonperfusion With Monthly Anti–Vascular Endothelial Growth Factor Therapy for Proliferative Diabetic Retinopathy. *Retina* **2018**, *38*, e74–e75. [CrossRef]
35. Levin, A.M.; Rusu, I.; Orlin, A.; Gupta, M.P.; Coombs, P.; D'Amico, D.J.; Kiss, S. Retinal reperfusion in diabetic retinopathy following treatment with anti-VEGF intravitreal injections. *Clin. Ophthalmol.* **2017**, *11*, 193–200. [CrossRef]
36. Leicht, S.F.; Kernt, M.; Neubauer, A.; Wolf, A.; Oliveira, C.M.; Ulbig, M.; Haritoglou, C. Microaneurysm Turnover in Diabetic Retinopathy Assessed by Automated RetmarkerDR Image Analysis—Potential Role as Biomarker of Response to Ranibizumab Treatment. *Ophthalmologica* **2014**, *231*, 198–203. [CrossRef]
37. Ip, M.S.; Zhang, J.; Ehrlich, J.S. The Clinical Importance of Changes in Diabetic Retinopathy Severity Score. *Ophthalmology* **2017**, *124*, 596–603. [CrossRef]
38. Dhoot, D.S.; Baker, K.; Saroj, N.; Vitti, R.; Berliner, A.J.; Metzig, C.; Thompson, D.; Singh, R.P. Baseline Factors Affecting Changes in Diabetic Retinopathy Severity Scale Score After Intravitreal Aflibercept or Laser for Diabetic Macular Edema. *Ophthalmology* **2018**, *125*, 51–56. [CrossRef]
39. Yu, H.J.; Ehlers, J.P.; Sevgi, D.D.; Hach, J.; O'Connell, M.; Reese, J.L.; Srivastava, S.K.; Wykoff, C.C. Real-Time Photographic- and Fluorescein Angiographic-Guided Management of Diabetic Retinopathy: Randomized PRIME Trial Outcomes. *Am. J. Ophthalmol.* **2021**, *226*, 126–136. [CrossRef]
40. Wykoff, C.C. *PANORAMA: A Phase 3, Double-Masked, Randomized Study of the Efficacy and Safety of Aflibercept in Patients with Moderately Severe to Severe NPDR*; Retina Consultants Houston: Houston, TX, USA, 2020.
41. Study of the Efficacy and Safety of Intravitreal (IVT) Aflibercept for the Improvement of Moderately Severe to Severe Non-Proliferative Diabetic Retinopathy (NPDR) (PANORAMA). 2020. Available online: https://clinicaltrials.gov/ct2/show/NCT027 18326 (accessed on 5 June 2021).
42. Fahrmann, E.R.; Adkins, L.; Driscoll, H.K. Modification of the Association Between Severe Hypoglycemia and Ischemic Heart Disease by Surrogates of Vascular Damage Severity in Type 1 Diabetes During 30 Years of Follow-up in the DCCT/EDIC Study. *Diabetes Care* **2021**, *44*, 1–8. [CrossRef]
43. Kloecker, D.E.; Khunti, K.; Davies, M.J.; Pitocco, D.; Zaccardi, F. Microvascular Disease and Risk of Cardiovascular Events and Death from Intensive Treatment in Type 2 Diabetes. *Mayo Clin. Proc.* **2021**, *96*, 1458–1469. [CrossRef]
44. Kramer, C.K.; Rodrigues, T.C.; Canani, L.H.; Gross, J.L.; Azevedo, M.J. Diabetic Retinopathy Predicts All-Cause Mortality and Cardiovascular Events in Both Type 1 and 2 Diabetes: Meta-analysis of observational studies. *Diabetes Care* **2011**, *34*, 1238–1244. [CrossRef]
45. Zhu, X.-R.; Zhang, Y.-P.; Bai, L.; Zhang, X.-L.; Zhou, J.-B.; Yang, J.-K. Prediction of risk of diabetic retinopathy for all-cause mortality, stroke and heart failure. *Medicine* **2017**, *96*, e5894. [CrossRef] [PubMed]

46. Guo, V.Y.; Cao, B.; Wu, X.; Lee, J.J.W.; Zee, B.C.-Y. Prospective Association between Diabetic Retinopathy and Cardiovascular Disease—A Systematic Review and Meta-analysis of Cohort Studies. *J. Stroke Cerebrovasc. Dis.* **2016**, *25*, 1688–1695. [CrossRef] [PubMed]
47. Sadda, S.R.; Nittala, M.G.; Taweebanjongsin, W.; Verma, A.; Velaga, S.B.; Alagorie, A.R.; Sears, C.M.; Silva, P.S.; Aiello, L.P. Quantitative Assessment of the Severity of Diabetic Retinopathy. *Am. J. Ophthalmol.* **2020**, *218*, 342–352. [CrossRef]

Journal of
*Personalized
Medicine*

MDPI

Article

Association between Retinal Thickness Variability and Visual Acuity Outcome during Maintenance Therapy Using Intravitreal Anti-Vascular Endothelial Growth Factor Agents for Neovascular Age-Related Macular Degeneration

Timothy Y. Y. Lai [1,2,*] and Ricky Y. K. Lai [2]

[1] Department of Ophthalmology & Visual Sciences, The Chinese University of Hong Kong, Central Ave, Shatin, New Territories, Hong Kong, China
[2] 2010 Retina & Macula Centre, Tsim Sha Tsui, Kowloon, Hong Kong, China; lai_r@yahoo.com
* Correspondence: tyylai@cuhk.edu.hk

Abstract: Previous studies based on clinical trial data have demonstrated that greater fluctuations in retinal thickness during the course of intravitreal anti-vascular endothelial growth factor (anti-VEGF) therapy for neovascular age-related macular degeneration (nAMD) is associated with poorer visual acuity outcomes. However, it was unclear whether similar findings would be observed in real-world clinical settings. This study aimed to evaluate the association between retinal thickness variability and visual outcomes in eyes receiving anti-VEGF therapy for nAMD using pro re nata treatment regimen. A total of 64 eyes which received intravitreal anti-VEGF therapy (bevacizumab, ranibizumab or aflibercept) for the treatment of nAMD were evaluated. Variability in spectral-domain optical coherence tomography (OCT) central subfield thickness (CST) was calculated from the standard deviation (SD) values of all follow-up visits after three loading doses from month 3 to month 24. Eyes were divided into quartiles based on the OCT CST variability values and the mean best-corrected visual acuity values at 2 years were compared. At baseline, the mean $\pm$ SD logMAR visual acuity and CST were 0.59 ± 0.39 and 364 ± 113 μm, respectively. A significant correlation was found between CST variability and visual acuity at 2 years (Spearman's $\rho = 0.54$, $p < 0.0001$), indicating that eyes with lower CST variability had better visual acuity at 2 years. Eyes with the least CST variability were associated with the highest mean visual acuity improvement at 2 years (quartile 1: +9.7 letters, quartile 2: +1.1 letters, quartile 3: −2.5 letters, quartile 4: −9.5 letters; $p = 0.018$). No significant difference in the number of anti-VEGF injections was found between the four CST variability quartile groups ($p = 0.21$). These findings showed that eyes undergoing anti-VEGF therapy for nAMD with more stable OCT CST variability during the follow-up period were associated with better visual outcomes. Clinicians should consider adopting treatment strategies to reduce CST variability during the treatment course for nAMD.

Keywords: neovascular age-related macular degeneration; anti-VEGF therapy; optical coherence tomography; retinal thickness; visual acuity; variability

Citation: Lai, T.Y.Y.; Lai, R.Y.K. Association between Retinal Thickness Variability and Visual Acuity Outcome during Maintenance Therapy Using Intravitreal Anti-Vascular Endothelial Growth Factor Agents for Neovascular Age-Related Macular Degeneration. *J. Pers. Med.* **2021**, *11*, 1024. https://doi.org/10.3390/jpm11101024

Academic Editors: Peter D. Westenskow and Andreas Ebneter

Received: 25 September 2021
Accepted: 11 October 2021
Published: 14 October 2021

Publisher's Note: MDPI stays neutral with regard to jurisdictional claims in published maps and institutional affiliations.

1. Introduction

Neovascular age-related macular degeneration (nAMD) is one of the most common causes of visual impairment in adults aged 50 years or older, especially in high-income regions [1]. Intravitreal anti-vascular endothelial growth factor (anti-VEGF) therapy is currently the standard-of-care treatment for patients with nAMD, and anti-VEGF therapy has been shown to be highly effective in reducing blindness due to nAMD [2–5]. Various treatment regimens using anti-VEGF therapy for nAMD have been adopted; these include proactive strategies such as fixed dosing and treat-and-extend, as well as more reactive approaches such as pro re nata (PRN) dosing and observe-and-plan [6–10]. Several studies and treatment guidelines have suggested that proactive treatment strategies such as fixed dosing and treat-and-extend dosing strategies might result in better visual acuity outcomes than PRN dosing when treating patients with nAMD [8–12]. However, a recent systematic review has demonstrated that after adjusting for the number of intravitreal anti-VEGF injections, neither the treatment dosing regimen adopted, nor the anti-VEGF agent used, were significant predictors for visual acuity changes [13]. Nonetheless, other factors such as age, anatomical status of the retina, including the presence or absence of subretinal and/or intraretinal fluids, optical coherence tomography (OCT) central macular thickness, and macular morphology have been implicated as important prognostic factors in determining visual outcomes [14–16].

Post hoc analyses of data from several clinical trials have evaluated the potential influence of OCT retinal thickness variation during the course of anti-VEGF therapy on the visual acuity outcomes of patients with nAMD [17,18]. The findings demonstrated that eyes which had greater fluctuation in the OCT retinal thickness during the course of anti-VEGF therapy had poorer visual acuity outcomes than eyes which had less fluctuation. However, most of these results were based on phase 3 clinical trial data with protocol-driven treatment, and it is unclear whether similar findings will be observed in the real-world clinical settings. The aim of our study was to evaluate the association between OCT retinal thickness fluctuation and visual acuity outcomes in patients receiving intravitreal anti-VEGF therapy in the real-world clinic setting using a personalized PRN dosing regimen.

2. Materials and Methods

2.1. Study Design

This was a retrospective study of consecutive eyes which received intravitreal anti-VEGF therapy from October 2012 to March 2018 for nAMD in the 2010 Retina and Macula Centre, Hong Kong. The inclusion criteria included the following: eyes which received three initial loading doses of intravitreal anti-VEGF agents at monthly intervals followed by additional PRN treatment; treatment-naïve eye with no prior treatment for nAMD; follow-up duration of at least 24 months after the first dose of intravitreal anti-VEGF injection; and availability of 3 or more OCT central subfield thickness (CST) values during the follow-up period. A treatment-agnostic approach was adapted for the study to include patients treated with any of the three available anti-VEGF agents (bevacizumab [Roche, Basel, Switzerland]; ranibizumab [Novartis, Basel, Switzerland]; and aflibercept [Bayer, Leverkusen, Germany]) because all three intravitreal anti-VEGF agents result in significant visual gain after treatment, as demonstrated in previous studies [17]. The study was conducted in accordance with the Declaration of Helsinki.

2.2. Optical Coherence Tomography Imaging and Measurements

Spectral-domain OCT scans were obtained using with the Cirrus HD-OCT 4000 (Carl Zeiss Meditec, Dublin, CA, USA). The "Macular Cube 512 × 128" scan protocol was used to obtain the OCT images of the macula, covering a retinal area of 6.0 × 6.0 mm. The OCT CST value of the individual OCT B-scan was obtained using the automated software and is defined as the distance between the middle of the retinal pigment epithelium and the internal limiting membrane at the central 1 mm subfield.

2.3. Statistical Analysis

Data were entered into computer spreadsheet software (Microsoft Excel for Mac version 16.52, Microsoft Corp, Redmond, WA, USA) and statistical analyses were carried out using a statistical module (StatPlus:mac Pro version 5.9.80, AnalystSoft Inc., Walnut, CA, USA) running within the spreadsheet software. OCT CST variability values were calculated from the standard deviation (SD) of all the available OCT CST values of the eye at each follow-up visit after the 3 loading doses of intravitreal anti-VEGF injection from month 3 to month 24. Eyes were then categorized into four quartiles of OCT CST variability, ranging from the group with the lowest variability values to the group with the highest variability values. Correlation analysis between the OCT CST variability values and the best-corrected visual acuity (BCVA) at 2 years was performed using the non-parametric Spearman's rank test. The mean change in BCVA and the mean number of intravitreal anti-VEGF injections were compared between the 4 OCT CST variability quartile groups using one-way ANOVA tests. A *p*-value of ≤ 0.05 was considered as statistically significant.

3. Results

3.1. Baseline Patient Demographics

A total of 64 eyes of 62 patients were included in the study. The mean ± SD age of the patients at baseline was 75.3 ± 9.4 years (range, 50 to 91 years). All patients were of Chinese ethnicity and there were 38 (61.3%) males and 24 (38.7%) females. At baseline, the mean ± SD logMAR BCVA was 0.59 ± 0.39 (range, 0.0 to 1.3) and the mean ± SD OCT CST was 364 ± 113 μm (range, 210 to 665 μm).

3.2. Correlation of OCT CST Variability Values and BCVA at 2 Years

The correlation between OCT CST variability values and the logMAR BCVA at 2 years is shown in Figure 1. There was a significant correlation between OCT CST variability values and the logMAR BCVA at 2 years (Spearman's $\rho = 0.54$, $p < 0.0001$). This indicates that eyes which had less variability in OCT CST values during the 2-year treatment period had better visual acuity at 2 years.

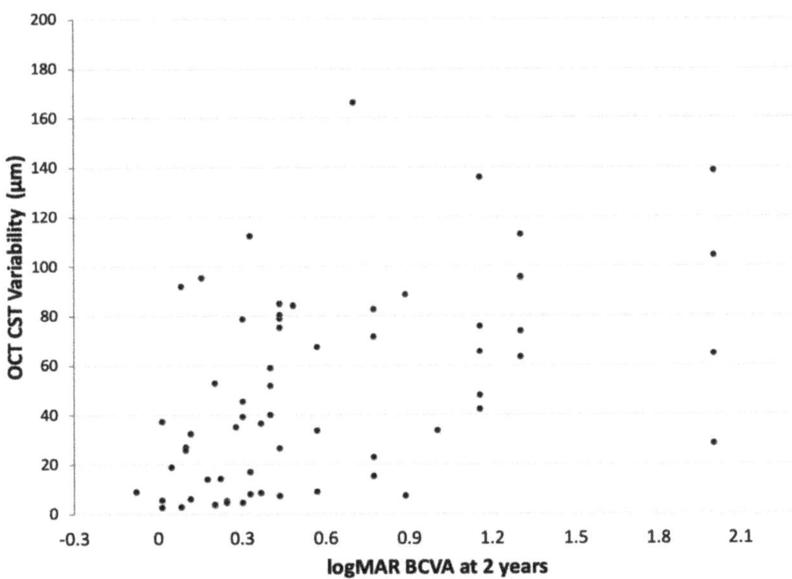

Figure 1. Correlation analysis between OCT CST variability and logMAR BCVA at 2 years.

3.3. Correlation of OCT CST Variability Quartiles and Change in BCVA at 2 Years

The OCT CST variability values for the four quartile groups are listed in Table 1. The lowest quartile (quartile 1) group had OCT CST variability values of <14.4 μm, whereas the highest quartile group (quartile 4) had OCT CST variability values of ≥ 78.7 μm.

The mean BCVA changes at 2 years for the four OCT CST variability quartile groups are displayed in Figure 2. There was a significant difference in the mean BCVA change between the four OCT CST variability groups ($p = 0.018$). The lowest two quartiles of CST variability had BCVA gains of 9.7 letters (quartile 1) and 1.1 letters (quartile 2), whereas the highest two quartile groups of OCT CST variability had BCVA losses of 2.5 letters (quartile 3) and 9.5 letters (quartile 4).

Table 1. OCT CST variability values of the 4 quartile groups during the study period.

Quartile 1 (Least Variability)	Quartile 2	Quartile 3	Quartile 4 (Greatest Variability)
<14.4 μm	14.4–39.4 μm	39.4–78.7 μm	≥78.7 μm

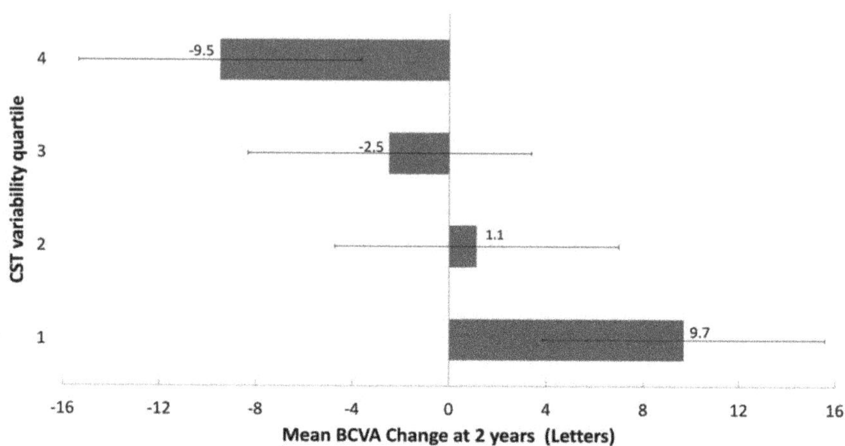

Figure 2. Mean BCVA change at 2 years for the CST variability quartile groups (error bars represent the 95% confidence intervals).

3.4. Anti-VEGF Injections and OCT CST Variability Quartiles

Various intravitreal anti-VEGF agents were used in the study, including aflibercept alone in 22 (34.3%) eyes, bevacizumab alone in 17 (26.6%) eyes, ranibizumab alone in 8 (12.5%) eyes, and a combination of the two agents in 17 (26.6%) eyes. The median number of intravitreal anti-VEGF injections during the 24 months was 7 (interquartile range [IQR], 4 to 10). The median number of injections in each OCT CST variability quartile group was 3 (IQR, 3 to 6.5) for quartile 1, 7.5 (IQR, 5.5 to 11.5) for quartile 2, 7.5 (IQR, 5.5 to 9.5) for quartile 3 and 7 (IQR, 3 to 10) for quartile 4. No significant difference in the number of injections was found between the four quartile groups (one-way ANOVA, $p = 0.23$).

4. Discussion

In this study, we utilized a personalized treatment approach of a PRN anti-VEGF therapy dosing regimen for treating nAMD patients in order to minimize the treatment burden in terms of the number of injections and drug costs associated with anti-VEGF therapy. In many Asian countries, anti-VEGF therapy is frequently self-financed or is only partially subsided, with a cap to the maximum number of anti-VEGF injections that can be reimbursed [19,20]. Therefore, patients often choose to receive a reactive PRN dosing regimen rather than more proactive treatment approaches, such as fixed dosing or treat-and-extend regimens. It is well known that the visual acuity of a substantial proportion of patients with nAMD could still deteriorate to before-treatment levels despite receiving regular anti-VEGF therapy, especially in the long term [21]. Our study demonstrated that eyes with greater variation in OCT retinal thickness during intravitreal anti-VEGF treatment for nAMD had worse visual outcomes in terms of both final visual acuity and mean change in BCVA at 2 years compared with eyes which had smaller variation in retinal thickness. Our findings based on a real-world clinical setting using a personalized PRN treatment regimen are similar to those observed in the CATT and IVAN studies, which used protocol-driven treatment regimens in clinical trial settings [17]. With the use of an artificial intelligence algorithm to analyze the volumes of macular fluid compartments,

Chakravarthy et al. also demonstrated that greater variations in the volume of retinal fluids were associated with worse visual acuity outcomes at 2 years [18].

Several reasons have been postulated to account for the worse visual outcomes in eyes which had greater variation in retinal thickness during anti-VEGF therapy for nAMD. A possible cause might be related to the increased development of fibrosis and macular atrophy [17]. Evans et al. found that the risks of developing fibrosis and macular atrophy increased with greater variation in OCT retinal thickness in eyes receiving anti-VEGF therapy for nAMD, with the highest quartile of retinal thickness variability having odds ratios of 1.95 and 2.10 in developing fibrosis and geographic atrophy, respectively [17]. Similar associations were observed regardless of whether the eye was allocated to the continuous or to the discontinuous PRN treatment groups. Another possible reason for the association between greater OCT variability and worse visual outcome might be due to possible undertreatment, because real-world studies have shown that better visual acuity outcomes might be associated with a higher number of anti-VEGF injections [22,23]. However, as observed in our study, the number of injections was similar between the four OCT CST variability quartile groups. Patients in the lowest two OCT CST variability quartiles had visual acuity gains which could be achieved with around three to four intravitreal anti-VEGF injections per year, and were similar to the highest two quartile groups. In the CATT and IVAN studies, a higher number of anti-VEGF injections was found to be associated with an increased likelihood of having greater OCT retinal thickness variability [17]. Therefore, the association between greater retinal thickness variability and worse visual acuity outcome is likely to be independent of the degree of anti-VEGF treatment.

The findings in our study suggest that when treating patients with nAMD, clinicians should aim to adopt treatment strategies which can minimize variation in OCT retinal thickness. This may be achieved by adopting a more objective individualized treat-and-extend treatment approach using OCT and visual acuity findings to increase the chance of maintaining a dry macula. A form of automated modified treat-and-extend protocol is being evaluated in the faricimab personalized treatment interval arm in the LUCERNE and TENAYA studies for nAMD [24,25]. Another potential option for minimizing the variation in retinal thickness is to utilize newer therapeutic agents which have stronger ability to resolve macular fluids or have increased treatment durability [24–26]. For example, brolucizumab has been shown to result in a significantly greater proportion of eyes with fluid resolution compared with aflibercept; thus, it might be useful in achieving less fluid fluctuation over the course of anti-VEGF therapy [27].

The main limitations of our current study include the retrospective nature of the study and the small number of eyes included. Due to the retrospective nature of the study, sample size and power calculations were not performed. We also only performed a qualitative assessment of the OCT findings based only on CST values; other measures such as volumes of various macular fluid compartments, fibrosis and macular atrophy were not evaluated. Nonetheless, our findings in the real-world clinical setting showed that greater variation in OCT retinal thickness following loading doses of anti-VEGF therapy for nAMD is associated with worse visual acuity outcomes at 2 years. Further studies to determine the optimal treatment regimen which can reduce or minimize retinal thickness variation to improve the treatment outcomes are warranted.

Author Contributions: Conceptualization, T.Y.Y.L.; methodology, T.Y.Y.L. and R.Y.K.L.; formal analysis, T.Y.Y.L.; investigation, R.Y.K.L.; resources, T.Y.Y.L.; data curation, T.Y.Y.L. and R.Y.K.L.; writing—original draft preparation, T.Y.Y.L.; writing—review and editing, T.Y.Y.L. and R.Y.K.L.; visualization, T.Y.Y.L. and R.Y.K.L. All authors have read and agreed to the published version of the manuscript.

Funding: This research received no external funding.

Institutional Review Board Statement: Ethical review and approval were waived for this study, due to the retrospective analysis of anonymized data.

Informed Consent Statement: Patient consent was waived due to the retrospective analysis of anonymized data.

Data Availability Statement: The data presented in this study are available on request from the corresponding author. The data are not publicly available due to privacy.

Acknowledgments: The authors would like to thank Adrian Lai and Joshua Lai for assisting in the OCT data collection.

Conflicts of Interest: T.Y.Y.L. reports grants from Bayer (Leverkusen, Germany), Chengdu Kanghong Biotechnology (Chengdu, China), Novartis (Basel, Switzerland), and Roche (Basel, Switzerland), and has received honoraria for consultancy and lecture fees from Bayer (Leverkusen, Germany), Boehringer Ingelheim (Ingelheim am Rhein, Germany), Novartis (Basel, Switzerland), and Roche (Basel, Switzerland). R.Y.K.L. declares no conflict of interest.

References

1. Flaxman, S.R.; Bourne, R.; Resnikoff, S.; Ackland, P.; Braithwaite, T.; Cicinelli, M.V.; Das, A.; Jonas, J.B.; Keeffe, J.; Kempen, J.H.; et al. Global causes of blindness and distance vision impairment 1990–2020: A systematic review and meta-analysis. *Lancet Glob. Health* **2017**, *5*, e1221–e1234. [CrossRef]

2. Adamis, A.P.; Brittain, C.J.; Dandekar, A.; Hopkins, J.J. Building on the success of anti-vascular endothelial growth factor therapy: A vision for the next decade. *Eye* **2020**, *34*, 1966–1972. [CrossRef] [PubMed]

3. Sloan, F.A.; Hanrahan, B.W. The effects of technological advances on outcomes for elderly persons with exudative age-related macular degeneration. *JAMA Ophthalmol.* **2014**, *132*, 456–463. [CrossRef] [PubMed]

4. Finger, R.P.; Daien, V.; Eldem, B.M.; Talks, J.S.; Korobelnik, J.F.; Mitchell, P.; Sakamoto, T.; Wong, T.Y.; Pantiri, K.; Carrasco, J. Anti-vascular endothelial growth factor in neovascular age-related macular degeneration-a systematic review of the impact of anti-VEGF on patient outcomes and healthcare systems. *BMC Ophthalmol.* **2020**, *20*, 294. [CrossRef] [PubMed]

5. Heath Jeffery, R.C.; Mukhtar, S.A.; Lopez, D.; Preen, D.B.; McAllister, I.L.; Mackey, D.A.; Morlet, N.; Morgan, W.H.; Chen, F.K. Incidence of newly registered blindness from age-related macular degeneration in Australia over a 21-Year Period: 1996–2016. *Asia Pac. J. Ophthalmol.* **2021**, *10*, 442–449. [CrossRef]

6. Mantel, I. Optimizing the anti-VEGF treatment strategy for neovascular age-related macular degeneration: From clinical trials to real-life requirements. *Transl. Vis. Sci. Technol.* **2015**, *4*, 6. [CrossRef]

7. García-Layana, A.; Figueroa, M.S.; Arias, L.; Araiz, J.; Ruiz-Moreno, J.M.; García-Arumí, J.; Gómez-Ulla, F.; López-Gálvez, M.I.; Cabrera-López, F.; García-Campos, J.M.; et al. Individualized therapy with ranibizumab in wet age-related macular degeneration. *J. Ophthalmol.* **2015**, *2015*, 412903. [CrossRef]

8. Garweg, J.G.; Niderprim, S.A.; Russ, H.M.; Pfister, I.B. Comparison of strategies of treatment with ranibizumab in newly-diagnosed cases of neovascular age-related macular degeneration. *J. Ocul. Pharmacol. Ther.* **2017**, *33*, 773–778. [CrossRef]

9. Hatz, K.; Prünte, C. Treat and extend versus pro re nata regimens of ranibizumab in neovascular age-related macular degeneration: A comparative 12 month study. *Acta Opphthalmol.* **2017**, *95*, e67–e72. [CrossRef]

10. Chaikitmongkol, V.; Sagong, M.; Lai, T.Y.Y.; Tan, G.S.W.; Fariza, N.; Ohji, M.; Mitchell, P.; Yang, C.H.; Ruamviboonsuk, P.; Wong, I.; et al. Treat-and-extend regimens for the management of neovascular age-related macular degeneration and polypoidal choroidal vasculopathy: Consensus and recommendations from the Asia-Pacific Vitreo-Retina Society. *Asia Pac. J. Ophthalmol.* **2021**, in press.

11. Okada, M.; Kandasamy, R.; Chong, E.W.; McGuiness, M.; Guymer, R.H. The treat-and-extend injection regimen versus alternate dosing strategies in age-related macular degeneration: A systematic review and meta-analysis. *Am. J. Ophthalmol.* **2018**, *192*, 184–197. [CrossRef] [PubMed]

12. Ross, A.H.; Downey, L.; Devonport, H.; Gale, R.P.; Kotagiri, A.; Mahmood, S.; Mehta, H.; Narendran, N.; Patel, P.J.; Parmar, N.; et al. Recommendations by a UK expert panel on an aflibercept treat-and-extend pathway for the treatment of neovascular age-related macular degeneration. *Eye* **2020**, *34*, 1825–1834. [CrossRef] [PubMed]

13. Spaide, R.F. Antivascular endothelial growth factor dosing and expected acuity outcome at 1 year. *Retina* **2021**, *41*, 1153–1163.

14. Phadikar, P.; Saxena, S.; Ruia, S.; Lai, T.Y.; Meyer, C.H.; Eliott, D. The potential of spectral domain optical coherence tomography imaging based retinal biomarkers. *Int. J. Retin. Vitr.* **2017**, *3*, 1. [CrossRef]

15. Zhang, X.; Lai, T.Y.Y. Baseline predictors of visual acuity outcome in patients with wet age-related macular degeneration. *BioMed Res. Int.* **2018**, *2018*, 9640131. [CrossRef]

16. Nguyen, V.; Puzo, M.; Sanchez-Monroy, J.; Gabrielle, P.H.; Garcher, C.C.; Baudin, F.; Wolff, B.; Castelnovo, L.; Michel, G.; O'Toole, L.; et al. Association between anatomical and clinical outcomes of neovascular age-related macular degeneration treated with antivascular endothelial growth factor. *Retina* **2021**, *41*, 1446–1454. [CrossRef] [PubMed]

17. Evans, R.N.; Reeves, B.C.; Maguire, M.G.; Martin, D.F.; Muldrew, A.; Peto, T.; Rogers, C.; Chakravarthy, U. Associations of variation in retinal thickness with visual acuity and anatomic outcomes in eyes with neovascular age-related macular degeneration lesions treated with anti-vascular endothelial growth factor agents. *JAMA Ophthalmol.* **2020**, *138*, 1043–1051. [CrossRef]

18. Kaiser, P.K.; Wykoff, C.C.; Singh, R.P.; Khanani, A.M.; Do, D.V.; Patel, H.; Patel, N. Retinal fluid and thickness as measures of disease activity in neovascular age-related macular degeneration. *Retina* **2021**, *41*, 1579–1586. [CrossRef]
19. Eldem, B.; Lai, T.Y.Y.; Ngah, N.F.; Vote, B.; Yu, H.G.; Fabre, A.; Backer, A.; Clunas, N.J. An analysis of ranibizumab treatment and visual outcomes in real-world settings: The UNCOVER study. *Graefes Arch. Clin. Exp. Ophthalmol.* **2018**, *256*, 963–973. [CrossRef]
20. Lai, T.Y.Y.; Cheung, C.M.G.; Mieler, W.F. Ophthalmic application of anti-VEGF therapy. *Asia Pac. J. Ophthalmol.* **2017**, *6*, 479–480. [CrossRef]
21. Chandra, S.; Arpa, C.; Menon, D.; Khalid, H.; Hamilton, R.; Nicholson, L.; Pal, B.; Fasolo, S.; Hykin, P.; Keane, P.A.; et al. Ten-year outcomes of antivascular endothelial growth factor therapy in neovascular age-related macular degeneration. *Eye* **2020**, *34*, 1888–1896. [CrossRef] [PubMed]
22. Ciulla, T.A.; Hussain, R.M.; Pollack, J.S.; Williams, D.F. Visual acuity outcomes and anti-vascular endothelial growth factor therapy intensity in neovascular age-related macular degeneration patients: A real-world analysis of 49 485 Eyes. *Ophthalmol. Retin.* **2020**, *4*, 19–30. [CrossRef] [PubMed]
23. Chandra, S.; Rasheed, R.; Menon, D.; Patrao, N.; Lamin, A.; Gurudas, S.; Balaskas, K.; Patel, P.J.; Ali, N.; Sivaprasad, S. Impact of injection frequency on 5-year real-world visual acuity outcomes of aflibercept therapy for neovascular age-related macular degeneration. *Eye* **2021**, *35*, 409–417. [CrossRef]
24. Nicolò, M.; Ferro Desideri, L.; Vagge, A.; Traverso, C.E. Faricimab: An investigational agent targeting the Tie-2/angiopoietin pathway and VEGF-A for the treatment of retinal diseases. *Expert Opin. Investig. Drugs* **2021**, *30*, 193–200. [CrossRef]
25. Khanani, A.M.; Russell, M.W.; Aziz, A.A.; Danzig, C.J.; Weng, C.Y.; Eichenbaum, D.A.; Singh, R.P. Angiopoietins as potential targets in management of retinal disease. *Clin. Ophthalmol.* **2021**, *15*, 3747–3755. [CrossRef]
26. Iglicki, M.; González, D.P.; Loewenstein, A.; Zur, D. Longer-acting treatments for neovascular age-related macular degeneration-present and future. *Eye* **2021**, *35*, 1111–1116. [CrossRef]
27. Dugel, P.U.; Singh, R.P.; Koh, A.; Ogura, Y.; Weissgerber, G.; Gedif, K.; Jaffe, G.J.; Tadayoni, R.; Schmidt-Erfurth, U.; Holz, F.G. HAWK and HARRIER: Ninety-Six-Week Outcomes from the Phase 3 Trials of Brolucizumab for Neovascular Age-Related Macular Degeneration. *Ophthalmology* **2021**, *128*, 89–99. [CrossRef] [PubMed]

Journal of
Personalized
Medicine

Article

Real-Time Diabetic Retinopathy Severity Score Level versus Ultra-Widefield Leakage Index-Guided Management of Diabetic Retinopathy: Two-Year Outcomes from the Randomized PRIME Trial

Hannah J. Yu [1], Justis P. Ehlers [2],*, Duriye Damla Sevgi [2], Margaret O'Connell [2], Jamie L. Reese [2], Sunil K. Srivastava [2] and Charles C. Wykoff [1,3],*

1   Retina Consultants of Texas, Retina Consultants of America, Houston, TX 77030, USA; hannahyu2019@gmail.com
2   Tony and Leona Campane Center for Excellence in Image-Guided Surgery and Advanced Imaging Research, Cole Eye Institute, Cleveland Clinic, 9500 Euclid Ave/i32, Cleveland, OH 44195, USA; sevgid@ccf.org (D.D.S.); oconnem6@ccf.org (M.O.); reesej3@ccf.org (J.L.R.); SRIVASS2@ccf.org (S.K.S.)
3   Blanton Eye Institute & Weill Cornell Medical College, Houston Methodist Hospital, 6560 Fannin St., Suite 750, Houston, TX 77030, USA
*   Correspondence: ehlersj@ccf.org (J.P.E.); charleswykoff@gmail.com (C.C.W.); Tel.: +1-(216)-636-0183 (J.P.E.); +1-(713)-524-3434 (C.C.W.); Fax: +1-(713)-795-4552 (C.C.W.)

Citation: Yu, H.J.; Ehlers, J.P.; Sevgi, D.D.; O'Connell, M.; Reese, J.L.; Srivastava, S.K.; Wykoff, C.C. Real-Time Diabetic Retinopathy Severity Score Level versus Ultra-Widefield Leakage Index-Guided Management of Diabetic Retinopathy: Two-Year Outcomes from the Randomized PRIME Trial. *J. Pers. Med.* **2021**, *11*, 885. https://doi.org/10.3390/jpm11090885

Academic Editors: Peter D. Westenskow and Andreas Ebneter

Received: 1 August 2021
Accepted: 28 August 2021
Published: 4 September 2021

Publisher's Note: MDPI stays neutral with regard to jurisdictional claims in published maps and institutional affiliations.

Abstract: The prospective PRIME trial applied real-time, objective imaging biomarkers to determine individualized retreatment needs with intravitreal aflibercept injections (IAI) among eyes with diabetic retinopathy (DR). 40 eyes with nonproliferative or proliferative DR without diabetic macular edema received monthly IAI until a DR severity scale (DRSS) level improvement of ≥ 2 steps was achieved. Eyes were randomized 1:1 to DRSS- or PLI- guided management. At the final 2-year visit, DRSS level was stable or improved compared to baseline in all eyes, and mean PLI decreased by 11% ($p = 0.73$) and 23.6% ($p = 0.25$) in the DRSS- and PLI-guided arms. In both arms, the percent of *pro re nata* (PRN) visits requiring IAI was significantly higher in year 2 versus 1 ($p < 0.0001$). The percent of PRN visits receiving IAI during year 1 was significantly correlated with the percent of PRN visits with IAI during year 2 ($p < 0.0001$). Through week 104, 77.4% of instances of DRSS level worsening in the DRSS-guided arm were preceded by or occurred alongside an increase of PLI. Overall, consistent IAI re-treatment interval requirements were observed longitudinally among individual patients. Additionally, PLI increases appeared to precede DRSS level worsening, highlighting PLI as a valuable biomarker in the management of DR.

Keywords: diabetic retinopathy; anti-vascular endothelial growth factor; diabetic retinopathy severity scale; panretinal leakage index

1. Introduction

Diabetic retinopathy (DR) is a leading cause of preventable vision loss [1]. In the setting of diabetic macular edema (DME) with visual loss, multiple studies have demonstrated the remarkable visual and anatomic value of consistent anti-vascular endothelial growth factor (VEGF) pharmacotherapy initiated early in the disease process [2–5]. Similarly, prospective studies involving eyes with DR without DME, both proliferative DR (PDR) [6–9] and nonproliferative DR (NPDR) [10,11], have repeatedly demonstrated clinically meaningful anatomic benefits with repeated anti-VEGF treatment compared to laser or observation. While fixed interval anti-VEGF dosing among a population has frequently been employed in clinical trials [2,3,10], this is rarely applied to routine clinical practice. Rather, specific examination and/or imaging-based biomarkers of disease activity are typically applied to guide dosing frequency using an individualized, patient-centric approach.

In the context of DME, the objective endpoint of anti-VEGF therapy is typically fluid resolution on optical coherence tomography (OCT), despite the imperfect correlation between visual acuity and OCT-based fluid status [12,13]. In comparison, OCT-based imaging is typically of limited utility in the management of eyes with DR without DME, because by definition they have no, or limited, central intraretinal or subretinal fluid. There is a need for objective, quantifiable and validated biomarkers of DR disease activity that can be utilized in clinical practice to guide dosing intervals among patients with DR being managed with anti-VEGF pharmacotherapy.

The Intravitreal Aflibercept as Indicated by Real-Time Objective Imaging to Achieve Diabetic Retinopathy Improvement (PRIME) trial was designed to evaluate the utilization of real-time assessments of DR severity levels on fundus photography and panretinal leakage index (PLI) on wide-field fluorescein angiography in guiding anti-VEGF retreatment decisions in the management of DR.

In the first year of the study, while no meaningful differences in visual or anatomic outcomes were observed between the two re-treatment strategies [14], results suggested that PLI worsening may precede DR severity worsening. In year 2 of the PRIME trial, re-treatment strategies remained consistent, but assessments were performed every other month instead of monthly. The purpose of the current work is to describe outcomes from the 2nd and final year of the PRIME trial.

2. Methods

The PRIME trial was a randomized, phase 2 clinical trial (ClinicalTrials.gov, NCT03531294). Institutional review board (IRB; Advarra IRB, Columbia, MA, USA) and ethics committee approval was obtained for this Health Insurance Portability and Accountability Act-compliant trial adhering to the tenets of the Declaration of Helsinki. Data were collected at the Retina Consultants of Texas (Houston and The Woodlands, TX, USA). All participants provided written informed consent prior to enrollment. Study methods were previously reported [14]. Briefly, 40 subjects with treatment-naïve diabetic retinopathy (DR) with a DR severity scale (DRSS) level of 47A to 71A as determined by the central reading center (CRC; Cole Eye Institute, Cleveland Clinic, Cleveland, OH, USA) were randomized 1:1 to a DRSS-guided or PLI-guided arm. Enrollment of eyes with PDR was limited to 50% of the total population. Subjects were excluded if they had spectral domain-OCT central subfield thickness (CST) of more than 320 μm in the study eye or had central DME causing loss of visual acuity (VA).

Subjects were assessed monthly (28 ± 7 days) through week 52; real-time CRC grading of DRSS and PLI was performed using Early Treatment Diabetic Retinopathy Study (ETDRS) 7-standard-field fundus photography imaging (FF4 fundus camera; Zeiss Meditec, Dublin, CA, USA) and ultra-widefield fluorescein angiography (UWFA; 200Tx device; Optos, Dunfermline, United Kingdom). The ETDRS scale [15] was used to grade DRSS (35 = mild NPDR; 43 = moderate NPDR; 47 = moderately severe NPDR; 53 = severe NPDR; 61 = mild PDR; 65 = moderate PDR; 71 and 75 = high-risk PDR; 81 and 85 = advanced PDR). All study eyes received monthly 2.0 mg intravitreal aflibercept injection (IAI) until a DRSS level improvement of ≥2 steps relative to baseline was achieved. PLI at the visit at which a DRSS level improvement of ≥2 steps was achieved was defined as the threshold leakage. For subjects randomized to the DRSS-guided arm, monthly re-treatment with IAI was initiated if a 1-step worsening of DRSS occurred compared to the best DRSS level achieved; treatment was stopped when the best or better DRSS level was again achieved. For subjects randomized to the PLI-guided arm, monthly re-treatment with IAI was initiated if PLI increased to 50% or higher of the difference between the baseline PLI and threshold PLI. Treatment was stopped when PLI decreased to the threshold value, or less. The PLI-guided protocol was optimized with multiple iterations during the early stage of the study, and the majority of assessments were performed using the protocol described above. Beginning at week 52, the interval between visits was lengthened to every other month (56 ± 14 days) and continued through the final visit at week 104, during which need for IAI re-treatment continued to be assessed using the pre-specified DRSS and PLI guidelines for each randomized arm.

Pro re nata (PRN, as needed) visits were defined as visits occurring after the initial ≥ 2 step DRSS improvement was made. PRN IAI were defined as IAI occurring at PRN visits.

The primary outcome measure of the PRIME trial was the incidence of DR-related adverse events among DR subjects receiving IAI. Secondary outcome measures included changes in DRSS level and PLI, number of IAI received, and changes in visual and anatomic measures. Observed data are reported; for subjects who did not complete the week 104 endpoint, corresponding data are included until withdrawal. Statistical analysis was performed using R version 3.5.2 (R Foundation, Vienna, Austria). One-sample and 2-sample paired and unpaired Student's *t*-tests were used to compare continuous variables, and Fisher's exact test was used to compare proportions. Pearson's correlation was used to assess linear relationships between variables.

3. Results

40 eyes from 40 subjects were enrolled and randomized in PRIME with a DRSS level of 47 to 71. As reported, baseline demographics appeared well-balanced between the DRSS- and PLI-guided arms [14]. 29 eyes (DRSS-guided = 13; PLI-guided = 16) entered into year 2 of the trial, and 25 eyes (DRSS-guided = 12; PLI-guided = 13), representing 62.5% of enrolled subjects, completed the week 104 visit. During year 2 of PRIME, 2 (6.9%) subjects died, 1 (3.4%) withdrew due to concerns regarding coronavirus disease 2019 (COVID-19), and 1 (3.4%) was loss to follow up (LTFU). The following data focuses on eyes that entered into year 2 of the PRIME trial. Among the 203 possible visits in year 2 for the 29 subjects who entered year 2, 163 (80.3%) were completed.

3.1. Re-Treatment Requirements in Year 2

Cumulatively through week 104, the DRSS-guided arm received a mean total 9.7 IAI (range, 6 to 17) and the PLI-guided arm received 9.8 IAI (range, 6 to 16; $p = 0.95$; Figure 1A). From week 52 through week 104, eyes received a mean 3.3 (range, 1 to 6) and 2.9 (range, 0 to 7) IAI ($p = 0.62$) in the DRSS-guided and PLI-guided arms respectively. In the DRSS-guided arm, 32.2% and 63.2% of PRN visits resulted in IAI re-treatment in year 1 and year 2, respectively ($p < 0.0001$); in the PLI-guided arm, 26.1% and 56.8% of PRN visits resulted in IAI re-treatment in year 1 and year 2 ($p < 0.0001$; Figure 1B). No significant differences were observed between the two arms related to the proportion of PRN visits resulting in IAI re-treatment in year 1 ($p = 0.45$) or year 2 ($p = 0.5$). Overall, after achieving the pre-defined threshold of ≥ 2-step DRSS level improvement with initial monthly dosing, a mean 5.9 (range, 3 to 11) and 5.4 (range, 3 to 11) PRN IAI were administered to the DRSS- and PLI-guided arms, respectively ($p = 0.65$), over a mean 13.8 ± 2.9 and 13.8 ± 2.7 PRN visits; yielding an average IAI re-treatment interval of every 3.5 months (range, 1.1 to 14). Consistent with this, eyes in the DRSS- and PLI-guided arms had a mean 4.5 (range, 2 to 8) and 4.4 (range, 2 to 12) months between their last monthly IAI and their first PRN IAI ($p = 0.93$); eyes with baseline NPDR and PDR had a mean 4.9 (range, 2 to 12) and 4.1 (range, 2 to 8) months between their last monthly IAI and their first PRN IAI ($p = 0.41$).

A relationship was identified among patient-level re-treatment requirements in the first and second years. Specifically, the percent of PRN IAI given during year 1 was significantly correlated with the percent of PRN IAI given during year 2 ($R = 0.73$, $p < 0.0001$; Figure 2A). In comparison, the absolute number of IAI needed to achieve the initial ≥ 2-step DRSS level improvement was not found to be correlated with the percent of PRN IAI given ($R = -0.07$, $p = 0.73$; Figure 2B).

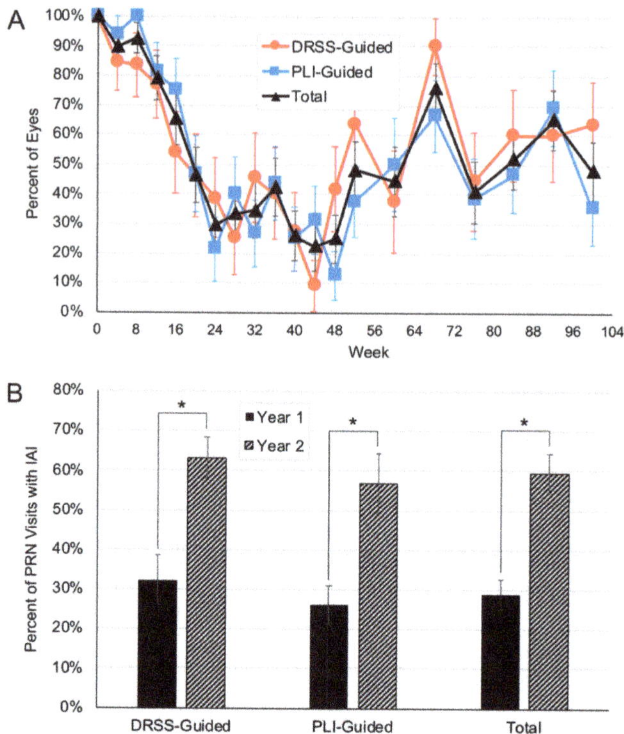

Figure 1. Intravitreal aflibercept injections (IAI) administered through week 104. (**A**) Percent of eyes treated at each visit in the diabetic retinopathy severity scale (DRSS)-guided and the panretinal leakage index (PLI)-guided arms. (**B**) Percent of *pro re nata* (PRN) visits with IAI administration in year 1 and year 2. In the DRSS-guided arm, 32.2% and 63.2% of PRN visits resulted in IAI administration in year 1 and year 2, respectively. In the PRN-guided arm, 26.1% and 56.8% of PRN visits resulted in IAI administration in year 1 and year 2, respectively. Asterisks indicate statistically significant differences between groups.

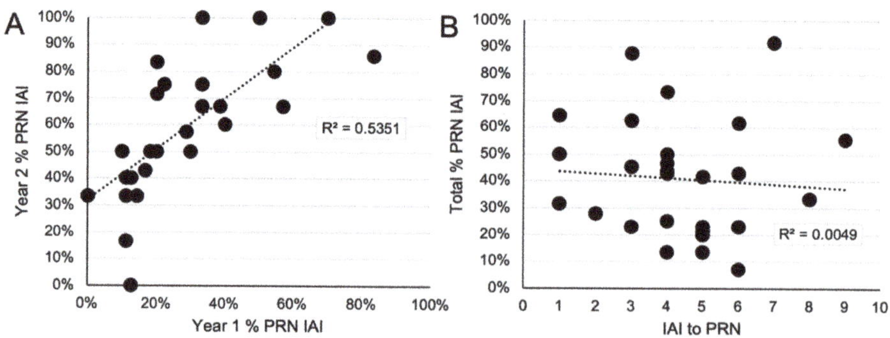

Figure 2. Linear correlations between intravitreal aflibercept injection (IAI) administrations. (**A**) The percent of *pro re nata* (PRN) visits resulting in IAI during year 1 versus the percent of PRN visits resulting in IAI during year 2. A significantly positive linear correlation was observed (R = 0.73, $p < 0.0001$; black dotted line). (**B**) The percent of IAI required to achieve a ≥2-step DRSS improvement versus the total percent of PRN visits requiring IAI through year 2. No significant correlation was observed (R = −0.07, $p = 0.73$; black dotted line).

3.2. DRSS and PLI Outcomes through Year 2

Through week 104, DRSS level stabilized or improved among 16% and 84% of all subjects who completed the week 104 visit (Figure 3) respectively; no subject experienced a worsening of DRSS level at week 104 compared to baseline. 6 (50%) and 10 (76.9%) eyes in the DRSS- and PLI-guided arms experienced a ≥2-step DRSS improvement at week 104 compared to baseline, respectively ($p = 0.23$). Through week 104, one (3.4%) subject never achieved ≥2-step DRSS improvement; this subject missed 12 of 21 visits (57.1%). In the DRSS-guided arm, 8 (61.5%) eyes had PDR at baseline; at week 104, 3 (25%) had PDR, 4 (50%) had NPDR, and 1 (12.5%) was LTFU (NPDR at last observed visit). In the PLI-guided arm, 7 (43.8%) eyes had PDR at baseline; at week 104, 0 had PDR, 4 (57.1%) had NPDR, and 3 (42.9%) were LTFU (2 were NPDR and 1 was PDR at last observed visit).

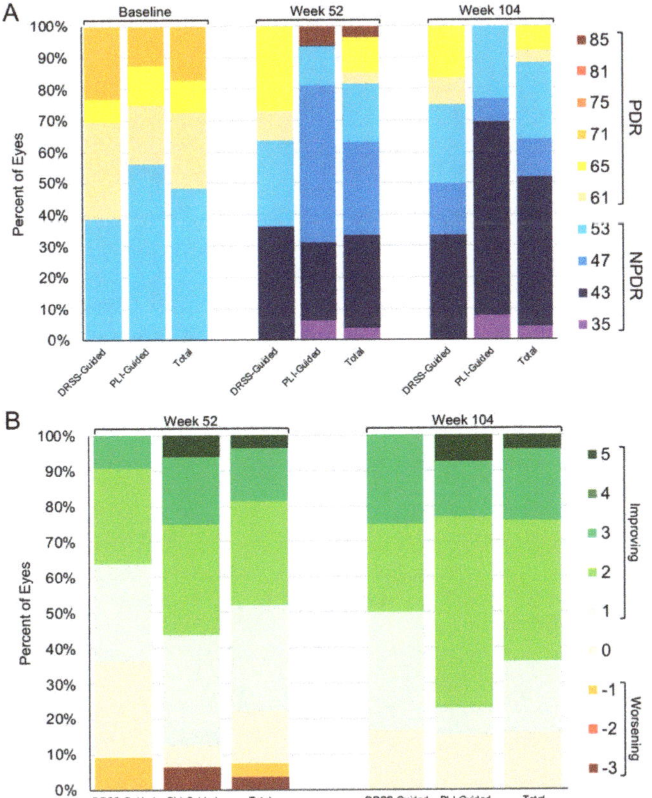

Figure 3. Diabetic retinopathy severity scale (DRSS) levels through week 104. (**A**) Absolute DRSS levels at baseline, week 52, and week 104 in the DRSS-guided and panretinal leakage index (PLI)-guided arms. (**B**) Change in DRSS level compared to baseline at week 52 and week 104 in the DRSS- and PLI-guided arms.

Through the end of the study, no significant differences were observed in change in PLI in either arm from baseline to week 104. Mean PLI decreased by 11% ($p = 0.73$) and 23.6% ($p = 0.25$) in the DRSS- and PLI-guided arms, respectively to an absolute PLI of 1.98% and 1.81% (Figure 4).

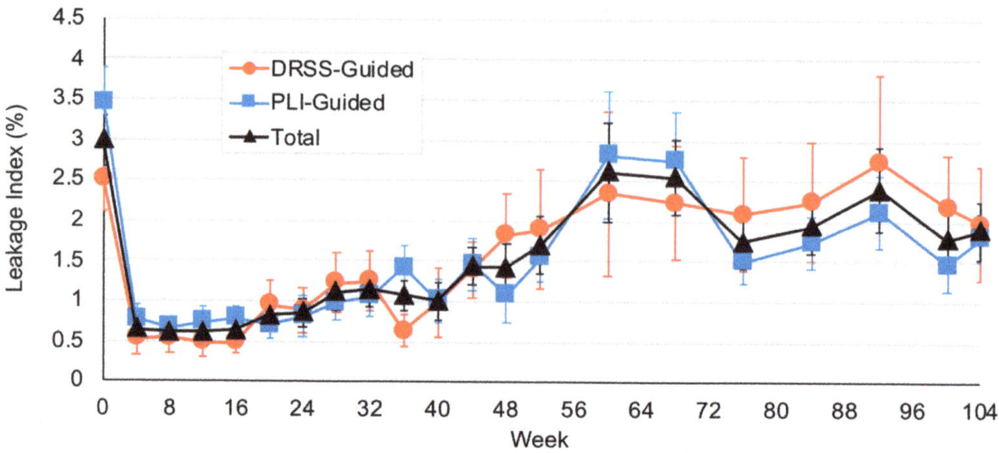

Figure 4. Panretinal leakage index (PLI) in through week 104. Mean PLI decreased by 11% (*p* = 0.73) and 23.6% (*p* = 0.25) compared to baseline in the diabetic retinopathy severity scale (DRSS)- and PLI-guided arms, respectively.

3.3. Indications for PLI as a Precursor to DRSS Worsening

Through week 52 of PRIME, increased PLI appeared to precede DRSS level worsening, a trend that continued through week 104. Overall, among subjects in the DRSS-guided arm who entered year 2 and achieved the initial $\geq$2-step DRSS improvement, 62 instances of $\geq$1-step DRSS worsening compared to their best DRSS were observed. 48 of these instances (77.4%) occurred simultaneously with, or were preceded by, a PLI indication for re-treatment in the visit immediately preceding the visit at which DRSS level worsening was observed. Inversely, among the same population, 57 instances of an increase in PLI above the PLI-retreatment threshold were observed; 48 of these cases (84.2%) occurred simultaneously with or were followed by a $\geq$1-step DRSS worsening compared to baseline in the visit immediately following the indication. Additionally, a total of 36 IAI re-treatment initiations according to DRSS were identified among this population, which occurred simultaneously with or were preceded by a PLI indication for re-treatment in 28 (77.8%) instances.

Two specific cases in the DRSS-guided arm highlight these increases in PLI at visits preceding DRSS level worsening. The first subject (subject 23) presented at baseline with a DRSS level of 53 and PLI of 4.87% (Figure 5A). The subject had a threshold leakage of 0.16% at week 24 and if the subject had been in the PLI-guided arm, they would have been re-treated at a PLI of 2.52%. Through week 104, the subject experienced 3 visits at which a $\geq$1-step DRSS worsening from the best-achieved DRSS was observed, weeks 40, 68, and 100; at the visit immediately preceding the visit or at the visit at which DRSS level worsened, a meaningful PLI increase was observed. At two of these visits, those prior to weeks 68 and 100, PLI increased above their PLI-retreatment threshold.

The second subject (subject 37) presented at baseline with a DRSS level of 53 and PLI of 0.85% (Figure 5B). The subject had a threshold leakage of 0.11% at week 16 and if the subject had been in the PLI-guided arm, they would have been re-treated at a PLI of 0.48%. Through week 104, the subject experienced 4 visits at which a $\geq$1-step DRSS worsening from the best-achieved DRSS was observed, weeks 36, 60, 68, and 92. At the visit immediately preceding each of these visits, an increase above their PLI-retreatment level was observed.

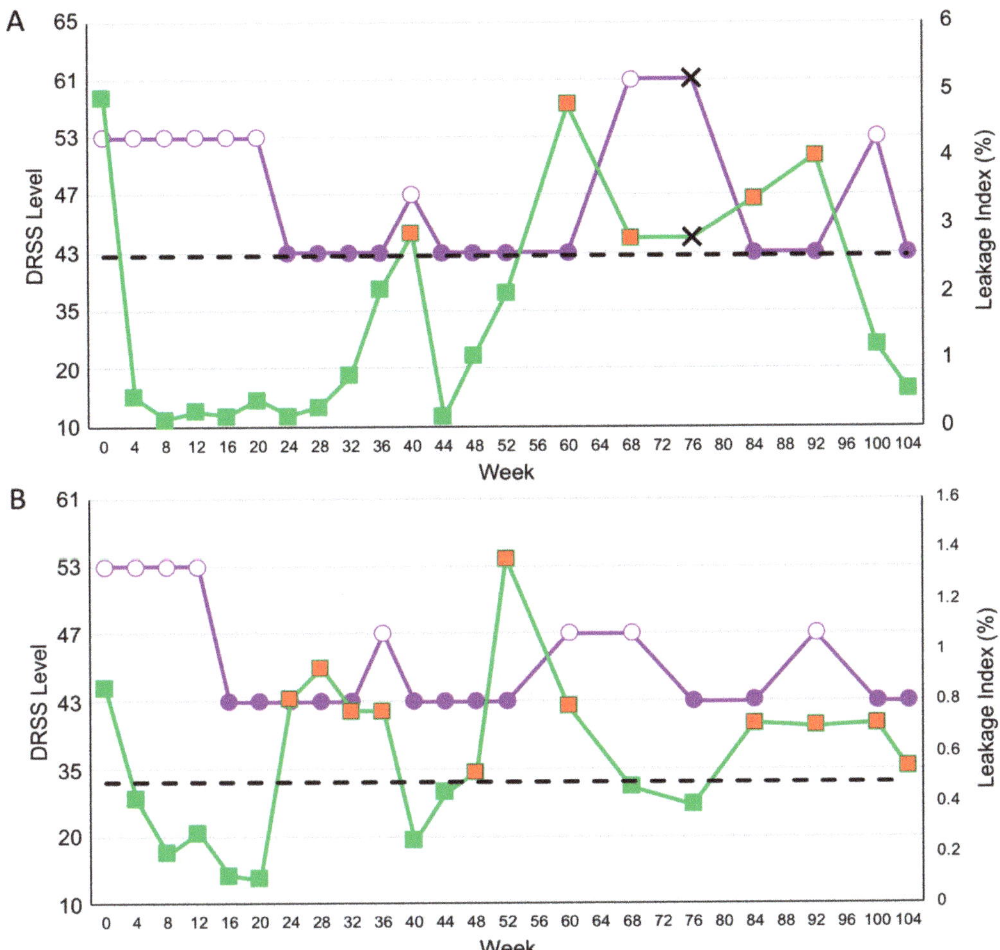

Figure 5. Subject cases from the diabetic retinopathy severity scale (DRSS)-guided arm through week 104. (**A**) Subject 23 and (**B**) subject 37 both demonstrated instances of ≥1-step DRSS worsening from the best-achieved DRSS, each of which was either preceded by or occurred alongside a meaningful panretinal leakage index (PLI) worsening. Purple indicates DRSS level; white-filled markers indicate a visit before the initial ≥2-step DRSS level improvement with initial monthly dosing or a visit with a DRSS worsening compared to best-achieved DRSS level; green indicates PLI level; red-filled markers indicate a worsening in PLI above the re-treatment level according to the PLI-guided protocol; black dotted lines indicate the re-treatment level as indicated by the PLI-guided protocol; black X markers indicate a missed visit.

8 of 17 PLI-guided eyes that completed the week 52 visit never experienced a ≥1-step worsening in DRSS level following the pre-defined initial ≥2-step improvement; 7 of these 8 eyes entered into year 2. Among these 7 eyes, one never experienced a ≥1-step DRSS level worsening through week 104; this eye received 6 IAI before initially achieving a ≥2-step DRSS level improvement and then received 3 PRN IAI during 13 PRN visits (20 months) through week 104. Of the remaining 6 eyes, one never regressed through W92 before becoming LTFU, after 4 IAI to ≥2-step DRSS level improvement and 5 PRN IAI during 14 PRN visits (19 months). One eye experienced DRSS worsening at week 60 after 6 months without PRN IAI, and another experienced DRSS worsening at week 68 after 14 months without PRN IAI; both of these eyes experienced an upward trend in PLI in

the months preceding the DRSS worsening. The final three eyes experienced a ≥1-step worsening in DRSS level after missing at least one visit and following 7, 7, and 14 months without PRN IAI, respectively; all three eyes experienced an increase in PLI alongside DRSS level worsening.

3.4. Visual and OCT-Based Anatomic Outcomes through Year 2

Through week 104, changes in visual and OCT-based anatomic outcomes were minor and similar between arms, as expected given the inclusion criteria. Among eyes that entered year 2, ETDRS BCVA increased in the DRSS- and PLI-guided arms, respectively, by 0.17 (95% CI, −2.7 to 3.4) and 0.38 (95% CI, −5.2 to 5.6) letters from baseline to week 104 to an absolute BCVA of 81.6 (approximate Snellen equivalent, 20/25; 95% CI, 82.1 to 91.0) and 86.5 (approximate Snellen equivalent, 20/20; 95% CI, 76.9 to 86.3) letters (Figure 6A); CST decreased in the DRSS- and PLI-guided arms, respectively, by −13.2 (95% CI, −39.4 to +13.0) and −5.5 (95% CI, 19.0 to +8.08) μm to an absolute CST of 254.4 (95% CI, 237.0 to 271.8) and 270.6 (95% CI, 255.9 to 285.3) μm (Figure 6B).

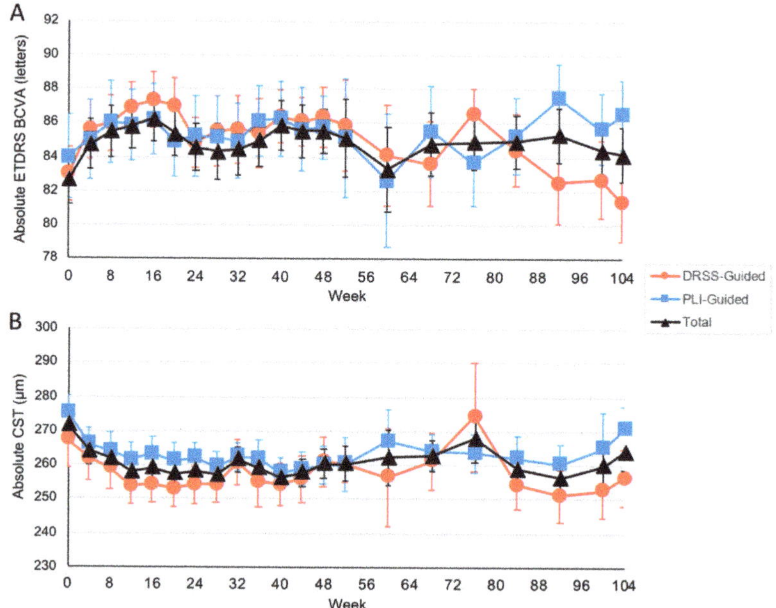

Figure 6. Visual and anatomic changes through week 104. (**A**) Mean absolute Early Treatment Diabetic Retinopathy Study (ETDRS) best-corrected visual acuity (BCVA) letters through week 104. (**B**) Mean absolute central subfield thickness (CST) through week 104. All error bars represent standard errors.

3.5. Adverse Events through Year 2

Ocular and systemic adverse events for year 2 are reported in Table 1. The incidence of DR-related adverse events was 3.4% among eyes entering year 2. No new safety signals or serious ocular adverse events occurred. There were no cases of center-involved DME development or new-onset neovascularization, and no subject received panretinal photocoagulation or vitrectomy.

Table 1. Ocular and Systemic Adverse Events in Year 2.

Ocular Adverse Events			
	DRSS-Guided	PLI-Guided	Total
Worsening of Cataracts	1	1	2
Worsening of PDR	1	0	1
Floaters	1	0	1
Flashes	1	0	1
Ocular Pain	1	0	1
Glaucoma	0	1	1
Cotton Wool Spots	0	1	1
Total # of AEs	5	3	8
Total # of Patients	4	3	7
Serious Systemic Adverse Events			
	DRSS-Guided	PLI-Guided	Total
Cerebrovascular Accident	0	1	1
Arthritic Hip Pain	0	1	1
Infection of Left Foot on Hallux	0	1	1
Bone Destruction of Right Foot	1	0	1
Pneumonia	1	1	2
COVID-19	0	1	1
Acute Chronic Renal Failure	0	1	1
Asthma	1	0	1
Worsening Anemia	1	0	1
Transient Ischemic Attack	1	0	1
Stage 2 Kidney Failure	1	0	1
Fatal Cardiovascular Disease/Diabetes	0	1	1
Fatal Cardiac Arrest	0	1	1
Total # of SAEs	6	8	14
Total # of Patients	4	6	10

DRSS = diabetic retinopathy severity scale; PLI = panretinal leakage index; PDR = proliferative diabetic retinopathy; AE = adverse event; COVID-19 = coronavirus disease 2019; SAE = serious adverse event; # = number.

Two deaths occurred in year 2 of PRIME attributed to cardiac arrest and cardiovascular disease/diabetes mellitus, respectively. Neither death was attributed to study drug or study procedure.

4. Discussion

The randomized PRIME trial explored the use of real-time DRSS and PLI assessments to determine re-treatment decisions for patients with DR without center-involved DME. Through week 104, all subjects in both the DRSS- and PLI-guided arms demonstrated either stable or improved DRSS compared to baseline, with all subjects requiring at least one PRN IAI during the trial, with a global mean of approximately one IAI injection required every 3.5 months.

Overall through week 104, the results of the PRIME trial were consistent with previous prospective trials reporting the need for consistent clinical follow-up and repeated anti-VEGF injections among DR patients due to the chronic, recurrent nature of the disease. While PRN trials have indicated that, once stabilized, some eyes may remain stable for long periods of time [7], cumulatively data indicate that the large majority of patients with DR will require re-treatment with anti-VEGF therapy [7,16,17], even in those with mild or apparent quiescent disease. In the current study, all patients required ongoing anti-VEGF therapy due to worsening of disease severity when treatment was discontinued. Nevertheless, in PRIME, the rate of progression of DR and the necessary re-treatment intervals were highly individualized between patients.

While overall outcomes were similar between the randomized arms in the current study, there were 3 important observations across the 2-year study.

First, in year 2, the percent of PRN visits that resulted in IAI re-treatment increased approximately two-fold in both arms. While this increase was related to the study design of reduced clinical visits in year 2, it is a clinically relevant observation. In clinical settings, patients are often transitioned to PRN regimens after a period of regular, fixed frequency dosing, and in doing so are often evaluated with less frequent visits. As demonstrated in the current dataset, among this population, when seeing patients less frequently with PRN dosing, a greater proportion of visits needing re-treatments is to be expected.

Second, a significant positive correlation was observed between the percent of PRN IAI required during year 1 and year 2. This observation is consistent with an understanding that while re-treatment intervals can vary widely among different patients receiving anti-VEGF pharmacotherapy for exudative retinal diseases, they appear to be remarkably consistent at the individual patient level. This has been elegantly demonstrated in a series of manuscripts by Fauser and colleagues in neovascular age-related macular degeneration; among this population, intravitreal VEGF suppression time after anti-VEGF treatment was intra-individually stable [18,19]. In the current series, the re-treatment intervals among patients were notably consistent longitudinally using either DRSS or PLI-guided re-treatment.

Third, PLI worsening often preceded or occurred alongside worsening in DRSS level, a phenomenon most obvious among the subjects randomized to PLI-guided re-treatment. Through week 104, 77% of instances of DRSS worsening were preceded or occurred simultaneously with a PLI indication for re-treatment, and 84% of instances of a PLI indication for re-treatment occurred alongside or before DRSS worsening, consistent with separate analyses correlating DRSS and PLI changes within patients [20]. Intuitively, this progression of disease severity aligns with current understanding of DR pathogenesis. As intraocular levels of anti-VEGF pharmacotherapy decrease in the vitreous following a bolus therapeutic injection, levels of pathologic VEGF concurrently rise, re-triggering breakdown of the blood-retina barrier leading to increased retinal vasculature permeability visualized by fluorescein leakage [21]. Subsequently, visible lesions of DR increase, including hemorrhages, leading to worsening of DR severity as observed by fundus photography [22]. Current observations suggest that PLI may prove to be a valuable biomarker for clinicians making retreatment decisions in DR patients—rising PLI levels may be indicative of a need for re-treatment with anti-VEGF therapy.

These results suggest that, at least in the context of bolus anti-VEGF monotherapy among eyes with DR without DME, regular clinical assessments and re-treatments are needed indefinitely in order to optimize outcomes. Sustained anti-VEGF therapies with next-generation approaches such as implantable devices [23], more durable pharmaceutical agents [24], and gene therapies [25,26] may be valuable to this patient population. Furthermore, treatments targeting additional mechanisms of action may also bring value to this space in the near future. For example, investigations into manipulating the Tie-2 and kallikrein systems for additive benefit among patients with DR and DME are ongoing [27,28].

Strengths of the PRIME trial include its randomized study design and its use of real-time, objective measurements of DRSS and PLI to guide re-treatment protocols. The key limitation of PRIME is the small sample size. However, given its pilot design, the results of the PRIME trial may be able to guide future, larger prospective studies evaluating biomarkers for re-treatment decisions. Another limitation is the modified version of pure DRSS grading used in the current study [14]. Additionally, there was a notably high rate of LTFU in the current study, with only 29 of 40 enrolled subjects entering year 2, and 25 subjects completing the week 104 endpoint. LTFU is a well-known challenge among patients with DR, as noted in year 1 of the PRIME trial [14] and in multiple other studies [29,30], despite the need for close clinical follow-up among this population. In the current trial, some cases of LTFU may have been due to the length of trial visits and the need for UWFA, FP, and multiple other images at each study visit. This challenge of LTFU was also exacerbated by the impact of COVID-19 during this study period. Finally, the current prospective study only considered the use of anti-VEGF management with aflibercept and did not include laser treatment options. Laser, especially pan-retinal photocoagulation (PRP), can be an excellent clinical tool either as monotherapy or in combination with anti-VEGF pharmacotherapy for the management of more advanced stages of DR.

While the 2-year results of PRIME suggest a meaningful and clinically relevant relationship between PLI and DRSS levels, with PLI increases typically preceding DRSS level worsening, the application of this to clinical practice is challenging. First, while UWFA provides remarkable insight into disease burden and severity not appreciable with fundus photography or ophthalmic examination, it is time-consuming and invasive, and therefore often not practical for all routine clinical visits; while OCT angiography (OCTA), can elegantly visualize the retinal vasculature with fast, non-invasive scan acquisition, it is not currently capable of measuring leakage index. Furthermore, PLI is not readily assessable using current commercially available technologies, and specific software systems would be required to measure and apply this biomarker in clinical practice. Lastly, the DRSS scale is not currently directly clinically applicable due to the strict methodology required to accurately grade imaging; it is our belief that advances in imaging and machine learning technology may allow for this scale to eventually be used clinically.

In conclusion, subjects enrolled in the PRIME trial were able to maintain or improve their DRSS levels through 2 years of PRN re-treatment according to real-time DRSS or PLI measurements. The lengthening of time between assessments in year 2 resulted in an increase in the proportion of visits requiring PRN IAI with subjects being treated an average of once every 3.5 months to maintain DR severity level improvements. Consistent with observations in year 1, worsening of PLI often preceded DRSS level worsening, indicating that PLI may be a valuable biomarker to guide management strategies in trials and clinical practice.

Author Contributions: Conceptualization, C.C.W. and J.P.E.; Methodology, C.C.W. and J.P.E.; Software: D.D.S., M.O., J.L.R., S.K.S.; Formal Analysis, H.J.Y.; Writing—Original Draft Preparation, H.J.Y.; Writing—Review & Editing, H.J.Y., C.C.W., J.P.E.; Supervision, C.C.W. and J.P.E. All authors have read and agreed to the published version of the manuscript.

Funding: Supported by Regeneron Pharmaceuticals, which had no role in the design or conduct of this research.

Institutional Review Board Statement: The study was conducted according to the guidelines of the Declaration of Helsinki, and a pproved by the Institutional Review Board/Ethics Committee of Advarra IRB (Protocol VGFTe-DR-1822, Approved 14 May 2018).

Informed Consent Statement: Informed consent was obtained from all subjects involved in the study.

Data Availability Statement: The data presented in this study are available on request from the corresponding author.

Conflicts of Interest: Authors H.J.Y., D.D.S., M.O., and J.L.R. report no conflict of interest; J.P.E.: Adverum (C), Aerpio (R,C), Alcon (R,C), Allegro (C), Allergan (C, R), Boehringer-Ingelheim (R), Genentech/Roche (R,C), Leica (C, P), Novartis (R,C), Thrombogenics/Oxurion (R,C), Regeneron (R, C), Santen (C), Stealth (C), Zeiss (C); S.K.S.: Abbvie (C), Allergan (R,C), Eyepoint (R,C), Eyevensys (R,C), Gilead (C), Leica (P), Novartis (C), Regeneron (R,C), Santen (R), Zeiss (C); C.C.W: Adverum (R, C), Aerie Pharmaceuticals (R C), Aldeyra (R), Alimera Sciences (R), Allergan (R,C), Allgenesis (C), Amgen (R), Apellis (R,C), Arrowhead Pharmaceuticals (C), Asclepix (R), Bausch + Lomb (C), Bayer (R, C), Bionic Vision Technologies (C), Boehringer Ingelheim (R), Chengdu Kanghong Biotechnologies (R,C), Clearside Biomedical (R,C), Eyepoint Pharmaceuticals (C), Gemini (R), Genetech (R,C), Graybug Vision (R), Gyroscope (R,C), IONIS Pharmaceutical (R), iRENIX (R), IVERIC Bio (R,C), Janssen (C), Kato Pharmaceuticals (C), Kodiak Sciences (R,C), LMRI (R), Long Bridge Medical (C), Neurotech Pharmaceuticals (R), NGM Biopharmaceuticals (R,C), Novartis (R,C), OccuRx (C), Ocular Therapeutix (C), ONL Therapeutics (C,O), Opthea Limited (C), Oxurion (R,C), Palatin

J. Pers. Med. **2021**, *11*, 885

(C), PolyPhotonix (C,O), Recens Medical (R,C,O), Regeneron (R,C), RegenXBio (R,C), Roche (R,C), SamChunDang Pharm (R), Surrozen (C), Taiwan Liposome Company (R), Takeda (C), Verana (C), Visgenx (O), Vitranu (C), Xbrane BioPharma (R).

References

1. Kempen, J.H.; O'Colmain, B.J.; Leske, M.C.; Haffner, S.M.; Klein, R.; Moss, S.E.; Taylor, H.R.; Hamman, R.F.; Eye Diseases Prevalence Research Group. The Prevalence of Diabetic Retinopathy among Adults in the United States. *Arch. Ophthalmol.* **2004**, *122*, 552–563. [CrossRef]
2. Brown, D.M.; Nguyen, Q.D.; Marcus, D.M.; Boyer, D.S.; Patel, S.; Feiner, L.; Schlottmann, P.G.; Rundle, A.C.; Zhang, J.; Rubio, R.G.; et al. Long-Term Outcomes of Ranibizumab Therapy for Diabetic Macular Edema: The 36-Month Results from Two Phase III Trials: RISE and RIDE. *Ophthalmology* **2013**, *120*, 2013–2022. [CrossRef]
3. Brown, D.M.; Schmidt-Erfurth, U.; Do, D.V.; Holz, F.G.; Boyer, D.S.; Midena, E.; Heier, J.S.; Terasaki, H.; Kaiser, P.K.; Marcus, D.M.; et al. Intravitreal Aflibercept for Diabetic Macular Edema: 100-Week Results from the VISTA and VIVID Studies. *Ophthalmology* **2015**, *122*, 2044–2052. [CrossRef] [PubMed]
4. Wykoff, C.C.; Marcus, D.M.; Midena, E.; Korobelnik, J.-F.; Saroj, N.; Gibson, A.; Vitti, R.; Berliner, A.J.; Williams Liu, Z.; Zeitz, O.; et al. Intravitreal Aflibercept Injection in Eyes With Substantial Vision Loss After Laser Photocoagulation for Diabetic Macular Edema: Subanalysis of the VISTA and VIVID Randomized Clinical Trials. *JAMA Ophthalmol.* **2017**, *135*, 107–114. [CrossRef] [PubMed]
5. Payne, J.F.; Wykoff, C.C.; Clark, W.L.; Bruce, B.B.; Boyer, D.S.; Brown, D.M.; TREX-DME Study Group. Randomized Trial of Treat and Extend Ranibizumab With and Without Navigated Laser Versus Monthly Dosing for Diabetic Macular Edema: TREX-DME 2-Year Outcomes. *Am. J. Ophthalmol.* **2019**, *202*, 91–99. [CrossRef]
6. Writing Committee for the Diabetic Retinopathy Clinical Research Network; Gross, J.G.; Glassman, A.R.; Jampol, L.M.; Inusah, S.; Aiello, L.P.; Antoszyk, A.N.; Baker, C.W.; Berger, B.B.; Bressler, N.M.; et al. Panretinal Photocoagulation vs. Intravitreous Ranibizumab for Proliferative Diabetic Retinopathy: A Randomized Clinical Trial. *JAMA* **2015**, *314*, 2137. [CrossRef] [PubMed]
7. Gross, J.G.; Glassman, A.R.; Liu, D.; Sun, J.K.; Antoszyk, A.N.; Baker, C.W.; Bressler, N.M.; Elman, M.J.; Ferris, F.L.; Gardner, T.W.; et al. Five-Year Outcomes of Panretinal Photocoagulation vs Intravitreous Ranibizumab for Proliferative Diabetic Retinopathy: A Randomized Clinical Trial. *JAMA Ophthalmol.* **2018**, *136*, 1138. [CrossRef] [PubMed]
8. Wykoff, C.C.; Nittala, M.G.; Zhou, B.; Fan, W.; Velaga, S.B.; Lampen, S.I.R.; Rusakevich, A.M.; Ehlers, J.P.; Babiuch, A.; Brown, D.M.; et al. Intravitreal Aflibercept for Retinal Nonperfusion in Proliferative Diabetic Retinopathy: Outcomes from the Randomized RECOVERY Trial. *Ophthalmol. Retina* **2019**, *3*, 1076–1086. [CrossRef] [PubMed]
9. Sivaprasad, S.; Prevost, A.T.; Vasconcelos, J.C.; Riddell, A.; Murphy, C.; Kelly, J.; Bainbridge, J.; Tudor-Edwards, R.; Hopkins, D.; Hykin, P.; et al. Clinical Efficacy of Intravitreal Aflibercept versus Panretinal Photocoagulation for Best Corrected Visual Acuity in Patients with Proliferative Diabetic Retinopathy at 52 Weeks (CLARITY): A Multicentre, Single-Blinded, Randomised, Controlled, Phase 2b, Non-Inferiority Trial. *Lancet* **2017**, *389*, 2193–2203. [CrossRef] [PubMed]
10. Brown, D.M.; Wykoff, C.C.; Boyer, D.; Heier, J.S.; Clark, W.L.; Emanuelli, A.; Higgins, P.M.; Singer, M.; Weinreich, D.M.; Yancopoulos, G.D.; et al. Evaluation of Intravitreal Aflibercept for the Treatment of Severe Nonproliferative Diabetic Retinopathy: Results From the PANORAMA Randomized Clinical Trial. *JAMA Ophthalmol.* **2021**. [CrossRef]
11. Maturi, R.K.; Glassman, A.R.; Josic, K.; Antoszyk, A.N.; Blodi, B.A.; Jampol, L.M.; Marcus, D.M.; Martin, D.F.; Melia, M.; Salehi-Had, H.; et al. Effect of Intravitreous Anti–Vascular Endothelial Growth Factor vs Sham Treatment for Prevention of Vision-Threatening Complications of Diabetic Retinopathy: The Protocol W Randomized Clinical Trial. *JAMA Ophthalmol.* **2021**. [CrossRef] [PubMed]
12. Ou, W.C.; Brown, D.M.; Payne, J.F.; Wykoff, C.C. Relationship Between Visual Acuity and Retinal Thickness During Anti-Vascular Endothelial Growth Factor Therapy for Retinal Diseases. *Am. J. Ophthalmol.* **2017**, *180*, 8–17. [CrossRef] [PubMed]
13. Diabetic Retinopathy Clinical Research Network; Browning, D.J.; Glassman, A.R.; Aiello, L.P.; Beck, R.W.; Brown, D.M.; Fong, D.S.; Bressler, N.M.; Danis, R.P.; Kinyoun, J.L.; et al. Relationship between Optical Coherence Tomography-Measured Central Retinal Thickness and Visual Acuity in Diabetic Macular Edema. *Ophthalmology* **2007**, *114*, 525–536. [CrossRef] [PubMed]
14. Yu, H.J.; Ehlers, J.P.; Sevgi, D.D.; Hach, J.; O'Connell, M.; Reese, J.L.; Srivastava, S.K.; Wykoff, C.C. Real-Time Photographic- and Fluorescein Angiographic-Guided Management of Diabetic Retinopathy: Randomized PRIME Trial Outcomes. *Am. J. Ophthalmol.* **2021**, *226*, 126–136. [CrossRef]
15. Early Treatment Diabetic Retinopathy Study Research Group Fundus Photographic Risk Factors for Progression of Diabetic Retinopathy. *Ophthalmology* **1991**, *98*, 823–833. [CrossRef]
16. Sun, J.K.; Wang, P.-W.; Taylor, S.; Haskova, Z. Durability of Diabetic Retinopathy Improvement with As-Needed Ranibizumab: Open-Label Extension of RIDE and RISE Studies. *Ophthalmology* **2019**, *126*, 712–720. [CrossRef] [PubMed]
17. Wykoff, C.C.; Ou, W.C.; Khurana, R.N.; Brown, D.M.; Lloyd Clark, W.; Boyer, D.S.; ENDURANCE Study Group. Long-Term Outcomes with as-Needed Aflibercept in Diabetic Macular Oedema: 2-Year Outcomes of the ENDURANCE Extension Study. *Br. J. Ophthalmol.* **2018**, *102*, 631–636. [CrossRef]
18. Muether, P.S.; Hermann, M.M.; Viebahn, U.; Kirchhof, B.; Fauser, S. Vascular Endothelial Growth Factor in Patients with Exudative Age-Related Macular Degeneration Treated with Ranibizumab. *Ophthalmology* **2012**, *119*, 2082–2086. [CrossRef]

19. Muether, P.S.; Hermann, M.M.; Dröge, K.; Kirchhof, B.; Fauser, S. Long-Term Stability of Vascular Endothelial Growth Factor Suppression Time under Ranibizumab Treatment in Age-Related Macular Degeneration. *Am. J. Ophthalmol.* **2013**, *156*, 989–993.e2. [CrossRef]

20. Ehlers, J.P.; Jiang, A.C.; Boss, J.D.; Hu, M.; Figueiredo, N.; Babiuch, A.; Talcott, K.; Sharma, S.; Hach, J.; Le, T.; et al. Quantitative Ultra-Widefield Angiography and Diabetic Retinopathy Severity: An Assessment of Panretinal Leakage Index, Ischemic Index and Microaneurysm Count. *Ophthalmology* **2019**, *126*, 1527–1532. [CrossRef]

21. Lechner, J.; O'Leary, O.E.; Stitt, A.W. The Pathology Associated with Diabetic Retinopathy. *Vis. Res.* **2017**, *139*, 7–14. [CrossRef] [PubMed]

22. Ip, M.S.; Zhang, J.; Ehrlich, J.S. The Clinical Importance of Changes in Diabetic Retinopathy Severity Score. *Ophthalmology* **2017**, *124*, 596–603. [CrossRef] [PubMed]

23. Campochiaro, P.A.; Marcus, D.M.; Awh, C.C.; Regillo, C.; Adamis, A.P.; Bantseev, V.; Chiang, Y.; Ehrlich, J.S.; Erickson, S.; Hanley, W.D.; et al. The Port Delivery System with Ranibizumab for Neovascular Age-Related Macular Degeneration: Results from the Randomized Phase 2 Ladder Clinical Trial. *Ophthalmology* **2019**, *126*, 1141–1154. [CrossRef]

24. A Trial to Evaluate the Efficacy, Durability, and Safety of KSI-301 Compared to Aflibercept in Participants with Diabetic Macular Edema (DME) (GLEAM). Available online: https://clinicaltrials.gov/ct2/show/NCT04611152 (accessed on 30 June 2021).

25. RGX-314 Gene Therapy Administered in the Suprachoroidal Space for Participants with Diabetic Retinopathy (DR) without Center Involved-Diabetic Macular Edema (CI-DME) (ALTITUDE). Available online: https://clinicaltrials.gov/ct2/show/NCT04567550 (accessed on 30 June 2021).

26. ADVM-022 Intravitreal Gene Therapy for DME (INFINITY). Available online: https://www.clinicaltrials.gov/ct2/show/NCT04418427 (accessed on 30 June 2021).

27. Heier, J.S.; Singh, R.P.; Wykoff, C.C.; Csaky, K.G.; Lai, T.Y.Y.; Loewenstein, A.; Schlottmann, P.G.; Paris, L.P.; Westenskow, P.D.; Quezada-Ruiz, C. The Angiopoietin/Tie Pathway in Retinal Vascular Diseases: A Review. *Retina* **2021**, *41*, 1–19. [CrossRef]

28. New Phase III Data Show Roche's Faricimab Is the First Investigational Injectable Eye Medicine to Extend Time between Treatments up to Four Months in Two Leading Causes of Vision Loss, Potentially Reducing Treatment Burden for Patients. Available online: https://www.roche.com/media/releases/med-cor-2021-02-12.htm (accessed on 30 June 2021).

29. Suresh, R.; Yu, H.; Thoveson, A.; Swisher, J.; Apolinario, M.; Zhou, B.; Shah, A.R.; Fish, R.H.; Wykoff, C.C. Loss to Follow-Up Among Patients with Proliferative Diabetic Retinopathy in Clinical Practice. *Am. J. Ophthalmol.* **2020**, S0002939420301112. [CrossRef] [PubMed]

30. Zhou, B.; Mitchell, T.C.; Rusakevich, A.M.; Brown, D.M.; Wykoff, C.C. Noncompliance in Prospective Retina Clinical Trials: Analysis of Factors Predicting Loss to Follow-Up. *Am. J. Ophthalmol.* **2020**, *210*, 86–96. [CrossRef]

Journal of
Personalized
Medicine

Article

Characterization of Risk Profiles for Diabetic Retinopathy Progression

José Cunha-Vaz [1,2,3,*] and Luís Mendes [1]

[1] AIBILI—Association for Innovation and Biomedical Research on Light and Image, 3000-548 Coimbra, Portugal; lgmendes@aibili.pt
[2] Coimbra Institute for Clinical and Biomedical Research (iCBR), Faculty of Medicine, University of Coimbra, 3000-548 Coimbra, Portugal
[3] Center for Innovative Biomedicine and Biotechnology (CIBB), University of Coimbra, 3000-548 Coimbra, Portugal
* Correspondence: cunhavaz@aibili.pt; Tel.: +351-239-480-136

Abstract: Diabetic retinopathy (DR) is a frequent complication of diabetes and, through its vision-threatening complications, i.e., macular edema and proliferative retinopathy, may lead to blindness. It is, therefore, of major relevance to identify the presence of retinopathy in diabetic patients and, when present, to identify the eyes that have the greatest risk of progression and greatest potential to benefit from treatment. In the present paper, we suggest the development of a simple to use alternative to the Early Treatment Diabetic Retinopathy Study (ETDRS) grading system, establishing disease severity as a necessary step to further evaluate and categorize the different risk factors involved in the progression of diabetic retinopathy. It needs to be validated against the ETDRS classification and, ideally, should be able to be performed automatically using data directly from the examination equipment without the influence of subjective individual interpretation. We performed the characterization of 105 eyes from 105 patients previously classified by ETDRS level by a Reading Centre using a set of rules generated by a decision tree having as possible inputs a set of metrics automatically extracted from Swept-source Optical Coherence Tomography (SS-OCTA) and Spectral Domain- OCT (SD-OCT) measured at different localizations of the retina. When the most relevant metrics were used to derive the rules to perform the organization of the full pathological dataset, taking into account the different ETDRS grades, a global accuracy equal to 0.8 was obtained. In summary, it is now possible to envision an automated classification of DR progression using noninvasive methods of examination, OCT, and SS-OCTA. Using this classification to establish the severity grade of DR, at the time of the ophthalmological examination, it is then possible to identify the risk of progression in severity and the development of vision-threatening complications based on the predominant phenotype.

Keywords: diabetic retinopathy; ETDRS classification; biomarkers; visual prognosis; phenotypes; personalized medicine

Citation: Cunha-Vaz, J.; Mendes, L. Characterization of Risk Profiles for Diabetic Retinopathy Progression. *J. Pers. Med.* **2021**, *11*, 826. https://doi.org/10.3390/jpm11080826

Academic Editors: Peter D. Westenskow and Andreas Ebneter

Received: 21 July 2021
Accepted: 20 August 2021
Published: 23 August 2021

1. Introduction

Diabetic retinopathy (DR) is a frequent complication of diabetes and, through its vision-threatening complications, i.e., macular edema and proliferative retinopathy, may lead to blindness. Diabetes is now regarded as a global epidemic. It is estimated that by 2045 there will be 629 million people worldwide affected by diabetes. Considering that a third of people with diabetes have signs of diabetic retinopathy, with 10% developing vision-threatening retinopathy, it is clearly one of the leading causes of blindness in working-age people [1].

The presence of nonproliferative retinopathy is identified by microvascular changes that are typically asymptomatic. Nonproliferative retinopathy progresses silently, without vision loss from mild to moderate to severe stages. However, the progression of nonproliferative retinopathy to vision-threatening stages, proliferative retinopathy, and macular edema with vision loss vary from individual to individual. The cumulative occurrence of rates of progression from mild nonproliferative to vision-threatening complications has been determined to be in the order of 14–16% [2,3]. However, when there is moderate to severe retinopathy then progression to complications is in the order of 58% [2]. In any case, predicting which people with nonproliferative retinopathy are at a high risk for progression to vision loss remains a challenge.

It is, therefore, of major relevance to identify the presence of retinopathy in diabetic patients and, when present, to identify the eyes that have the greatest risk of progression and greatest potential to benefit from treatment.

2. Phenotypes of Diabetic Retinopathy Progression

When looking at the initial stages of DR, it is said that it is present when microaneurysms and small hemorrhages appear on ophthalmoscopic examination. On histopathological examination, the first vascular changes occur in the small vessels in the form of vasoregression with endothelial proliferation, pericyte damage, and the development of microaneurysms. Characteristically these initial lesions are focal and located at the posterior pole of the retina. With the progression of the disease, the capillaries of the arterial side of the retinal circulation show increased vasoregression. With cell loss and closure, the number of microaneurysms increases, and the areas of capillary closure enlarge. As the areas of capillary closure enlarge, they are seen to be crossed by the remaining enlarged capillaries, which appear to act as arteriovenous shunts, receiving the blood directly from the surrounding closed capillary net [4,5]. Recent clinical studies using optical coherence tomography angiography (OCTA) show that capillary closure occurs very early in diabetic retinal disease and is initiated in the macula [6,7]. Later on, as the disease progresses with remodeling of the retinal circulation and an altered retinal blood flow distribution, probably through preferential arteriovenous preferential channels, capillary closure develops also in more peripheral regions of the retina [8,9].

Using the examination methods now available, it may be stated that the earliest alterations that may be detected clinically in the retina in diabetes are alterations in the neurosensory retinal function, breakdown of the blood–retinal barrier, and capillary closure. These alterations can be detected before ophthalmoscopic signs of DR are visible in preclinical retinopathy [6,10].

Hyperglycemia appears to be sufficient to initiate the development of DR, as revealed by the development of retinopathy in animals experimentally made hyperglycemic. However, the observation that not all individuals with poor metabolic control develop advanced stages of DR suggests that other factors, such as environmental factors and genetic predispositions, are likely to determine individual susceptibility to the disease.

Diabetic retinopathy has been generally considered to be a microvascular complication of diabetes, limiting the diagnostic and therapeutic focus to the vascular system. However, recent evidence has been accumulating, suggesting that DR involves the neuronal as well as the vascular compartments.

The neurosensory retina has recently been shown to be altered very early in diabetes and may function as the trigger for microcirculatory changes [11]. Together with reduced corneal nerve sensation and impaired autonomic innervation of the pupil, altered function of the retinal indicates that diabetes causes denervation of multiple sensory inputs to the eye. There are, therefore, strong arguments for diabetic retina neuropathy. However, it is now clear that the micro-vascular changes occur to a different degree in different patients [6]. In fact, the direct link between neurodegeneration and microvascular disease may occur predominantly in the initial stages of the retinal disease, when different individuals respond differently to the neurodegenerative changes [12].

Regarding systemic factors, it is recognized that the duration of diabetes and the level of metabolic control determine the development of DR. However, these risk factors do not explain the great variability that characterizes the evolution and rate of progression of retinopathy in different diabetic individuals. There are many diabetic patients who, after many years with diabetes, never develop sight-threatening retinal changes, whereas other patients progress rapidly. This is a message of major relevance when dealing with type 2 diabetes.

It is now accepted that only a subset of individuals with diabetes who develop retinal changes is expected to progress to advanced retinopathy stages and is at risk of losing functional vision during their lifetime.

We have identified three major phenotypes of DR progression: One, the neurogenerative phenotype characterized by slow progression, where neurodegeneration is the only identified alteration and the retinal changes may be only a manifestation of the systemic neuropathy; a second one, the leaky phenotype characterized by the added occurrence of edema resulting from the breakdown of the blood–retinal barrier which may occur at any time in the disease progression, even in the absence of relevant microvascular pathology and, finally, a third one, the ischemic phenotype, identified by increased microaneurysm turnover and the presence of active microvascular lesions. In a series of follow-up studies of 2 and 5 years, in eyes with minimal retinopathy, the first phenotype was identified in 40% of the eyes with any evidence of retinopathy and only rarely progressed to sight-threatening complications, whereas the second phenotype representing approximately 30% of the eyes with minimal retinopathy showed a relatively high risk for development of mild and manageable macular edema, and the third representing the remaining 30% of the eyes showed the higher risk for development of both clinically significant macular edema and proliferative retinopathy [3].

A fundamental characteristic of DR is that its progression varies in different individuals, and the development of vision-threatening complications occurs only in a few individuals. The activity of the disease and its progression varies from patient to patient, making identification of biomarkers of progression of DR to vision-threatening complications a major need.

Treatment options presently available are limited. In those receiving treatment, therapy is largely limited to pan-retinal photocoagulation and anti-Vascular Endothelial Growth Factor (VEGF) intravitreal injections [13]. Pan-retinal photocoagulation is, however, usually deferred until nonproliferative retinopathy becomes severe and is associated with some degree of decline in visual function. Although there is no effective treatment that prevents nonproliferative diabetic retinopathy progression, the changes occurring in the retina may regress, indicating that there is a real opportunity for drugs or other treatments to stop retinopathy progression [6]. Treatments directed at the initial stages of nonproliferative diabetic retinopathy, particularly treatments that can be administered outside of a clinical setting, such as oral or topical formulations, are particularly desirable [14].

In the present paper, we review the ETDRS current nonproliferative retinopathy classification to grade retinopathy severity of the ischemic phenotype and suggest the development of a simple to use alternative to the ETDRS grading system, which establishes disease severity as a necessary step to further evaluate and categorize the different risk factors involved in the progression of diabetic retinopathy.

The final objective is to identify in an individual patient its risk profile in order to establish a personalized program of care with timely intervention, and it is realized that an automated and simple to use a severity grading system of DR is a necessary step that will facilitate the identification of risk: markers of DR progression.

3. The Early Treatment Diabetic Retinopathy Study and Classification of Diabetic Retinopathy Severity

The Early Treatment for Diabetic Retinopathy Study (ETDRS) disease severity scale is based on a modified version of the Airlie House classification system. The ETDRS classification is made using stereoscopic color fundus photography obtained from seven standard fields (30 degrees) and is the reference standard for grading diabetic retinopathy severity (ETDRS Report 10). More recently, digital evaluations were demonstrated to be comparable, and these have replaced stereoscopic color photographs [15].

An alternative simplified scale, the International Clinical Diabetic Retinopathy Disease severity scale, was developed with the main objective of facilitating communication between all levels of healthcare provider, but did not replace the ETDRS classification, because the severity grades needed the classic ETDRS grading process to be substantiated and verified. It is relevant that this classification chose to separate the grading of two phenotypes, the ischemic phenotype based in the ETDRS classification and the edema of the leaky phenotype [16].

The ETDRS classification is definitely the validated gold standard for diabetic retinopathy severity classification but has major limitations; it is time-consuming and not practical for daily clinical use. There is, therefore, a clear and established need for an alternative method that can perform at least as well as the ETDRS classification and can be used in clinical practice.

An alternative ETDRS classification should be able to discriminate reliably the different severity stages of diabetic retinopathy, as well as the present ETDRS classification. It needs to be validated against the ETDRS classification and, ideally, should be able to be performed automatically using data directly from the examination equipment without the influence of subjective individual interpretation.

The final goal, again, would be to use this framework on which risk factors and specific rates of progression could be identified to be associated with increased risk of progression.

4. Automated Alternative ETDRS Classification

The ETDRS classification is based on the identification of a series of lesions, mostly resulting from microvascular disease, and based on their distribution in the seven fundus image fields collected (Table 1). Different grades of severity are identified by the number of quadrants of the retina showing specific lesions, such as microaneurysms and hemorrhages, intraretinal vascular abnormalities, venous loops, hard exudates, and cotton wool spots, all of them resulting from microvascular pathology (Table 2).

Table 1. ETDRS Final Scale. Adapted from ETDRS Report No. 12 [17].

	ETDRS Final Scale
Level	**Definition**
10	Microaneurysms (MAs) and other characteristics absent
20	Define presence of Mas and other characteristics absent
35 A	Definite presence of venous loops in 1 field
35 B	Questionable soft exudates, Intraretinal Microvascular Abnormality (IrMA), or venous beading
35 C	Presence of Hemorrhage
35 D	Definite presence of hemorrhage in 1–5 field
35 E	Moderately severe hemorrhages in 1 field
35 F	Definite presence soft exudates in 1 field
43 A	Moderately severe hemorrhages in 4–5 fields or severe hemorrhages in 1 field
43 B	Definite presence of IrMA in 1–3 fields
47 A	Both 43 A and 43 B definitions
47 B	Definite presence of IrMA in 4–5 fields
47 C	Severe hemorrhages in 2–3 fields
47 D	Definite venous beading in 1 field
53 A	≥ 2 level 47 definition
53 B	Severe hemorrhages in 4–5 fields
53 C	Moderately severe presence of IrMA in 1 field
53 D	Definite presence of venous beading in 2–3 fields
53 D	Two or more level 53 definitions

Table 2. Basis for an alternative Diabetic Retinopathy classification.

Lesion Type (Fundus Photos)	Description	Alternative Metric	Examination
Microaneurysms and dot hemorrhages	Capillary wall outpouching (adjacent to capillary closure)	Capillary closure metrics: 1. Vessel density 2. Intercapillary spaces 3. Foveal avascular zone (FAZ) metrics	Optical Coherence Tomography (Angiography (OCT-A)
Venous beading Venous loops	Shunt vessels (?), tortuosity of venula (Adjacent to capillary closure)		
Intraretinal Microvascular Abnormalities (IrMA)	Remodeling new capillary "buds" (Adjacent to capillary closure)	Measurements performed at: 1. Perifovea 2. Mid-periphery 3. Number of quadrants involved	
Cotton wool spots	Site of ischemia, poor perfusion	Capillary closure metrics: 1. Vessel density 2. Intercapillary spaces Number of Quadrants	OCTA
Hard Exudates	Lipid and lipoproteins deposits	Central retinal thickness	Optical coherence tomography (OCT)
Neurodegenerative changes		Ganglion cell + Inner Plexiform layers thinning	

However, at least three main disease pathways occur in diabetic retinal disease: neurodegeneration, edema, and ischemia [6,18]. A forward-looking approach should be to develop a quantitative assessment of these three different disease pathways. Our group used a non-invasive, multimodal approach to evaluate and quantify the relative relevance of these disease pathways in eyes with nonproliferative diabetic retinopathy using OCT and OCTA. Neurodegeneration was identified by the thinning of the retinal tissue (ganglion cell layers plus inner plexiform layer—GCL + IPL), this metric comparing favorably with thinning of the retinal nerve fiber layer (RNFL) because of robustness of measurements. Retinal edema was characterized by increases in retinal thickness. Finally, ischemia was identified by microvascular metrics using OCTA.

This and a series of other studies have confirmed the relevance of vessel density metrics obtained with SD-OCTA and SS-OCTA, independently of the observer, to identify microvascular pathology occurring in the diabetic retina and their value in discriminating different ETDRS grades [6,7,10,19].

We propose that lesion distribution, which is taken into consideration in the ETDRS classification and has also been confirmed with OCTA, should be taken into account when using automated evaluation with OCTA [20,21].

Involvement of retinal mid-peripheral and peripheral regions, well demonstrated in studies with widefield fluorescein angiography, have confirmed that mid-peripheral and peripheral retinal changes need to be considered and may be essential for determining retinopathy progression, at least in the more advanced stages of DR [9,14,22]. Widefield OCTA using swept source OCTA enables the determination of the microvascular metrics over large fields of the retina [21]. Our group has been able to show that retinal capillary closure, quantified by vessel density metrics using SS-OCTA, can identify the more severe stages of nonproliferative diabetic retinopathy and discriminate them from the initial stages of NPDR. A combination of acquisition protocols, using SS-OCTA, allows discrimination between eyes with mild NPDR (ETDRS 20–35) and eyes with moderate-to-severe NPDR (ETDRS grades 43–53).

This study also showed that taking into account the lesions in different quadrants contributes to discriminating levels of retinopathy severity.

We have now tested a composite of features that can be identified and obtained directly from imaging equipment available commercially, such as OCT and OCTA, which discriminate the different ETDRS severity grades. Initial results are promising and indicate that it is possible to replace the laborious and complicated ETDRS grading based on the interpretation of fundus images by a series of metrics obtained directly from OCT and OCTA equipment (Table 3).

Table 3. Alternative Diabetic Retinopathy grading. Automated collection of parameters.

Alternative DR Grading		Perifovea	Mid-Periphery
1. Skeletonized vessel density SCP, DCP, FR	Capillary closure	✓*	✓*
2. FAZ area and circularity	Capillary closure	✓	
3. Thinning of GCL	Neurodegeneration	✓*	✓*

DCP: Deep Capillary Plexus; SCP: Superficial Capillary Plexus; FR: Full Retina; FAZ: Foveal Avascular Zone; GCL: Ganglion Cell Layer; *: number of quadrants with lesions.

We performed the characterization of 105 eyes from 105 patients, previously classified by ETDRS level by the Coimbra Ophthalmology Reading Centre (CORC) and the object of a previous report [8] using a set of rules generated by a decision tree, having as possible inputs a set of metrics automatically extracted from SS-OCTA and SS-OCT measured at different localization of the retina (Figure 1). Data quality was checked as described in the previous work [8]. The values of the metrics were computed on the ARI network (https://arinetworkhub.com/, accessed on 1 August 2021) using the algorithms provided by the manufacture: ETDRS Retina Thickness version 0.1 (metrics related to the thickness of GCL) and the Density Quantification version 0.3.5 (metrics related with VD, PD, and FAZ).

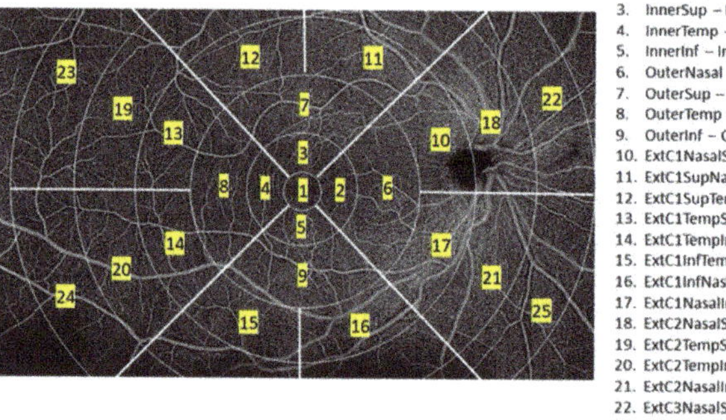

Figure 1. Localization of the 25 subregions that were used to decompose the 15 mm × 9 mm region acquired for each eye using the Zeiss PLEXT Elite 9000 (Carl Zeiss Meditec, Dublin, CA, USA). For each subregion, the mean value of the VD and GCL was measured. Aggregate measures were also used. The values of the inner ring (2,3,4,5), outer ring (6,7,8,9), inner circle (1,2,3,4,5), outer circle (1,2,3,4,5,6,7,8,9), and rings of the extended circles C1 (10,11,12,13,14,15,16), C2 (18,19,20,21), C3 (22, 23,24,25) were also used.

We started by organizing the data on rules derived from a CART (Classification and Regression Trees) decision tree between consecutive ETDRS levels. The rules were generated using the CART algorithm implemented on the python package Scikit-learn (https://scikit-learn.org/ accessed on 21 July 2021) version 0.23.2. The most important metrics included in the rules derived from the data extracted from healthy eyes (28 eyes) and eyes belonging to the ETDRS 10–20 group (34 eyes) were associated with changes in the FAZ, VD in perifovea, and thinning of the GCL + IPL localized in the mid-periphery. When the task was to organize data belonging to ETDRS 10–20 group (34 eyes) and ETDRS 35 (39 eyes), the most relevant discriminating features were associated with the VD in the central retina, extending to more quadrants and GCL thinning in the retinal mid-periphery. Between the ETDRS 35 (39 eyes) and ETDRS 43–47 (24 eyes), analysis of the most important features suggested that the VD changes were extended to the retinal mid-periphery and involved the deep capillary plexus predominately. Figures 2 and 3 show the rules derived when the task was to organize the data comparing ETDRS 10–20 with ETDRS 35 and ETDRS 35 with ETDRS 43–47, respectively. When the most relevant metrics were used to derive the rules to perform the organization of the full pathological dataset, taking into account the different ETDRS grades, a global accuracy equal to 0.8 was obtained.

```
|----- GCL_ExtC3   <= 34.2
|      |----- GCL_ExtC1SupTemp <= 37.6
|      |      |----- #eyesPerClass: [0, 2] assignedClass: ETDRS 35
|      |----- GCL_ExtC1SupTemp >  37.6
|      |      |----- #eyesPerClass: [15, 0] assignedClass: ETDRS 10-20
|----- GCL_ExtC3   >  34.2
|      |----- VD_InnerRing_SCP<= 28.5
|      |      |----- GCL_ExtC3TempInf <= 29.2
|      |      |      |----- #eyesPerClass: [4, 0] assignedClass: ETDRS 10-20
|      |      |----- GCL_ExtC3TempInf >  29.2
|      |      |      |----- VD_OuterNasal_SCP  <= 33.7
|      |      |      |      |----- VD_OuterNasal_DCP<= 14.4
|      |      |      |      |      |----- VD_CircleRing_DCP <= 10.7
|      |      |      |      |      |  |----- #eyesPerClass: [0, 4] assignedClass: ETDRS 35
|      |      |      |      |      |----- VD_CircleRing_DCP >  10.7
|      |      |      |      |      |  |----- #eyesPerClass: [4, 0] assignedClass: ETDRS 10-20
|      |      |      |      |----- VD_OuterNasal_DCP>  14.4
|      |      |      |      |      |----- GCL_ExtC1InfTemp <= 45.0
|      |      |      |      |      |  |----- #eyesPerClass: [0, 32] assignedClass: ETDRS 35
|      |      |      |      |      |----- GCL_ExtC1InfTemp >  45.0
|      |      |      |      |      |  |----- #eyesPerClass: [1, 0] assignedClass: ETDRS 10-20
|      |      |      |----- VD_OuterNasal_SCP  >  33.7
|      |      |      |      |----- VD_ExtC2TempInf_DCP<= 8.7
|      |      |      |      |      |----- #eyesPerClass: [0, 1] assignedClass: ETDRS 35
|      |      |      |      |----- VD_ExtC2TempInf_DCP>  8.7
|      |      |      |      |      |----- #eyesPerClass: [3, 0] assignedClass: ETDRS 10-20
|      |----- VD_InnerRing_SCP>  28.5
|      |      |----- #eyesPerClass: [7, 0] assignedClass: ETDRS 10-20
```

Figure 2. Rules derived using a CART decision tree when the task was to organize the values of the OCT and OCTA metrics extracted from the dataset comparing eyes with ETDRS 10–20 and ETDRS 35 levels. At each leaf, the number (#) of eyes selected by that rule and the most popular class (the assigned class) are presented.

An automated ETDRS classification using the data derived from OCTA and OCT equipment is, therefore, within reach. The limitations are mainly associated with the need for good quality images, which can originate from data sets with smaller sample sizes [8]. It is now necessary to collect more data using the same approach in order to apply machine learning techniques to create and validate a model to perform the ETDRS classification automatically.

The identification of IrMAs by OCTA is also considered another important step to further characterize the severity of disease before the development of PDR. It is relevant that other authors have recently attempted similar approaches [23,24].

```
|----- VD_OuterNasal_SCP  <= 34.1
|      |----- VD_OuterTemp_SCP <= 17.1
|      |      |----- VD_ExtC2NasalInf_DCP<= 4.9
|      |      |      |----- VD_InnerNasal_SCP <= 6.2
|      |      |      |      |----- #eyesPerClass: [0, 2] assignedClass: ETDRS 43-47
|      |      |      |----- VD_InnerNasal_SCP >  6.2
|      |      |      |      |----- #eyesPerClass: [5, 0] assignedClass: ETDRS 35
|      |      |----- VD_ExtC2NasalInf_DCP>  4.9
|      |      |      |----- GCL_ExtC1NasalSup<= 36.3
|      |      |      |      |----- #eyesPerClass: [1, 0] assignedClass: ETDRS 35
|      |      |      |----- GCL_ExtC1NasalSup> 36.3
|      |      |      |      |----- #eyesPerClass: [0, 9] assignedClass: ETDRS 43-47
|      |----- VD_OuterTemp_SCP >  17.1
|      |      |----- VD_ExtC1NasalInf_SCP<= 39.3
|      |      |      |----- GCL_ExtC2TempInf <= 32.0
|      |      |      |      |----- #eyesPerClass: [0, 1] assignedClass: ETDRS 43-47
|      |      |      |----- GCL_ExtC2TempInf >  32.0
|      |      |      |      |----- #eyesPerClass: [32, 0] assignedClass: ETDRS 35
|      |      |----- VD_ExtC1NasalInf_SCP>  39.3
|      |      |      |----- ExtendedC1TemporalInferiorSuper <= 19.5
|      |      |      |      |----- #eyesPerClass: [0, 2] assignedClass: ETDRS 43-47
|      |      |      |----- ExtendedC1TemporalInferiorSuper >  19.5
|      |      |      |      |----- #eyesPerClass: [1, 0] assignedClass: ETDRS 35
|----- VD_OuterNasal_SCP  >  34.1
|      |----- #eyesPerClass: [0, 10] assignedClass: ETDRS 43-47
```

Figure 3. Rules derived using a CART decision tree when the task was to organize the values of the OCT and OCTA metrics extracted from the dataset when comparing with ETDRS 35 and ETDRS 43–47 level having into account the ETDRS level. At each leaf, the number (#) of eyes selected by that rule and the most popular class (the assigned class) are presented.

In order to establish the progression and risk of progression, it is crucial to be able to compare examinations performed at regular intervals, and this is possible with OCT and OCTA using standardized equipment and using information given directly from the equipment.

There is now evidence supporting the concept that DR initiates in the macular area by a deficient vascular response to abnormal neuronal toxicity caused by hyperglycemia, progressing later to the entire retina. There is initial swelling of the central retina, demanding a vascular response that is not adequate in some patients [12]. The eyes that are not able to respond adequately develop progressive capillary closure, initially around the FAZ and perifovea. This capillary closure increases as the disease progresses, with the development of collateral vascular channels and capillary closure extending to the entire retina, including mid-periphery and periphery. This capillary closure is accompanied by progressive thinning of the retina and a neovascular response that finally dominates the picture and leads to full-blown proliferative retinopathy.

5. Risk Profiles of DR Progression

In two recent follow-up studies performed by our group in different patient cohorts, one performed during a 5-year period, using OCT, and another for a period of 3-years, using both OCT and OCTA, we verified that the risk of retinopathy progression, identified by ETDRS gradings, is different between different individuals with type 2 diabetes and that ocular imaging risk markers are stronger predictors of progression to vision-threatening complications, macular edema, and proliferative retinopathy than systemic markers of metabolic control (Table 4) [23]. Furthermore, the progression of DR severity in the initial stages of DR, determined by 2-or-more-step worsening of ETDRS severity score, is associated with microvascular disease progression identified by increased microaneurysm turnover obtained from fundus photography and decreased vessel density obtained from OCTA examination [24,25].

Table 4. The five most relevant features obtained when a set of rules generated by a decision tree having as possible inputs a set of metrics automatically extracted from SS-OCTA-OCT was used to organize the data. Three different sets of rules were obtained after applying the CART method to eyes with consecutive ETDRS levels.

Control vs. ETDRS 10–20	ETDRS 10–20 vs. ETDRS 35	ETDRS 35 vs. ETDRS 43–47
Faz_Circularity	GCL_ExtC3	VD_OuterNasal_SCP
VD_CSF_Retina	VD_InnerRing_SCP	VD_OuterTemp_SCP
VD_ExtC3_DCP	GCL_ExtC3TempInf	VD_ExtC2NasInf_Deep
GCL_ExtC2NasSup	VD_OutCircle_DCP	VD_InnerNas_SCP
VD_ExtC2NasalInf_DCP	VD_EXTC1InfTemp_SCP	VD_ExtC1NasalInf_SCP

Identifying people with the greatest risk of progression and greatest potential to benefit from treatment is clearly an important goal that appears to be now within reach using these new available methods of examination, OCT and OCTA.

Diabetic patients followed annually with OCTA and OCT show that metrics to evaluate vessel density (fovea and mid-periphery using SS-OCTA), FAZ metrics (using OCTA) to identify the ischemic phenotype, together with OCT evaluation of neurodegenerative changes (GCL + IPL thinning) can be used to identify different risk profiles of DR progression. The identification of the ischemic phenotype and leaky phenotype allows the identification of the eyes that are at risk of developing vision-threatening complications, macular edema, or proliferative retinopathy.

The results of a three-year longitudinal study using OCT and OCTA metrics show that this goal is even more relevant when it is realized that the alterations occurring in the diabetic retina are reversible until relatively late in the disease progress and vary widely between different individuals [3]. In this three-year follow-up study, we observed that there is, indeed, marked variability in the progression of the retinal microvascular and neurogenerative changes occurring in T2 diabetes. Some patients show steady and progressive worsening, whereas others show a variable course and evidence of reversibility of their changes. These observations offer two important messages. First, the reversibility of capillary closure opens the door for early intervention with the possibility of stopping and delaying disease progression. Second, each patient should be followed closely, and risk factors should be considered to determine a specific risk profile for that patient [25].

In summary, it is now possible to envision an automated classification of DR progression using noninvasive methods of examination, OCT, and SS-OCTA. Using this classification to establish the severity grade of DR, at the time of the ophthalmological examination, it is then possible to identify the risk of progression in severity and the development of vision-threatening complications based on the predominant phenotype: ischemic, leaky, or neurodegenerative.

6. Directions for Future Studies

Current nonproliferative retinopathy classifications, such as ETDRS grading, are limited by their complexity and difficulty to use in daily clinical practice. However, multimodal imaging using OCT and OCTA is offering automated alternatives based on information that can be obtained directly from the equipment and is expected to replace the laborious and complicated ETDRS classification in the near future.

Different individuals with T2 diabetes and with the same severity degree of retinopathy show different rates of progression, both in disease severity and development of vision-threatening complications. An immediate and attainable goal will be the identification of the risk profile of each individual in order to achieve personalized monitoring of diabetic retinopathy progression and identify new tools for early intervention.

Author Contributions: J.C.-V. and L.M. collected and analyzed data, reviewed, wrote, and edited the manuscript. J.C.-V. is the guarantor of this work and, as such, had full access to all the data in the study and takes responsibility for the integrity of the data and the accuracy of the data analysis. All authors have read and agreed to the published version of the manuscript.

Funding: This work was supported by AIBILI and by COMPETE Portugal2020 and by the Fundação para a Ciência e Tecnologia (02/SAICT/2017–032412) under the project FILTER (Framework to Develop and Validate Automated Image Analysis Systems for Early Diagnosis and Treatment of Eyes at Risk in Blinding Age-Related Diseases).

Institutional Review Board Statement: The tenets of the Declaration of Helsinki were followed, and approval was obtained from the local Institutional Ethical Review Board, AIBILI'S Ethics Committee for Health, with the code 030/2015/AIBILI/CE.

Informed Consent Statement: Written informed consent was obtained by each participant agreeing to participate in the study.

Data Availability Statement: Data will be available upon request to the correspondent author.

Conflicts of Interest: J.C.-V. reports grants from Carl Zeiss Meditec (Dublin, CA, USA) and is consultant for Alimera Sciences (Alpharetta, GA, USA), Allergan (Dublin, Ireland), Bayer (Leverkusen, Germany), Gene Signal (Lausanne, Switzerland), Novartis (Basel, Switzerland), Pfizer (New York, NY, USA), Precision Ocular Ltd. (Oxford, UK), Roche (Basel, Switzerland), Sanofi-Aventis (Paris, France), Vifor Pharma (Glattbrugg, Switzerland), and Carl Zeiss Meditec (Dublin, CA, USA). L.G.M. declares no conflicts of interest.

References

1. IDF. *IDF Diabetes Atlas*, 9th ed.; IDF: Brussels, Belgium, 2019.
2. Sato, Y.; Lee, Z.; Hayashi, Y. Subclassification of preproliferative diabetic retinopathy and glycemic control: Relationship between mean hemoglobin A1C value and development of proliferative diabetic retinopathy. *Jpn. J. Ophthalmol.* **2001**, *45*, 523–527. [CrossRef]
3. Marques, I.P.; Madeira, M.H.; Messias, A.L.; Santos, T.; Martinho, A.C.-V.; Figueira, J.; Cunha-Vaz, J. Retinopathy phenotypes in type 2 diabetes with different risks for macular edema and proliferative retinopathy. *J. Clin. Med.* **2020**, *9*, 1433. [CrossRef]
4. Cogan, D.G.; Kuwabara, T. Capillary Shunts in the Pathogenesis of Diabetic Retinopathy. *Diabetes* **1963**, *12*, 293–300. [CrossRef] [PubMed]
5. Cunha-Vaz, J.G. Pathophysiology of diabetic retinopathy. *Br. J. Ophthalmol.* **1978**, *62*, 351–355. [CrossRef]
6. Marques, I.P.; Alves, D.; Santos, T.; Mendes, L.; Santos, A.R.; Lobo, C.; Durbin, M.; Cunha-Vaz, J. Multimodal Imaging of the Initial Stages of Diabetic Retinopathy: Different Disease Pathways in Different Patients. *Diabetes* **2019**, *68*, 648–653. [CrossRef] [PubMed]
7. Durbin, M.K.; An, L.; Shemonski, N.D.; Soares, M.; Santos, T.; Lopes, M.; Neves, C.; Cunha-Vaz, J. Quantification of Retinal Microvascular Density in Optical Coherence Tomographic Angiography Images in Diabetic Retinopathy. *JAMA Ophthalmol.* **2017**, *135*, 370. [CrossRef]
8. Santos, T.; Warren, L.H.; Santos, A.R.; Marques, I.P.; Kubach, S.; Mendes, L.G.; De Sisternes, L.; Madeira, M.H.; Durbin, M.; Cunha-Vaz, J.G. Swept-source OCTA quantification of capillary closure predicts ETDRS severity staging of NPDR. *Br. J. Ophthalmol.* **2020**. [CrossRef]
9. Silva, P.S.; Dela Cruz, A.J.; Ledesma, M.G.; Van Hemert, J.; Radwan, A.; Cavallerano, J.D.; Aiello, L.M.; Sun, J.K.; Aiello, L.P. Diabetic retinopathy severity and peripheral lesions are associated with nonperfusion on ultrawide field angiography. *Ophthalmology* **2015**, *122*, 2465–2472. [CrossRef] [PubMed]
10. Marques, I.P.; Alves, D.; Santos, T.; Mendes, L.; Lobo, C.; Santos, A.R.; Durbin, M.; Cunha-Vaz, J. Characterization of Disease Progression in the Initial Stages of Retinopathy in Type 2 Diabetes: A 2-Year Longitudinal Study. *Investig. Opthalmol. Vis. Sci.* **2020**, *61*, 20. [CrossRef]
11. Sohn, E.H.; Van Dijk, H.W.; Jiao, C.; Kok, P.H.B.; Jeong, W.; Demirkaya, N.; Garmager, A.; Wit, F.; Kucukevcilioglu, M.; Van Velthoven, M.E.J.; et al. Retinal neurodegeneration may precede microvascular changes characteristic of diabetic retinopathy in diabetes mellitus. *Proc. Natl. Acad. Sci. USA* **2016**, *113*, E2655–E2664. [CrossRef]
12. Ludovico, J.; Bernardes, R.; Pires, I.; Figueira, J.; Lobo, C.; Cunha-Vaz, J. Alterations of retinal capillary blood flow in preclinical retinopathy in subjects with type 2 diabetes. *Graefe's Arch. Clin. Exp. Ophthalmol.* **2003**, *241*, 181–186. [CrossRef]
13. Royle, P.; Mistry, H.; Auguste, P.; Shyangdan, D.; Freeman, K.; Lois, N.; Waugh, N. Pan-retinal photocoagulation and other forms of laser treatment and drug therapies for non-proliferative diabetic retinopathy: Systematic review and economic evaluation. *Health Technol. Assess.* **2015**, *19*, 1–247. [CrossRef]
14. Sivaprasad, S.; Pearce, E. The unmet need for better risk stratification of non-proliferative diabetic retinopathy. *Diabet. Med.* **2019**, *36*, 424–433. [CrossRef]

15. Hubbard, L.D.; Sun, W.; Cleary, P.A.; Danis, R.P.; Hainsworth, D.P.; Peng, Q.; Susman, R.A.; Aiello, L.P.; Davis, M.D. Comparison of digital and film grading of diabetic retinopathy severity in the diabetes control and complications trial/epidemiology of diabetes interventions and complications study. *Arch. Ophthalmol.* **2011**, *129*, 718–726. [CrossRef]

16. Wilkinson, C.P.; Ferris, F.L.; Klein, R.E.; Lee, P.P.; Agardh, C.D.; Davis, M.; Dills, D.; Kampik, A.; Pararajasegaram, R.; Verdaguer, J.T.; et al. Proposed international clinical diabetic retinopathy and diabetic macular edema disease severity scales. *Ophthalmology* **2003**, *110*, 1677–1682. [CrossRef]

17. ETDRS. Grading diabetic retinopathy from stereoscopic color fundus photographs–an extension of the modified Airlie House classification. ETDRS report number 10. *Ophthalmology* **1991**, *98*, 786–806. [CrossRef]

18. Cunha-Vaz, J.; Ribeiro, L.; Lobo, C. Phenotypes and biomarkers of diabetic retinopathy. *Prog. Retin. Eye Res.* **2014**, *41*, 90–111. [CrossRef] [PubMed]

19. Nesper, P.L.; Roberts, P.K.; Onishi, A.C.; Chai, H.; Liu, L.; Jampol, L.M.; Fawzi, A.A. Quantifying Microvascular Abnormalities with Increasing Severity of Diabetic Retinopathy Using Optical Coherence Tomography Angiography. *Invest. Ophthalmol. Vis. Sci.* **2017**, *58*, BIO307–BIO315. [CrossRef] [PubMed]

20. Hove, M.N.; Kristensen, J.K.; Lauritzen, T.; Bek, T. The relationships between risk factors and the distribution of retinopathy lesions in type 2 diabetes. *Acta Ophthalmol. Scand.* **2006**, *84*, 619–623. [CrossRef] [PubMed]

21. Santos, A.R.; Mendes, L.; Madeira, M.H.; Marques, I.P.; Tavares, D.; Figueira, J.; Lobo, C.; Cunha-Vaz, J.G. Microaneurysm Turnover in mild Non-Proliferative Diabetic Retinopathy is Associated with Progression and Development of Vision-threatening Complications: A 5-year longitudinal study. *J. Clin. Med.* **2021**, *10*, 2142. [CrossRef]

22. Wang, F.P.; Saraf, S.S.; Zhang, Q.; Wang, R.K.; Rezaei, K.A. Ultra-Widefield Protocol Enhances Automated Classification of Diabetic Retinopathy Severity with OCT Angiography. *Ophthalmol. Retin.* **2020**, *4*, 415–424. [CrossRef]

23. Li, X.; Xie, J.; Zhang, L.; Cui, Y.; Zhang, G.; Chen, X.; Wang, J.; Zhang, A.; Huang, T.; Meng, Q. Identifying microvascular and neural parameters related to the severity of diabetic retinopathy using optical coherence tomography angiography. *Investig. Ophthalmol. Vis. Sci.* **2020**, *61*, 39. [CrossRef]

24. Custo Greig, E.; Brigell, M.; Cao, F.; Levine, E.S.; Peters, K.; Moult, E.M.; Fujimoto, J.G.; Waheed, N.K. Macular and Peripapillary Optical Coherence Tomography Angiography Metrics Predict Progression in Diabetic Retinopathy: A Sub-analysis of TIME-2b Study Data. *Am. J. Ophthalmol.* **2020**, *219*, 66–76. [CrossRef]

25. Martinho, A.C.-V.; Marques, I.P.; Messias, A.L.; Santos, T.; Madeira, M.H.; Sousa, D.C.; Lobo, C.; Cunha-Vaz, J. Ocular and Systemic Risk Markers for Development of Macular Edema and Proliferative Retinopathy in Type 2 Diabetes: A 5-Year Longitudinal Study. *Diabetes Care* **2021**, *44*, e12–e14. [CrossRef]

Journal of *Personalized Medicine*

Article

Glycemic Gap as a Useful Surrogate Marker for Glucose Variability and Progression of Diabetic Retinopathy

Shi-Chue Hsing [1], Chin Lin [2,3,4], Jiann-Torng Chen [5], Yi-Hao Chen [5] and Wen-Hui Fang [6,*]

[1] National Defense Medical Center, Department of Internal Medicine, Tri-Service General Hospital, Taipei 11490, Taiwan; lars0121@gmail.com
[2] National Defense Medical Center, Graduate Institute of Life Sciences, Taipei 11490, Taiwan; xup6fup0629@gmail.com
[3] National Defense Medical Center, School of Medicine, Taipei 11490, Taiwan
[4] National Defense Medical Center, School of Public Health, Taipei 11490, Taiwan
[5] National Defense Medical Center, Department of Ophthalmology, Tri-Service General Hospital, Taipei 11490, Taiwan; jt66chen@gmail.com (J.-T.C.); doc30879@mail.ndmctsgh.edu.tw (Y.-H.C.)
[6] National Defense Medical Center, Department of Family and Community Medicine, Tri-Service General Hospital, No 161, Min-Chun E. Rd., Sec. 6, Neihu, Taipei 11490, Taiwan
*   Correspondence: rumaf.fang@gmail.com; Tel.: +886-2-87923311 (ext. 12322); Fax: +886-2-66012632

Abstract: (1) Background: Recent studies have reported that the glucose variability (GV), irrespective of glycosylated hemoglobin (HbA1c), could be an additional risk factor for the development of diabetic retinopathy (DR). However, measurements for GV, such as continuous glucose monitoring (CGM) and fasting plasma glucose (FPG) variability, are expensive and time consuming. (2) Methods: This present study aims to explore the correlation between the glycemic gap as a measurement of GV, and DR. In total, 2565 patients were included in this study. We evaluated the effect of the different types of glycemic gaps on DR progression. (3) Results: We found that the area under the curve (AUC) values of both the glycemic gap and negative glycemic gap showed an association with DR progression. (4) Conclusions: On eliminating the possible influences of chronic blood glucose controls, the results show that GV has deleterious effects that are associated with the progression of DR. The glycemic gap is a simple measurement of GV, and the predictive value of the negative glycemic gap in DR progression shows that GV and treatment-related hypoglycemia may cause the development of DR. Individual treatment goals with a reasonable HbA1c and minimal glucose fluctuations may help in preventing DR.

Keywords: diabetic retinopathy; type 2 diabetes; progression; glycemic gap; glucose variability

Citation: Hsing, S.-C.; Lin, C.; Chen, J.-T.; Chen, Y.-H.; Fang, W.-H. Glycemic Gap as a Useful Surrogate Marker for Glucose Variability and Progression of Diabetic Retinopathy. *J. Pers. Med.* **2021**, *11*, 799. https://doi.org/10.3390/jpm11080799

Academic Editors: Peter D. Westenskow and Andreas Ebneter

Received: 22 July 2021
Accepted: 14 August 2021
Published: 16 August 2021

Publisher's Note: MDPI stays neutral with regard to jurisdictional claims in published maps and institutional affiliations.

1. Introduction

Diabetic retinopathy (DR) has been identified as one of the major microvascular complications of diabetes mellitus (DM). The common risk factors for DR include poor glycemic control, disease duration, and systolic hypertension. Glycosylated hemoglobin (HbA1c), which indicates an average glucose level over the previous 2–3 months, is the strongest marker of glycemic control [1,2]. However, studies have found a U-shaped association between the HbA1c and mortality in patients, and tight glucose control did not improve healthcare outcomes [3,4]. These findings have suggested that aside from HbA1c, other factors of glycemic control may be related to development of diabetes-related complications.

Recently, studies reported that glucose variability (GV), irrespective of HbA1c, could be an additional risk factor for the development of DR. These measurements for short- and long-term GV, including HbA1c variability, continuous glucose monitoring (CGM), and fasting plasma glucose (FPG) variability were associated with the risk of DR in patients with type 2 DM [5–8].

High HbA1c standard deviations were reported as a predictor for DR development in two longitudinal studies with a large sample size among Asians with type 2 DM and long follow-up period [9,10]. In addition, the HbA1c standard deviation may play a greater role in DR development than the mean HbA1c when the HbA1c variation magnitude is wide. However, a wide standard deviation only indicates that the values are spread out over a wider range without directionality. We could not explain that the effect was caused by hyperglycemia due to poor control or iatrogenic hypoglycemia due to an intensive HbA1c treatment goal.

In a previous study, the glycemic gap, which is calculated from the glucose level minus the HbA1c-derived average glucose, was used to evaluate stress-induced hyperglycemia (SIH). The results suggested that the glycemic gap was associated with disease severity and adverse clinical outcomes in diabetic patients with liver abscesses, pneumonia, acute myocardial infarction, acute ischemic stroke, chronic obstructive lung disease, necrotizing fasciitis, and in-hospital cardiac arrest [11–13].

We hypothesized that GV is associated with an increased risk of developing DR in type 2 DM patients, and that the glycemic gap could be a simple measurement for GV. The aims of the present study were to explore the correlation between the glycemic gap as a measurement for GV, and DR, and to evaluate the effect of a negative/positive magnitude of GV in the development of DR. In addition, we sought to justify the use of the glycemic gap as a tool to reflect individual patterns of glycemic control beyond the HbA1c level.

2. Materials and Methods

2.1. Population

A tertiary referral medical center in northern Taiwan provided the research data from 12 October 2012 to 11 September 2018. This study was approved by the Ethics Committee of Tri-Service General Hospital (IRB NO 2-105-05-073). Diabetic patients with more than 2 fundus color photography tests were included in this study. We defined time 0 (the beginning of the follow-up period) as the first eye examination. Initially, there were 5974 potential cases. However, those without type 2 DM were excluded. In this study, type 2 DM is defined as having a prescription for insulin or an oral antidiabetic and 1 of the following conditions: (1) at least two International Classification of Diseases (ICD) codes of type 2 DM (ICD9: 250 and ICD10: E11) at least 6 months from the start of the study; (2) at least 2 records of $\geq$126 mg/dL of blood glucose before meals (ante cibum, AC) at least 6 months from the start of the study; and (3) at least 2 records of $\geq$6.5% HbA1c at least 6 months from the start of the study. Moreover, patients without HbA1c and glucose AC results within 14 days before the start of the study were also excluded. Ultimately, 2642 patients with 14,409 test results were included for analysis (Figure 1).

2.2. Measurements and Variables

DR was graded based on the International Clinical Diabetic Retinopathy Disease Severity Scale [14] as follows: (0) no apparent retinopathy; (1) mild non-proliferative DR (NPDR): microaneurysms only; (2) moderate NPDR: any finding of microaneurysms, dot and blot hemorrhages, hard exudates, or cotton wool spots but less than severe NPDR; (3) severe NPDR: intraretinal hemorrhages ($\geq$20 in each of the 4 quadrants), definite venous beading in 2 quadrants, or intraretinal microvascular abnormalities in 1 quadrant but no signs of proliferative retinopathy; and (4) proliferative DR (PDR): 1 or more findings of neovascularization and vitreous or preretinal hemorrhages. The end of the follow-up period was determined by the following: (1) the change in the grade of DR or (2) the end of the last fundus color photography if there was no progression of DR. The patients with PDR at the start of the study were excluded, leaving 2565 patients to be further analyzed.

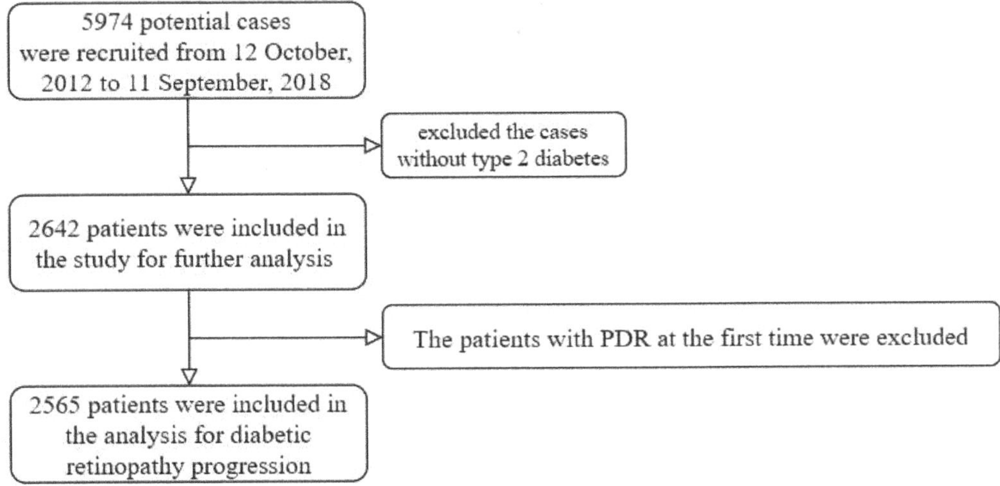

Figure 1. Flowchart of the enrolled patients.

The expected glucose AC was calculated by the following equation: $(28.7 \times HbA1c) - 46.7$ [15]. The difference between the observed and expected glucose AC was referred to as the glycemic gap in this study. If there were several glucose AC tests within 14 days before the start of the study, the latest result was selected. To further explore the effect of the amplitude of glycemic excursion poor control and excessive control, we defined 2 additional variables as the following: (1) positive glycemic gap—the maximum of a series of glycemic gap within 14 days, and the negative number was defined as 0 in this calculation; (2) negative glycemic gap—the minimum of a series of glycemic gap within 14 days, and the positive number was defined as 0 in this calculation. We also collected other laboratory data from our electronic health records within 30 days before the start of the study: total cholesterol, low-density lipoprotein (LDL), high-density lipoprotein (HDL), triglyceride, creatinine, uric acid, hemoglobin, white blood cell (WBC), platelet count, neutrophils, lymphocytes, albumin, and total bilirubin. The complete blood count was obtained using the automated XN-9000 hematology analyzer (Sysmex Co., Kobe, Japan). HbA1c was tested by the automated Glycated Hemoglobin Analyzer, HLC-723 G11 (Tosoh Bioscience Inc., Tokyo, Japan). An AU5800 Series Clinical Chemistry Analyzer (Beckman Coulter Inc., Brea, CA, USA) analysis system was used to detect chemistry assays. The missing rate of mentioned laboratory data was less than 30%. We used multiple imputation to impute the missing values.

The other demographic characteristics and comorbidities were collected from our electronic health records. The collected patients' characteristics included their gender, age, systolic blood pressure (SBP), and diastolic blood pressure (DBP). Comorbidities, including hypertension, lipidemia, ischemic heart disease, heart failure, chronic obstructive pulmonary disease (COPD), stroke, and diabetic neuropathy, were recorded after a subsequent chart review according to the ICD 9 and ICD 10 codes.

2.3. Deep Learning Model for Grading DR

Due to the numerous fundus color photography results, it was not possible to record them one by one. Therefore, we used a software we have developed previously to grade DR [16]. The deep learning model was based on the convolutional neural network. Kaggle, a coding website [17] with 35,126 images with corresponding DR grade, provided fundus color photography for the deep learning model. The model architecture was based on a 50-layer SE-ResNeXt [18]. The public score was 0.837 in a test with 53,576 images, which was seventh on the leaderboard, while the private score was 0.841, which was third. This model was applied to our images. Each test was conducted in both eyes, and the severity was determined by the result of the eye with more severe DR.

2.4. Statistical Methods

We presented the characteristics as means and standard deviations, numbers of patients, or percentages, where appropriate. They were then compared using either analysis of variance or the chi-squared test, as appropriate. A p-value < 0.05 was considered significant. Statistical analyses were conducted using R software version 3.4.3.

The primary analysis evaluated the effect of the different types of glycemic gaps in DR progression. Kaplan–Meier curves were obtained to evaluate the progression between the participants with higher glycemic gaps and those with lower glycemic gaps. The log-rank test was used to test the statistical significance, and the cutoff value of the glycemic gap was decided by the median. All variables were evaluated on their effect on DR progression using the univariate Cox proportional hazards model. We used the multivariable Cox proportional hazards model to adjust the potential confounding factors. The selection of the adjusted variables was based on the significance of the univariate analysis. Because the proportional hazards assumption of the Cox proportional hazards model might have been violated in our data, we also used the time-dependent receiver operating characteristic curve to present the effect of the different types of glycemic gaps with respect to time.

We performed secondary analyses with stratified analysis. Because the initial conditions of the different patients varied in real-world clinical practice, the initial grade of DR and the baseline HbA1c were used as the stratified variables. We only presented the analysis of the most significant type of glycemic gap in DR progression. The Cox proportional hazards model was used for the interaction analysis, and the adjusted variables were the same with the primary analysis.

3. Results

Table 1 shows the baseline characteristics of 2565 patients according to DR. In total, 1046 patients were classified as no DR, 480 patients as mild NPDR, 757 patients as moderate NPDR, and 282 patients as severe NPDR. The DR severity at the beginning of our study was significantly associated with younger age, higher creatinine level, higher HbA1c, lower albumin levels, higher uric acid levels, and more hypertensive events.

The conditional inference tree is defined as a method to minimize the information entropy from the classification of the characteristics used. Thus, it also can be used to find the most significant cutoff value of interesting predictors. We used a conditional inference tree to predict the DR progression using the glycemic gaps. Finally, an optimal cutoff value of 45 mg/dL was determined to minimize the information entropy.

Considering the baseline glycemic control, we divided the patients into three groups equally based on HbA1c level for the subgroup analysis based on the HbA1c level. Table 2 presents the basic characteristics of the patients with different glycemic controls (by HbA1c). The subjects were divided into Q1 (HbA1c < 6.8%), Q2 (HbA1c = 6.8–8.3%), and Q3 (HbA1c > 8.4%). Poorer glycemic control (higher HbA1c) was related to the DR severity, younger age, lower albumin level, and higher uric acid level.

Table 3 presents the risk of DR progression through the end of the study. The initial DR grade, glycemic control, glycemic gap, and negative glycemic gap were found to have statistical significance in HR (p < 0.05), both in the crude and adjusted model. The cut points of glycemic gap (45 mg/dL) and negative glycemic gap (45 mg/dL) were associated with significantly higher HR for DR progression when compared with a gap of <45 mg/dL, both in the crude and adjusted model.

Table 1. The baseline characteristics of patients according to diabetic retinopathy.

	Initial Grade of DR				*p*-Value
	No DR *n* = 1046	**Mild NPDR** *n* = 480	**Moderate NPDR** *n* = 757	**Severe NPDR** *n* = 282	
Basic characteristics					
Gender					0.201
Female	506(48.4%)	215(44.8%)	329(43.5%)	129(45.7%)	
Male	540(51.6%)	265(55.2%)	428(56.5%)	153(54.3%)	
Age (years)	63.76 ± 12.99	63.07 ± 12.26	59.59 ± 11.11	57.15 ± 11.71	<0.001
SBP (mmHg)	139.06 ± 20.30	140.37 ± 23.46	141.14 ± 21.70	141.83 ± 23.19	0.116
DBP (mmHg)	78.84 ± 11.98	78.66 ± 11.73	79.87 ± 12.21	80.62 ± 12.25	0.046
Comorbidity					
Hypertension	389(37.2%)	211(44.0%)	319(42.1%)	128(45.4%)	0.014
Lipidemia	318(30.4%)	140(29.2%)	225(29.7%)	87(30.9%)	0.948
Ischemic heart disease	222(21.2%)	125(26.0%)	169(22.3%)	69(24.5%)	0.180
Heart failure	54(5.2%)	34(7.1%)	67(8.9%)	30(10.6%)	0.002
COPD	38(3.6%)	21(4.4%)	12(1.6%)	2(0.7%)	0.001
Stroke	145(13.9%)	66(13.8%)	90(11.9%)	36(12.8%)	0.635
Diabetic neuropathy	67(6.4%)	29(6.0%)	75(9.9%)	19(6.7%)	0.019
Laboratory test					
HbA1c (%)	7.63 ± 1.91	7.91 ± 1.91	8.20 ± 2.03	8.40 ± 2.16	<0.001
Triglyceride (mg/dL)	152.89 ± 104.63	154.97 ± 118.99	158.71 ± 118.46	151.34 ± 88.38	0.671
Total cholesterol (mg/dL)	170.82 ± 40.39	167.83 ± 38.99	173.07 ± 43.54	172.47 ± 44.15	0.169
LDL cholesterol (mg/dL)	99.07 ± 33.42	98.40 ± 31.41	100.16 ± 35.22	100.33 ± 34.48	0.771
HDL cholesterol (mg/dL)	46.83 ± 13.34	46.13 ± 12.95	45.99 ± 12.90	46.65 ± 13.08	0.537
Creatinine (mg/dL)	1.37 ± 1.75	1.70 ± 2.17	1.65 ± 2.00	1.63 ± 2.01	0.002
ALT	24.94 ± 27.24	24.39 ± 31.69	22.94 ± 37.79	25.10 ± 61.13	0.677
Total bilirubin	0.64 ± 0.41	0.66 ± 0.61	0.63 ± 0.53	0.59 ± 0.35	0.200
WBC	7.40 ± 5.58	7.39 ± 2.58	7.28 ± 4.72	7.86 ± 7.21	0.444
PLT	213.06 ± 76.15	216.90 ± 74.50	221.20 ± 81.48	227.85 ± 84.12	0.019
Neutrophil	63.63 ± 11.29	64.49 ± 11.30	63.81 ± 10.73	64.29 ± 11.07	0.497
Lymphocyte	27.43 ± 10.33	26.56 ± 10.38	27.03 ± 9.85	26.72 ± 9.72	0.409
Uric acid (mg/dL)	5.88 ± 1.78	6.08 ± 1.85	6.10 ± 1.89	6.32 ± 1.81	0.001
Hb (g/dL)	12.95 ± 1.95	12.70 ± 2.01	12.59 ± 2.19	12.58 ± 2.07	0.001
Albumin (g/dL)	3.91 ± 0.53	3.88 ± 0.56	3.83 ± 0.59	3.76 ± 0.61	<0.001
Result					
Glycemic gap	49.54 ± 59.55	52.82 ± 68.25	59.75 ± 74.43	65.29 ± 94.93	0.001
Positive glycemic gap	46.43 ± 52.36	47.63 ± 40.13	52.90 ± 44.44	54.184 ± 44.18	0.008
Negative glycemic gap	45.72 ± 52.23	50.06 ± 67.47	56.70 ± 73.08	62.184 ± 93.49	<0.001

Testing by Fisher's exact test, Wilcoxon test, or Kruskal–Wallis test, respectively; SBP: systolic blood pressure; DBP: diastolic blood pressure; HbA1c: glycated hemoglobin; LDL: low-density lipoprotein; HDL: high-density lipoprotein; Hb: hemoglobin; ALT: alanine aminotransferase.

Table 2. The baseline characteristics of patients based on glycemic control. (divided into 3 groups equally based on HbA1c level at the time of enrollment).

	Into 3 Groups Equally Based on HbA1c Level			*p*-Value
	HbA1c Q1 (<6.8) *n* = 836	**HbA1c Q2** (6.8−8.3) *n* = 854	**HbA1c Q3** (>8.3) *n* = 875	
Basic characteristics				
DR severity				<0.001
No DR	394(47.1%)	377(44.1%)	275(31.4%)	
Mild NPDR	150(17.9%)	161(18.9%)	169(19.3%)	
Moderate NPDR	217(26.0%)	232(27.2%)	308(35.2%)	
Severe NPDR	75(9.0%)	84(9.8%)	123(14.1%)	
Gender				0.377
Female	368(44.0%)	398(46.6%)	413(47.2%)	
Male	468(56.0%)	456(53.4%)	462(52.8%)	
Age (years)	62.12 ± 12.45	63.62 ± 11.87	59.34 ± 12.52	<0.001
SBP (mmHg)	139.00 ± 20.39	140.85 ± 21.13	140.78 ± 23.30	0.140
DBP (mmHg)	78.44 ± 11.53	78.72 ± 11.41	80.72 ± 12.98	<0.001
Comorbidity				
Hypertension	345(41.3%)	369(43.2%)	333(38.1%)	0.088
Lipidemia	237(28.3%)	255(29.9%)	278(31.8%)	0.301
Ischemic heart disease	189(22.6%)	204(23.9%)	192(21.9%)	0.62
Heart failure	52(6.2%)	57(6.7%)	76(8.7%)	0.109
COPD	18(2.2%)	29(3.4%)	26(3.0%)	0.296
Stroke	126(15.1%)	106(12.4%)	105(12.0%)	0.127
Diabetic neuropathy	42(5.0%)	49(5.7%)	99(11.3%)	<0.001
Laboratory test				
Triglyceride (mg/dL)	132.54 ± 73.00	142.32 ± 91.21	188.31 ± 143.38	<0.001
Total cholesterol (mg/dL)	164.75 ± 39.84	166.52 ± 37.60	181.65 ± 44.64	<0.001
LDL cholesterol (mg/dL)	95.05 ± 33.14	96.30 ± 31.20	106.59 ± 35.41	<0.001
HDL cholesterol (mg/dL)	47.39 ± 13.04	47.05 ± 13.36	44.92 ± 12.81	<0.001
Creatinine (mg/dL)	1.69 ± 2.21	1.51 ± 1.84	1.43 ± 1.75	0.015
Uric acid (mg/dL)	6.07 ± 1.86	5.97 ± 1.78	6.06 ± 1.86	0.469
Hb (g/dL)	12.54 ± 2.17	12.70 ± 1.96	13.02 ± 2.00	<0.001
Albumin (g/dL)	3.87 ± 0.57	3.91 ± 0.56	3.82 ± 0.56	0.004
Result				
Glycemic gap	31.92 ± 51.45	47.39 ± 52.99	84.18 ± 88.80	<0.001
Positive glycemic gap	28.87 ± 26.18	44.48 ± 32.84	73.87 ± 61.59	<0.001
Negative glycemic gap	29.62 ± 50.92	45.44 ± 52.43	78.58 ± 83.07	<0.001

Testing by Fisher's exact test, Wilcoxon test, or Kruskal–Wallis test, respectively; SBP: systolic blood pressure; DBP: diastolic blood pressure; HbA1c: glycated hemoglobin; LDL: low-density lipoprotein; HDL: high-density lipoprotein; Hb: hemoglobin.

Table 3. The risk of diabetic retinopathy progression at the end of the study.

	Crude-HR (95% CI)	*p*-Value	Adjusted-HR (95% CI)	*p*-Value
Initial DR grade		<0.001		<0.001
No DR	1.00		1.00	
Mild NPDR	2.00 (1.72–2.32)	<0.001	1.98 (1.70–2.31)	<0.001
Moderate NPDR	1.82 (1.58–2.09)	<0.001	1.78 (1.54–2.04)	<0.001
Servere NPDR	0.87 (0.69–1.09)	0.223	0.86 (0.69–1.08)	0.191
Glycemic control		0.003		0.060
HbA1c Q1 (<6.8)	1.00		1.00	
HbA1c Q2 (6.8−8.3)	1.00 (0.86–1.15)	0.976	0.97 (0.84–1.12)	0.668
HbA1c Q3 (>8.3)	1.23 (1.07–1.41)	0.004	1.13 (0.98–1.31)	0.083
Glycemic gap		<0.001		0.001
<45 mg/dL	1.00		1.00	
>45 mg/dL	1.26 (1.13–1.42)	<0.001	1.22 (1.09–1.37)	0.001
Positive glycemic gap		0.065		0.076
<45 mg/dL	1.00		1.00	
>45 mg/dL	1.21 (0.99–1.49)	0.065	1.21 (0.98–1.48)	0.076
Negative glycemic gap		0.001		0.005
<45 mg/dL	1.00		1.00	
>45 mg/dL	1.22 (1.08–1.37)	0.001	1.18 (1.05–1.33)	0.005
Gender		0.019		0.033
Female	1.00		1.00	
Male	1.15 (1.02–1.29)	0.019	1.13 (1.01–1.27)	0.033
Age	1.00 (1.00–1.00)	0.962	1.00 (1.00–1.01)	0.945
Height	1.18 (0.92–1.51)	0.184	0.96 (0.71–1.30)	0.804
Weight	0.95 (0.74–1.21)	0.669	0.88 (0.50–1.57)	0.676
Systolic blood pressure	1.00 (1.00–1.00)	0.269	1.00 (1.00–1.00)	0.639
Diastolic blood pressure	1.00 (1.00–1.01)	0.214	1.00 (1.00–1.01)	0.300
Comorbidity				
Hypertension	1.14 (1.02–1.28)	0.027	1.07 (0.95–1.21)	0.253
Lipidemia	0.97 (0.85–1.10)	0.609	0.94 (0.82–1.07)	0.335
Ischemic heart disease	1.09 (0.96–1.25)	0.184	1.03 (0.89–1.19)	0.682
Stroke	1.10 (0.93–1.30)	0.249	1.07 (0.91–1.27)	0.415
Diabetic neuropathy	1.30 (1.07–1.59)	0.009	1.21 (0.99–1.49)	0.066
Heart failure	1.16 (0.95–1.43)	0.152	1.08 (0.87–1.34)	0.472
Laboratory test				
Triglyceride	1.00 (1.00–1.00)	0.488	1.00 (1.00–1.00)	0.602
Total cholesterol	1.00 (1.00–1.00)	0.483	1.00 (1.00–1.00)	0.420
Low-density lipoprotein	1.00 (1.00–1.00)	0.488	1.00 (1.00–1.00)	0.602
High-density lipoprotein	1.00 (1.00–1.00)	0.983	1.00 (1.00–1.01)	0.630
Hemoglobin	0.99 (0.97–1.02)	0.699	1.00 (0.97–1.03)	0.835

All result of adjusted-HR were adjusted by grade of diabetic retinopathy, gender, hypertension, and diabetic neuropathy. Crude-HR: crude hazard ratio; HR: hazard ratio; CI: confidence interval; HbA1c: glycated hemoglobin.

Figure 2 presents the time-dependent area under the curve (AUC). Compared with the positive glycemic gap, both the glycemic gap and negative glycemic gap showed large AUC values for predicting DR progression (Figure 2). This relationship (similar AUC association: glycemic gap > negative glycemic gap > positive glycemic gap) was found at almost every time point in all groups and in every HbA1c subgroup.

As shown in Figure 3, the Kaplan–Meier survival curve showed that the poor glycemic control group was more positively associated with DR progression as compared with the good and intermediate glycemic control groups. In addition, the survival curve did not reveal a difference between the good glycemic control and intermediate control groups. The Kaplan–Meier survival curve showed that a glycemic gap > 45 mg/dL was associated with a significantly higher risk for DR progression when compared with a gap of <45 mg/dL for both the glycemic gap and negative glycemic gap (log-rank test $p < 0.05$).

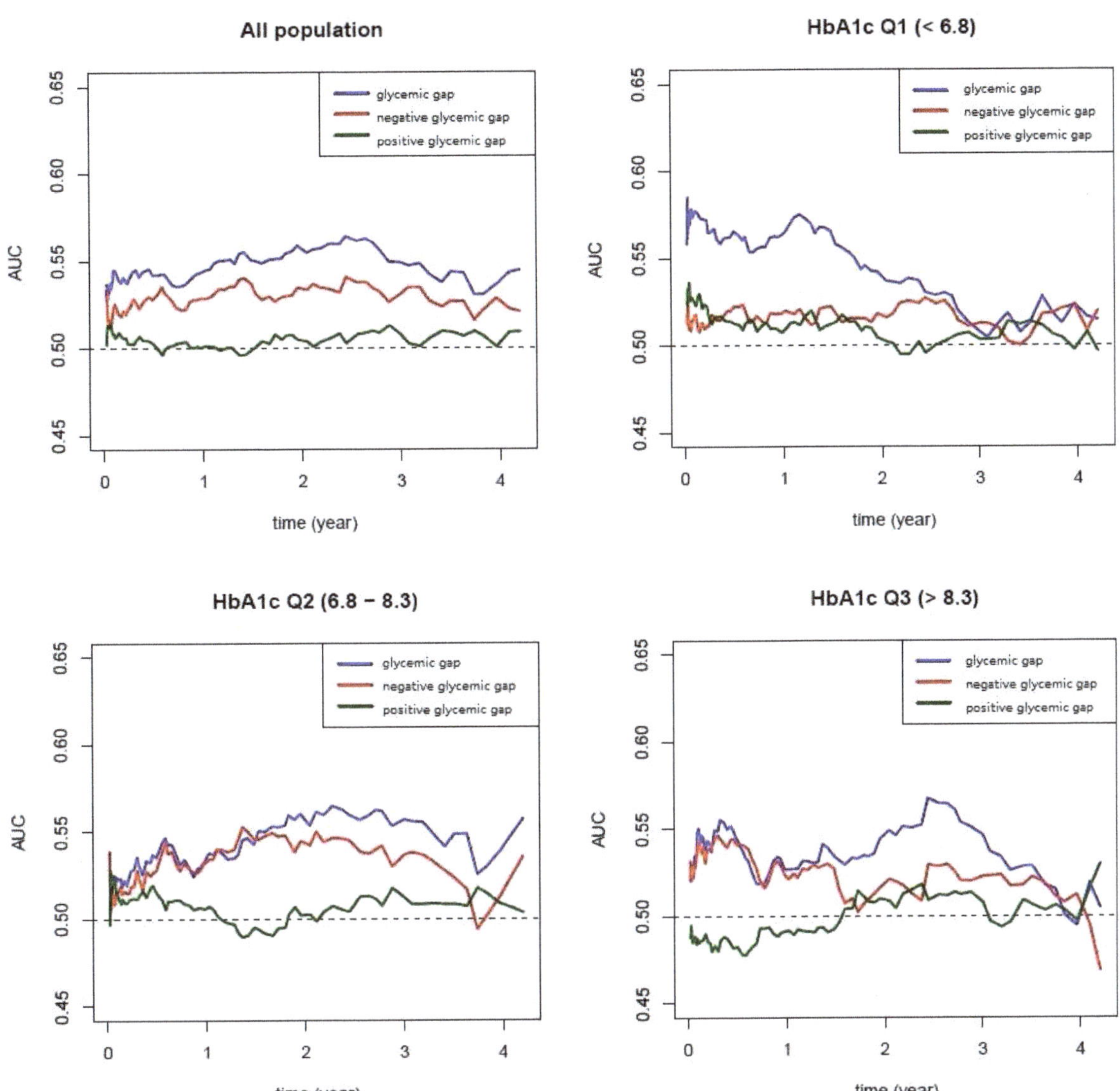

Figure 2. Time-dependent area under the curve (AUC): a similar association, in that the glycemic gap and negative glycemic gap both showed higher AUC for predicting diabetic retinopathy progression as compared to the positive glycemic gap, was found in all populations and each glycemic control group.

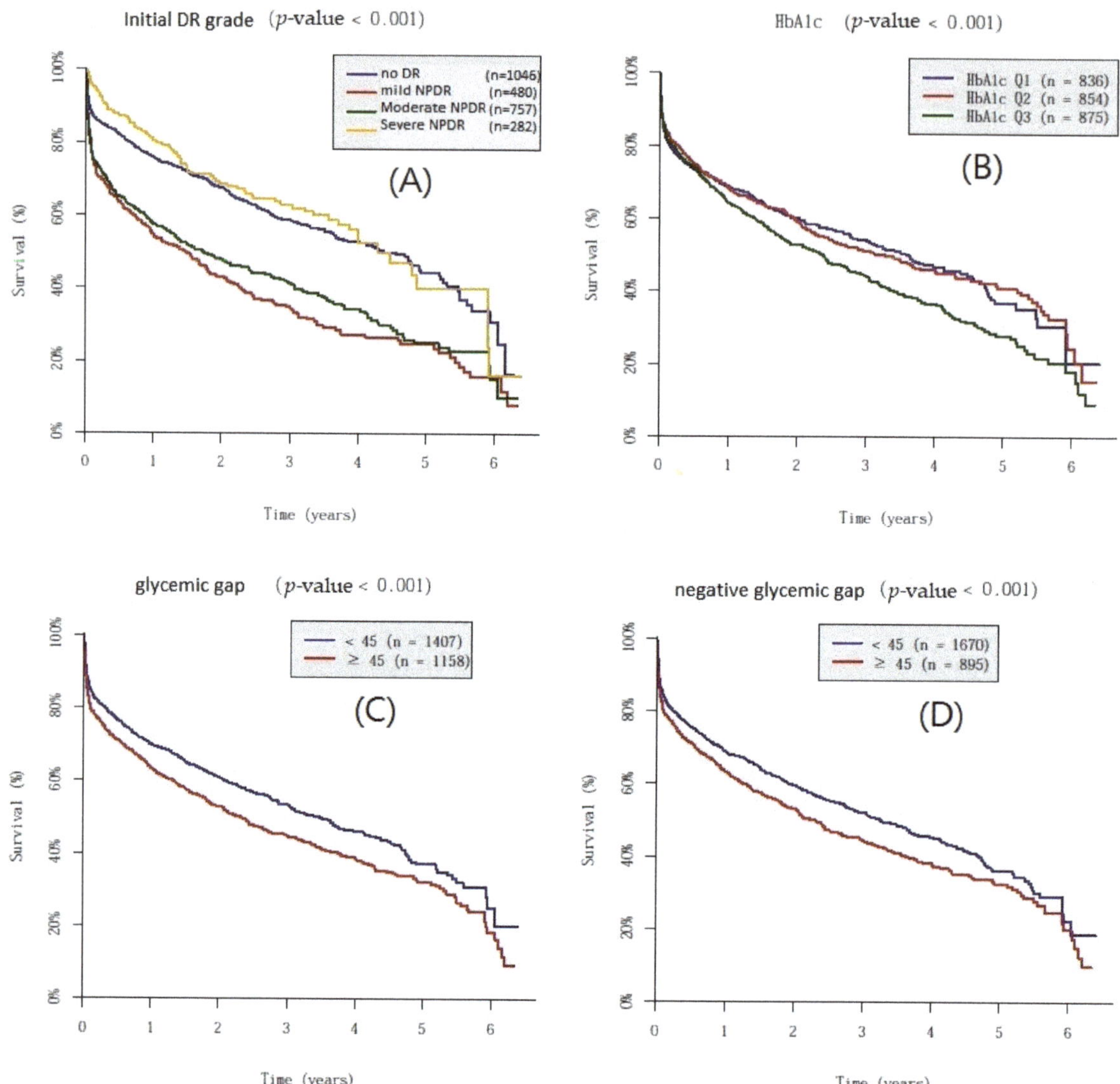

Figure 3. Kaplan–Meier survival curves. (**A**): Stepwise diabetic retinopathy progression with different initial grades of diabetic retinopathy. (**B**): A stepwise fashion of diabetic retinopathy progression with differing glycemic control. (**C**): A stepwise fashion of diabetic retinopathy progression with a glycemic gap > 45 mg/dL as a cutoff value. (**D**): A stepwise fashion of diabetic retinopathy progression with a negative glycemic gap > 45 mg/dL as a cutoff value.

As shown in Tables 4 and 5, subgroup analyses were performed to determine the potential moderating effect of glycemic control and initial DR grade on the relationships between the glycemic gap and DR progression. No significant values (*p*-value for interaction < 0.05) were noted in any of the subgroups. The relationships were determined to be stronger in the no DR, severe NPDR, and HbA1c Q2 groups.

Table 4. The risk of diabetic retinopathy progression in each subgroup at the end of the study; a negative glycemic gap >45 mg/dL was the cutoff value.

		Crude-HR (95% CI)	*p*-Value	Adjusted-HR (95% CI)	*p*-Value
Initial DR grade			0.164 (interaction)		0.129 (interaction)
No DR	<45 mg/dL	1.00		1.00	
	>45 mg/dL	1.30 (1.06–1.59)	0.012	1.30 (1.06–1.59)	0.013
Mild NPDR	<45 mg/dL	1.00		1.00	
	>45 mg/dL	1.03 (0.81–1.32)	0.790	1.01 (0.79–1.29)	0.952
Moderate NPDR	<45 mg/dL	1.00		1.00	
	>45 mg/dL	1.14 (0.93–1.40)	0.200	1.14 (0.93–1.39)	0.221
Servere NPDR	<45 mg/dL	1.00		1.00	
	>45 mg/dL	1.71 (1.13–2.57)	0.010	1.76 (1.16–2.66)	0.007
Glycemic control			0.709 (interaction)		0.686 (interaction)
HbA1c Q1 (<6.8)	<45 mg/dL	1.00		1.00	
	>45 mg/dL	1.25 (0.88–1.78)	0.217	1.23 (0.87–1.76)	0.244
HbA1c Q2(6.8−8.3)	<45 mg/dL	1.00		1.00	
	>45 mg/dL	1.22 (1.00–1.50)	0.053	1.21 (0.98–1.48)	0.073
HbA1c Q3 (>8.3)	<45 mg/dL	1.00		1.00	
	>45 mg/dL	1.11 (0.92–1.34)	0.296	1.10 (0.91–1.33)	0.339

All results of the adjusted-HRs were adjusted by grade of diabetic retinopathy, gender, hypertension, and diabetic neuropathy. Crude-HR: crude hazard ratio; HR: hazard ratio; CI: confidence interval.

Table 5. The risk of diabetic retinopathy progression in each subgroup at the end of the study; a glycemic gap > 45 mg/dL was the cutoff value.

		Crude-HR (95% CI)	*p*-Value	Adjuseted-HR(95% CI)	*p*-Value
Initial DR grade			0.374 (interaction)		0.336 (interaction)
No DR	<45 mg/dL	1.00		1.00	
	>45 mg/dL	1.28 (1.05–1.56)	0.014	1.26 (1.03–1.53)	0.023
Mild NPDR	<45 mg/dL	1.00		1.00	
	>45 mg/dL	1.07 (0.85–1.35)	0.565	1.06 (0.84–1.33)	0.647
Moderate NPDR	<45 mg/dL	1.00		1.00	
	>45 mg/dL	1.25 (1.03–1.53)	0.025	1.25 (1.03–1.53)	0.026
Servere NPDR	<45 mg/dL	1.00		1.00	
	>45 mg/dL	1.56 (1.03–2.36)	0.036	1.56 (1.02–2.36)	0.038
Glycemic control			0.787 (interaction)		0.806 (interaction)
HbA1c Q1 (<6.8)	<45 mg/dL	1.00		1.00	
	>45 mg/dL	1.15 (0.89–1.49)	0.277	1.15 (0.89–1.48)	0.300
HbA1c Q2(6.8−8.3)	<45 mg/dL	1.00		1.00	
	>45 mg/dL	1.28 (1.05–1.57)	0.016	1.26 (1.03–1.54)	0.025
HbA1c Q3 (>8.3)	<45 mg/dL	1.00		1.00	
	>45 mg/dL	1.20 (0.98–1.48)	0.079	1.19 (0.97–1.47)	0.099

All results of adjusted-HRs were adjusted by grade of diabetic retinopathy, gender, hypertension, and diabetic neuropathy. Crude-HR: crude hazard ratio; HR: hazard ratio; CI: confidence interval.

4. Discussion

The two major findings of the present study are as follows:

(1) The glycemic gap, by eliminating the possible influences of chronic blood glucose controls, supports the possible deteriorating effects of glucose fluctuation on the progression of DR.

(2) Both the positive and negative glucose gaps were associated with the progression of DR. Furthermore, a negative glucose gap revealed stronger AUC values in the progression of DR. The negative glycemic gap correlated with the progression of DR and indicated the negative effects of hypoglycemia episodes on the progression of DR.

To the best of our knowledge, this is the first study to examine the link between the glycemic gap and DR in type 2 DM patients. In addition, the glycemic gap may be a useful and simple tool for the assessment of GV. We suggest that the glycemic gap could be further studied as an assessment method for GV.

HbA1c has been well accepted as the gold standard for the evaluation of glycemic control, and an improvement in the HbA1c greatly reduces the risks of complications in patients with type 2 DM [19]. HbA1c-centered studies with aggressive treatment goals, such as ACCORD and VADT, have been performed. However, very low and high mean HbA1c values were associated with increased mortality in patients, revealing that tight glucose control did not improve healthcare outcomes [3,4]. Therefore, a reasonable A1C goal for adults is <7%; meanwhile, a lower glycemic goal of <7.5% for older adults was suggested in the practical guidelines [20]. This result was also true in our study. The survival curve showed that the poor glycemic control group had a significantly more rapid progression of DR, while the curve of the good glycemic control group was similar to that of the intermediate glycemic control group. Because the HbA1c only provides an average glucose level in the previous 3 months and does not reflect individual patterns of glycemic control, the development of diabetic complications is influenced not only by HbA1c but also by other factors.

In recent years, glucose fluctuation, which is also called as GV, was reported as a contributor to the development of diabetes-related complications irrespective of the HbA1c level [21,22]. DR is one of the major microvascular complications of DM. To date, numerous studies have reported that GV may also contribute to the development and progression of DR. These measurements of GV include HbA1c variability, CGM, and FPG variability.

Studies have provided evidence that long-term HbA1c variability predicted the risk of microvascular disease and DR in patients with type 2 DM [23–25]. In addition, the HbA1c variability was associated with macrovascular but not with microvascular diseases in type 2 DM patients, whereas short-term GV was associated with both macrovascular and microvascular events in the ADVANCE trial [26]. The HbA1c variability was associated with nephropathy but not with retinopathy in the RIACE Italian Multicenter Study [27]. One meta-analysis reported that DR was associated with the long-term HbA1c variability in type 1 DM but not in type 2 DM [28]. However, recent study with a long follow-up period reported that HbA1c variability played an important role in the development of DR when the HbA1c variability was high (>0.75% [10]; the cutoff value of HbA1c SD was 1.24 [9]). One longitudinal study in Taiwan that included 3152 type 2 DM patients who were followed-up for at least 2 years showed that HbA1c variability is an independent risk factor for DR. HbA1c variability may play a role in DR development when the mean value of HbA1c variability index is >0.75% [10]. The HbA1c standard deviation was associated with a risk of ≥3 steps of progression on the Early Treatment DR Study person scale as well as progression to PDR in a Korean retrospective study that had a follow-up period of more than 5 years [9]. The optimal cutoff value of the HbA1c SD was 1.24 for PDR and ≥3 steps of progression. GV may play a greater role in DR development when the magnitude of variation over a long period is large.

A Chinese cross-sectional study that assessed the 72 h CGM of 3262 type 2 DM patients found that the time in range (the percentage of time that the glucose range was within 70–180 mg/dL) and GV were associated with the prevalence of all stages of DR after adjusting for clinical risk factors [5,6]. Another Italian study that assessed the 72 h CGM reported that some parameters of GV were associated with DR, but its significance was lost in the multivariate analysis [29]. However, 33 type 1 diabetes mellitus and 35 type 2 diabetes mellitus cases were combined in the analysis to increase statistical power, thereby resulting in the heterogeneity of participants in the study. On the other hand, a meta-analysis revealed that the FPG variability is significantly associated with an increased risk of DR and all-cause mortality in patients with type 2 DM [8].

These studies have demonstrated that the deleterious effects of GV on DR have been underestimated. The parameters of GV have been identified to be important for the selection of the optimal treatment strategy and for the evaluation of the risk of DR.

However, the studies on GV mostly measured the value of dispersion without positive and negative directionality. We could not determine whether the association resulted from hyperglycemia with poor control or iatrogenic hypoglycemia due to an intensive HbA1c treatment goal. Further research should be performed to validate whether an HbA1c with a steady GV is a better treatment goal.

The glycemic gap was found to be a good parameter for SIH. In addition, it was revealed as a good predictor for disease severity and adverse clinical outcomes in diabetic patients with liver abscesses, pneumonia, acute myocardial infarction, acute ischemic stroke, chronic obstructive lung disease, necrotizing fasciitis, and in-hospital cardiac arrest [11–13]. The glycemic gap cutoff values of 45.26 [30], 42 [12], and 40 [31] were reported to predict adverse outcomes simultaneously. We then used the glycemic gap as a measurement for GV, and we also calculated the positive magnitude (positive glycemic gap) and negative magnitude (negative glycemic gap).

Compared with HbA1c variation, self-monitoring of blood glucose, or CGM, the glycemic gap is a simple and reliable measurement for GV in our study, and it can be performed by drawing blood only once. We found that both the glycemic gap and negative glycemic gap have AUC values that can significantly predict the progression of DR after adjustments; this was not true with the positive glycemic gap (Figure 2). In addition, greater AUC values were found in almost any time period despite dividing the patients into three subgroups based on the HbA1c level. This suggests that GV was independently related to DR regardless of the group. We have also found that a glycemic gap greater than the cutoff level of 45 mg/dL was associated with DR progression.

High GV increases the episodes of hypoglycemia. Hypoglycemia stimulates oxidative stress, endothelial dysfunction, and inflammatory reactions associated with microvascular changes [32–34]. Retinal Müller cells, which are identified as the major glial component of the retina, are closely associated with astrocytes, endothelial cells, and neurons, and play a part in the regulation of the blood–retinal barrier. In DM, it has been well established that Müller cells become activated [35–37]. An in vitro study of rats showed that the response of retinal Müller cells to glucose excursions was different under basal and high glucose concentrations. When the glucose concentration was normal, significant glial activation was detected not only in response to a constantly high glucose level but also to alternating low and high glucose levels. When the glucose concentration was high, a response was observed only upon exposure to a lower glucose concentration. This showed that the activation of retinal Müller cells occurs in response to glucose excursions [38]. This might explain the AUC values for predicting DR progression, as shown in Figure 3. This association was also supported by previous studies showing iatrogenic hypoglycemia due to low but oscillating HbA1c levels as contributing to the risk of DR progression [39].

Our study finding suggests that beyond the average glucose level, GV has different roles in the development of DR in type 2 DM. The worsening of DR might be affected by the episodes of hypoglycemia and sudden glycemic compensation caused by treatment [40]. A negative glycemic gap correlated with the progression of DR, which supported the possible deteriorating effects of hypoglycemic episodes on the progression of DR. In clinical practice, individualized treatments with an optimal balance between the mean HbA1c and GV are important to minimize the development of diabetes complications. An individualized treatment with an achievable goal of glycemic control and maintained GV may be similar to, or even better than, intensive goals leading to glycemic fluctuations. Future longitudinal trials and analysis with similar HbA1c levels and different GV will help validate and elucidate this effect.

The strengths of our study include the use of a longitudinal design, rather than a cross-sectional design, and a well-validated grading of DR. We also collected detailed glucose records while each patient's fundus color photography was being performed.

A few limitations of our study should be recognized. First, all patients were from a single center; thus, sampling bias and selection bias are deemed inevitable. Second, due to the longitudinal design, the results could not be used to establish a cause–effect relationship.

Third, the duration of DM in each patient, which was common risk factor for DR prevalence, was not mentioned in our study. Finally, the study population were from Taiwan; we cannot assure the generalizability of the results to other populations.

5. Conclusions

In conclusion, the glycemic gap, as a measurement for GV, is associated with DR progression in type 2 DM patients. The association between a negative glycemic gap and progression of DR indicated the negative effects of iatrogenic hypoglycemia. Our study results suggested that individual treatment goals with a reasonable HbA1c and stable GV may help to prevent DR. An achievable goal of glycemic control with a maintained GV may be better than intensive goals with wide glycemic fluctuations. Additional randomized prospective follow-up trials could help find an optimal balance between the mean HbA1c and GV.

Author Contributions: S.-C.H. contributed to the drafting and critical revision of the manuscript; C.L. contributed to algorithm development and static analysis; J.-T.C. and Y.-H.C. contributed towards confirming DR grading; W.-H.F. contributed to concept generation, methodology, data acquisition/interpretation, and the approval of the article′s authors. All authors have read and agreed to the published version of the manuscript.

Funding: The study was funded by the Ministry of Science and Technology, Taiwan (108-2314-B-016-001 to C. Lin, 109-2314-B-016-026 to C. Lin, 109-2314-B-016-021 to WH Fang), the National Science and Technology Development Fund Management Association, Taiwan (108-3111-Y-016-009 and 109-3111-Y-016-002 to C. Lin), and the Cheng Hsin General Hospital, Taiwan (CHNDMC-109-19 to C. Lin). The study was supported by funding from the Ministry of Science and Technology, Taiwan (MOST 108-2314-B-016-001 to C. Lin, MOST 109-2314-B-016-026 to C. Lin, MOST 109-2314-B-016-021 to WH Fang), the National Science and Technology Development Fund Management Association, Taiwan (MOST 108-3111-Y-016-009 and MOST 109-3111-Y-016-002 to C. Lin), and the Cheng Hsin General Hospital, Taiwan (CHNDMC-109-19 to C. Lin).

Institutional Review Board Statement: The study was conducted according to the guidelines of the Declaration of Helsinki, and approved by the Ethics Committee of Tri-Service General Hospital. Research ethics approval was given by the institutional review board without individual consent (IRB NO 2-105-05-073).

Informed Consent Statement: Patient consent was waived due to the retrospective nature of the study and the analysis used anonymous clinical data.

Data Availability Statement: Data available in a publicly accessible repository that does not issue DOIs. The data presented in this study are openly available in kaggle; reference number [17].

Conflicts of Interest: The authors declare no conflict of interest.

References

1. Diabetes Control; Complications Trial Research Group. The relationship of glycemic exposure (HbA1c) to the risk of development and progression of retinopathy in the diabetes control and complications trial. *Diabetes* **1995**, *44*, 968–983. [CrossRef]
2. Nathan, D.M.; McGee, P.; Steffes, M.W.; Lachin, J.M.; DCCT/EDIC Research Group. Relationship of glycated albumin to blood glucose and HbA1c values and to retinopathy, nephropathy, and cardiovascular outcomes in the DCCT/EDIC study. *Diabetes* **2014**, *63*, 282–290. [CrossRef] [PubMed]
3. Currie, C.J.; Peters, J.R.; Tynan, A.; Evans, M.; Heine, R.J.; Bracco, O.L.; Zagar, T.; Poole, C.D. Survival as a function of HbA1c in people with type 2 diabetes: A retrospective cohort study. *Lancet* **2010**, *375*, 481–489. [CrossRef]
4. Huang, E.S.; Liu, J.Y.; Moffet, H.H.; John, P.M.; Karter, A.J. Glycemic control, complications, and death in older diabetic patients: The diabetes and aging study. *Diabetes Care* **2011**, *34*, 1329–1336. [CrossRef]
5. Lu, J.; Ma, X.; Zhang, L.; Mo, Y.; Ying, L.; Lu, W.; Zhu, W.; Bao, Y.; Zhou, J. Glycemic variability assessed by continuous glucose monitoring and the risk of diabetic retinopathy in latent autoimmune diabetes of the adult and type 2 diabetes. *J. Diabetes Investig.* **2019**, *10*, 753–759. [CrossRef] [PubMed]
6. Lu, J.; Ma, X.; Zhou, J.; Zhang, L.; Mo, Y.; Ying, L.; Lu, W.; Zhu, W.; Bao, Y.; Vigersky, R.A.; et al. Association of time in range, as assessed by continuous glucose monitoring, with diabetic retinopathy in type 2 diabetes. *Diabetes Care* **2018**, *41*, 2370–2376. [CrossRef]

7. Škrha, J.; Šoupal, J.; Prázný, M. Glucose variability, HbA1c and microvascular complications. *Rev. Endocr. Metab. Disord.* **2016**, *17*, 103–110. [CrossRef] [PubMed]
8. Zhao, Q.; Zhou, F.; Zhang, Y.; Zhou, X.; Ying, C.J.D.R.; Practice, C. Fasting plasma glucose variability levels and risk of adverse outcomes among patients with type 2 diabetes: A systematic review and meta-analysis. *Diabetes Res. Clin. Pract.* **2019**, *148*, 23–31. [CrossRef]
9. Park, J.Y.; Hwang, J.H.; Kang, M.J.; Sim, H.E.; Kim, J.S.; Ko, K.S.J.R. Effects of glycemic variability on the progression of diabetic retinopathy among patients with type 2 diabetes. *Retina* **2021**, *41*, 1487–1495. [CrossRef]
10. Hu, J.; Hsu, H.; Yuan, X.; Lou, K.; Hsue, C.; Miller, J.D.; Lu, J.; Lee, Y.; Lou, Q. HbA1c variability as an independent predictor of diabetes retinopathy in patients with type 2 diabetes. *J. Endocrinol. Investig.* **2021**, *44*, 1229–1236. [CrossRef] [PubMed]
11. Chen, P.-C.; Tsai, S.-H.; Wang, J.-C.; Tzeng, Y.-S.; Wang, Y.-C.; Chu, C.-M.; Chu, S.-J.; Liao, W.-I. An elevated glycemic gap predicts adverse outcomes in diabetic patients with necrotizing fasciitis. *PLoS ONE* **2019**, *14*, e0223126. [CrossRef]
12. Liao, W.-I.; Lin, C.-S.; Lee, C.-H.; Wu, Y.-C.; Chang, W.-C.; Hsu, C.-W.; Wang, J.-C.; Tsai, S.-H. An elevated glycemic gap is associated with adverse outcomes in diabetic patients with acute myocardial infarction. *Sci. Rep.* **2016**, *6*, 27770. [CrossRef]
13. Liao, W.-I.; Sheu, W.H.-H.; Chang, W.-C.; Hsu, C.-W.; Chen, Y.-L.; Tsai, S.-H. An Elevated Gap between Admission and A1C-Derived Average Glucose Levels Is Associated with Adverse Outcomes in Diabetic Patients with Pyogenic Liver Abscess. *PLoS ONE* **2013**, *8*, e64476.
14. Wong, T.Y.; Sun, J.; Kawasaki, R.; Ruamviboonsuk, P.; Gupta, N.; Lansingh, V.C.; Maia, M.; Mathenge, W.; Moreker, S.; Muqit, M.M.; et al. Guidelines on diabetic eye care: The international council of ophthalmology recommendations for screening, follow-up, referral, and treatment based on resource settings. *Ophthalmology* **2018**, *125*, 1608–1622. [CrossRef] [PubMed]
15. Nathan, D.M.; Kuenen, J.; Borg, R.; Zheng, H.; Schoenfeld, D.; Heine, R.J. Translating the A1C assay into estimated average glucose values. *Diabetes Care* **2008**, *31*, 1473–1478. [CrossRef] [PubMed]
16. Hsing, S.-C.; Lee, C.-C.; Lin, C.; Chen, J.-T.; Chen, Y.-H.; Fang, W.-H. The Severity of Diabetic Retinopathy Is an Independent Factor for the Progression of Diabetic Nephropathy. *J. Clin. Med.* **2021**, *10*, 3. [CrossRef]
17. Available online: https://www.kaggle.com/c/diabetic-retinopathy-detection (accessed on 12 June 2020).
18. Hu, J.; Shen, L.; Sun, G. Squeeze-and-excitation networks. In Proceedings of the IEEE Conference on Computer Vision and Pattern Recognition, Salt Lake City, UT, USA, 18–23 June 2018; pp. 7132–7141.
19. Sherwani, S.I.; Khan, H.A.; Ekhzaimy, A.; Masood, A.; Sakharkar, M.K. Significance of HbA1c test in diagnosis and prognosis of diabetic patients. *Biomark. Insights* **2016**, *11*, BMI-S38440. [CrossRef]
20. Care, D. 6. Glycemic targets: Standards of medical care in diabetes—2019. *Diabetes Care* **2019**, *42*, S61–S70.
21. Bergenstal, R.M. Glycemic variability and diabetes complications: Does it matter? Simply put, there are better glycemic markers! *Diabetes Care* **2015**, *38*, 1615–1621. [CrossRef] [PubMed]
22. Hirsch, I.B. Glycemic variability and diabetes complications: Does it matter? Of course it does! *Diabetes Care* **2015**, *38*, 1610–1614. [CrossRef]
23. Cardoso, C.; Leite, N.; Moram, C.; Salles, G. Long-term visit-to-visit glycemic variability as predictor of micro-and macrovascular complications in patients with type 2 diabetes: The Rio de Janeiro Type 2 Diabetes Cohort Study. *Cardiovasc. Diabetol.* **2018**, *17*, 33. [CrossRef]
24. Park, S.P. Long-term HbA1c variability and the progression of diabetic retinopathy in patients with type 2 diabetes. *Investig. Ophthalmol. Vis. Sci.* **2019**, *60*, 1073.
25. Yang, C.-Y.; Su, P.-F.; Hung, J.-Y.; Ou, H.-T.; Kuo, S. Comparative predictive ability of visit-to-visit HbA1c variability measures for microvascular disease risk in type 2 diabetes. *Cardiovasc. Diabetol.* **2020**, *19*, 105. [CrossRef]
26. Hirakawa, Y.; Arima, H.; Zoungas, S.; Ninomiya, T.; Cooper, M.; Hamet, P.; Mancia, G.; Poulter, N.; Harrap, S.; Woodward, M.; et al. Impact of visit-to-visit glycemic variability on the risks of macrovascular and microvascular events and all-cause mortality in type 2 diabetes: The ADVANCE trial. *Diabetes Care* **2014**, *37*, 2359–2365. [CrossRef] [PubMed]
27. Penno, G.; Solini, A.; Bonora, E.; Fondelli, C.; Orsi, E.; Zerbini, G.; Morano, S.; Cavalot, F.; Lamacchia, O.; Laviola, L.; et al. HbA1c variability as an independent correlate of nephropathy, but not retinopathy, in patients with type 2 diabetes: The Renal Insufficiency And Cardiovascular Events (RIACE) Italian multicenter study. *Diabetes Care* **2013**, *36*, 2301–2310. [CrossRef] [PubMed]
28. Gorst, C.; Kwok, C.S.; Aslam, S.; Buchan, I.; Kontopantelis, E.; Myint, P.K.; Heatlie, G.; Loke, Y.; Rutter, M.K.; Mamas, M.A. Long-term glycemic variability and risk of adverse outcomes: A systematic review and meta-analysis. *Diabetes Care* **2015**, *38*, 2354–2369. [CrossRef]
29. Sartore, G.; Chilelli, N.C.; Burlina, S.; Lapolla, A. Association between glucose variability as assessed by continuous glucose monitoring (CGM) and diabetic retinopathy in type 1 and type 2 diabetes. *Acta Diabetol.* **2013**, *50*, 437–442. [CrossRef]
30. Donagaon, S.; Dharmalingam, M. Association between Glycemic Gap and adverse outcomes in critically ill patients with diabetes. *Indian J. Endocrinol. Metab.* **2018**, *22*, 208. [PubMed]
31. Chen, P.-C.; Liao, W.-I.; Wang, Y.-C.; Chang, W.-C.; Hsu, C.-W.; Chen, Y.-H.; Tsai, S.-H. An elevated glycemic gap is associated with adverse outcomes in diabetic patients with community-acquired pneumonia. *Medicine* **2015**, *94*, e1456. [CrossRef]
32. Ceriello, A.; Novials, A.; Ortega, E.; Canivell, S.; La Sala, L.; Pujadas, G.; Bucciarelli, L.; Rondinelli, M., Genovese, S. Vitamin C further improves the protective effect of glucagon-like peptide-1 on acute hypoglycemia-induced oxidative stress, inflammation, and endothelial dysfunction in type 1 diabetes. *Diabetes Care* **2013**, *36*, 4104–4108. [CrossRef]

33. Wang, J.; Alexanian, A.; Ying, R.; Kizhakekuttu, T.J.; Dharmashankar, K.; Vasquez-Vivar, J.; Gutterman, D.D.; Widlansky, M.E. Acute exposure to low glucose rapidly induces endothelial dysfunction and mitochondrial oxidative stress: Role for AMP kinase. *Arterioscler. Thromb. Vasc. Biol.* **2012**, *32*, 712–720. [CrossRef]

34. Yousefzade, G.; Nakhaee, A. Insulin-induced hypoglycemia and stress oxidative state in healthy people. *Acta Diabetol.* **2012**, *49*, 81–85. [CrossRef]

35. Gerhardinger, C.; Costa, M.B.; Coulombe, M.C.; Toth, I.; Hoehn, T.; Grosu, P. Expression of acute-phase response proteins in retinal Muller cells in diabetes. *Investig. Ophthalmol. Vis. Sci.* **2005**, *46*, 349–357. [CrossRef] [PubMed]

36. Kusner, L.L.; Sarthy, V.P.; Mohr, S. Nuclear translocation of glyceraldehyde-3-phosphate dehydrogenase: A role in high glucose-induced apoptosis in retinal Muller cells. *Investig. Ophthalmol. Vis. Sci.* **2004**, *45*, 1553–1561.

37. Puro, D.G. Diabetes-induced dysfunction of retinal Müller cells. *Trans. Am. Ophthalmol. Soc.* **2002**, *100*, 339. [PubMed]

38. Picconi, F.; Parravano, M.; Sciarretta, F.; Fulci, C.; Nali, M.; Frontoni, S.; Varano, M.; Caccuri, A.M. Activation of retinal Müller cells in response to glucose variability. *Endocrine* **2019**, *65*, 542–549. [CrossRef] [PubMed]

39. Cryer, P.E. Glycemic goals in diabetes: Trade-off between glycemic control and iatrogenic hypoglycemia. *Diabetes* **2014**, *63*, 2188–2195. [CrossRef]

40. FLAT-SUGAR Trial Investigators. Design of FLAT-SUGAR: Randomized trial of prandial insulin versus prandial GLP-1 receptor agonist together with basal insulin and metformin for high-risk type 2 diabetes. *Diabetes Care* **2015**, *38*, 1558–1566. [CrossRef]

Journal of
*Personalized
Medicine*

Article

Ethnic Disparities in the Development of Sight-Threatening Diabetic Retinopathy in a UK Multi-Ethnic Population with Diabetes: An Observational Cohort Study

Manjula D. Nugawela [1], Sarega Gurudas [1], A Toby Prevost [2], Rohini Mathur [3], John Robson [4], Wasim Hanif [5], Azeem Majeed [6] and Sobha Sivaprasad [1,7,*]

1   UCL Institute of Ophthalmology, 11-43 Bath Street, London EC1V 9EL, UK;
    manjula.nugawela@ucl.ac.uk (M.D.N.); sarega.gurudas.17@ucl.ac.uk (S.G.)
2   Department of Population Health Sciences, King's College London, London WC2R 2LS, UK;
    toby.1.prevost@kcl.ac.uk
3   London School of Hygiene & Tropical Medicine, Keppel Street, London WC1E 7HT, UK;
    rohini.mathur@lshtm.ac.uk
4   Institute of Population Health Sciences, Queen Mary University of London, London E1 4NS, UK;
    j.robson@qmul.ac.uk
5   Birmingham City School of Nursing and Midwifery, Westbourne Road, Birmingham B15 3TN, UK;
    hanif.wasim@uhb.nhs.uk
6   School of Public Health, Imperial College London, London SW7 2AZ, UK; a.majeed@imperial.ac.uk
7   Moorfields Eye Hospital NHS Foundation Trust, 162, City Road, London EC1V 2PD, UK
*   Correspondence: sobha.sivaprasad@nhs.net

Citation: Nugawela, M.D.;
Gurudas, S.; Prevost, A.T.; Mathur, R.;
Robson, J.; Hanif, W.; Majeed, A.;
Sivaprasad, S. Ethnic Disparities in
the Development of
Sight-Threatening Diabetic
Retinopathy in a UK Multi-Ethnic
Population with Diabetes: An
Observational Cohort Study. *J. Pers.
Med.* **2021**, *11*, 740. https://doi.org/
10.3390/jpm11080740

Academic Editors:
Peter D. Westenskow
and Andreas Ebneter

Received: 5 July 2021
Accepted: 26 July 2021
Published: 28 July 2021

Abstract: There is little data on ethnic differences in incidence of DR and sight threatening DR (STDR) in the United Kingdom. We aimed to determine ethnic differences in the development of DR and STDR and to identify risk factors of DR and STDR in people with incident or prevalent type II diabetes (T2DM). We used electronic primary care medical records of people registered with 134 general practices in East London during the period from January 2007–January 2017. There were 58,216 people with T2DM eligible to be included in the study. Among people with newly diagnosed T2DM, Indian, Pakistani and African ethnic groups showed an increased risk of DR with Africans having highest risk of STDR compared to White ethnic groups (HR: 1.36 95% CI 1.02–1.83). Among those with prevalent T2DM, Indian, Pakistani, Bangladeshi and Caribbean ethnic groups showed increased risk of DR and STDR with Indian having the highest risk of any DR (HR: 1.24 95% CI 1.16–1.32) and STDR (HR: 1.38 95% CI 1.17–1.63) compared with Whites after adjusting for all covariates considered. It is important to optimise prevention, screening and treatment options in these ethnic minority groups to avoid health inequalities in diabetes eye care.

Keywords: type 2 diabetes; retinopathy; ethnicity; general practice; risk factors

1. Introduction

Diabetic retinopathy (DR) is a major microvascular complication of diabetes [1,2]. DR can progress from no DR to sight threatening diabetic retinopathy (STDR) without any symptoms [3]. In the United Kingdom (UK), the prevalence of DR and STDR among people with type 2 diabetes (T2DM) is approximately 30% and 1.5%, respectively [4,5]. In absolute numbers, this equates to around 1.5 million people with DR and 140,000 with STDR. In the UK, the two largest minority ethnic groups are South Asian (7.5%) and Afro-Caribbean (3.3%) compared to the White ethnic groups who make up 86% of the population [6]. These ethnic minority groups are likely to develop diabetes from a younger age and have a higher prevalence of diabetes compared to the White counterparts [7–9]. Similarly, the prevalence of DR is higher in South Asian and Black ethnic groups in the UK and these findings are consistent with global literature [4,10–16]. These ethnic differences in prevalence of DR are often explained by earlier age of onset of diabetes, and suboptimal control of risk factors. However, there is little data on differences in the incidence of DR and STDR in these ethnic groups when compared to their White counterparts. Studies on incidence of DR and STDR are mainly reported from Clinical Practice Research Datalink (CPRD) where ethnicity records are missing in about a third of the people with diabetes or from regions that do not have sufficient ethnic minorities to study the incidence at a population level [4]. It is also unclear if any ethnic differences in incident DR and STDR are due to variations in control of modifiable systemic risk factors.

Duration of diabetes is one of the strongest risk factors of DR [17,18]. In the UK, all people with newly diagnosed diabetes are referred for annual DR screening within 3 months of the screening programme being notified of the diagnosis [19]. People with diabetes registered in the same practice have equal access to the same group of general practitioners and access to care is free in the National Health Service (NHS). The population in East London consists of a high proportion of people from ethnic minority groups. Therefore, it is a suitable population in which to evaluate whether ethnicity is an independent risk factor for incidence of DR and STDR.

The aim of this study was to determine ethnic differences in the development of DR and STDR, and to identify risk factors for DR/STDR development in people with newly diagnosed and prevalent T2DM at baseline. We focussed on the two main ethnic minority groups, South Asian and Black and then further analysed the sub-groups of Indian, Bangladeshi, Pakistani and Caribbean and African ethnic groups to identify the role of ethnicity as an independent risk factor for DR/STDR.

2. Materials and Methods

The Moorfields Research Management Committee approved the use of this fully anonymised UK dataset for model development and validation (SIVS1047) and further ethics approval was not required. Approval was also obtained from the Caldicott guardian of this anonymised dataset in Queen Mary University London (QMUL). Individual level patient consent was not obtained, as this was an observational study using de-identified data.

This cohort study was conducted using de-identified data from general practice electronic health records collected in three Clinical Commissioning Groups (CCGs) in East London, which include Newham, Tower Hamlets and City and Hackney. These data covered more than 98% of the GP-registered population in these CCGs. The data included demographic information, diagnoses, prescriptions, referrals, laboratory test results and clinical values. Diagnoses, symptoms and clinical values in this dataset were recorded using the Read code classification.

We included all adults with a diagnostic Read code for T2DM during the period 2007–2017 and aged 18+ at study entry. Study baseline was defined as the date of the initial T2DM diagnostic read code for each person within the study start and end dates. Start date was defined as the latest of study start date (January 2007), 12 months after the patient's current registration date, or the date the patient turned 18. Follow-up time end date is defined as the earliest date of transferring out of the practice, or latest data collection date or date of death or January 2017. People who were lost to follow up were censored at the date they left the study. People with Type 1 diabetes, gestational diabetes, other forms of diabetes were excluded.

People were classified as having newly diagnosed (incident) T2DM if their T2DM onset date was the same as their initial T2DM diagnosis date within the study start and end dates. If they had an earlier T2DM onset date than the initial T2DM diagnosis date within the study period, they were classified as people with prevalent T2DM. Self-reported ethnicity was identified using the relevant Office for National Statistic ethnic group classification in the electronic health record of each patient. Ethnicity was considered in two ways. First in three categories as White, South Asian and Black and then we further considered the subgroups of Indian, Pakistani, Bangladeshi, African and Caribbean. Chinese, other Asian, other Black and any other ethnic group were grouped as 'Other'. A small proportion of people with missing or unknown ethnicity (1.8%) were not included in the analysis. Socio-economic deprivation was categorised into quintiles of the Townsend score, which is made up of unemployment, overcrowding and car ownership variables with the most affluent in quintile 1 and most deprived in quintile 5.

The first record of any DR and any STDR were considered as the outcomes. Read codes for classification levels of DR severity levels included both the American Academy of Ophthalmology International Classification and UK National Screening Committee classification [3]. Those who had severe Non-Proliferative DR, Proliferative DR, grading classification for proliferative DR-R3 or grading classification for maculopathy-M1, were identified as people with STDR. A person's DR grade was defined as the DR severity level in the worst eye.

Covariates that are commonly found to be associated with DR and STDR according to existing literature were considered in this study [17,18], and these included age, gender, hyperglycaemia, systolic blood pressure (SBP), duration of diabetes, body mass index (BMI), total cholesterol, high density lipoprotein (HDL), eGFR, antidiabetic medication, history of cardiovascular disease (ischaemic heart disease, heart failure, stroke, peripheral vascular disease, cardiovascular death, Acute Myocardial Infarction, Bypass graft/Angioplasty, Angina Pectoris, Cardiac Arrhythmia, Major ECG abnormality, Silent Myocardial Infarction, Congestive Heart Failure, Transient Ischaemic Attack, Arterial Event requiring surgery), statin prescription and antihypertensive medication prescription. These covariates were measured at baseline for each individual and the closest record to the index date (± 6 months) was selected for clinical variables.

Sperate analysis were undertaken for the two groups of study population, people with newly diagnosed and prevalent diabetes, and then for the two outcomes DR and STDR. Univariable statistics were used to describe the baseline characteristics of the study population overall and by ethnic group. Kaplan-Meier Survival estimates with log-rank test were used to observe any differences in DR and STDR development over time by different ethnic groups. Multivariable Cox regression was then used to identify any association between the ethnic groups and development of any DR or STDR. Cox regression models were developed separately among people with newly diagnosed and prevalent diabetes for the two outcomes of DR and STDR resulting four models. All four models were adjusted for all covariates considered in the study, and the effects of each covariate was reported with hazard ratios (HR) with 95% confidence intervals (CI). Risk factors for development of DR were identified using the statistical significance of each covariate in each fully adjusted model ($p < 0.05$). We further analysed the ethnic differences in developing DR/STDR by stratifying the exposure variable into the more detailed ethnic categories of Indian, Pakistani, Bangladeshi and African and Caribbean. All statistical analyses were performed using Stata V.16.

3. Results

3.1. Overall Study Population

There were 71,406 people with a T2DM record within the study period who were registered with a GP practice for at least one year period. Of these, we excluded those who had background retinopathy ($n = 6926$) and sight threatening retinopathy ($n = 409$) before the baseline. People who had history of anti-diabetic treatment before their T2DM onset date ($n = 3318$) and people who were prescribed two antidiabetic drugs or insulin on the T2DM diagnosis date ($n = 1466$) were excluded and these were defined as coding errors. Of the remaining 59,287 people, ethnicity was recorded on 98% people and there were only 1071 (2%) who did not have their ethnicity recorded (unknown or missing) and these were also excluded, leaving 58,216 people eligible to be included in the study of whom 32,652 were people with newly diagnosed T2DM and 25,564 people with prevalent T2DM at the study baseline with mean duration of diabetes 7.4 (SD 6.1) years.

3.2. Baseline Characteristics of the Study Population

The following results are presented separately for groups of study population with newly diagnosed type 2 diabetes (Table 1) and prevalent T2DM at baseline (Table 2). Among people with newly diagnosed T2DM, we observed incidence rates of 82.8 and 4.2 events per 1000 person years for DR and STDR, respectively. Among people with known T2DM, the incidence rates for DR and STDR were 135.4 and 16.0 events per 1000 person years, respectively (Table 3).

Table 1. Baseline characteristics of the study population—People with newly diagnosed type 2 diabetes.

	Whole Population	White	South Asian	Black	Other
	N = 32,652	N = 8727	N = 15,291	N = 6639	N = 1995
>Age at T2DM diagnosis, years					
<45	10,136 (31.0)	1423 (16.3)	6536 (42.7)	1678 (25.3)	499 (25.0)
45 to <55	9352 (28.6)	2258 (25.9)	4329 (28.3)	2176 (32.8)	589 (29.5)
55 to <65	6789 (20.8)	2360 (27.0)	2530 (16.6)	1369 (20.6)	530 (26.6)
65 to <75	4091 (12.5)	1622 (18.6)	1269 (8.3)	947 (14.3)	253 (12.7)
75+	2284 (6.9)	1064 (12.2)	627 (4.1)	469 (7.1)	124 (6.2)
Gender					
Male	18,332 (56.1)	5064 (58.0)	8842 (57.8)	3381 (50.9)	1045 (52.4)
Female	14,320 (43.8)	3663 (41.9)	6449 (42.2)	3258 (49.1)	950 (47.7)
Townsend Score (quintiles)					
1 (most affluent)	6874 (21.0)	1484 (17.0)	4043 (26.4)	978 (14.7)	369 (18.5)
2	6795 (20.8)	1766 (20.2)	3159 (20.7)	1425 (21.5)	445 (22.3)
3	6409 (19.6)	1931 (22.1)	2708 (17.7)	1369 (20.6)	401 (20.1)
4	6335 (19.4)	1774 (20.3)	2807 (18.4)	1390 (20.9)	364 (18.3)
5 (most deprived)	6130 (18.8)	1739 (19.9)	2535 (16.6)	1448 (21.8)	408 (20.4)
Not recorded	109 (0.3)	33 (0.4)	39 (0.3)	29 (0.4)	8 (0.4)
Body Mass Index (kg/m^2)					
<18.5	321(1.0)	71 (0.8)	167 (1.1)	51 (0.8)	32 (1.6)
18.5 to <25	4627 (14.2)	686 (7.9)	3002 (19.6)	610 (9.2)	329 (16.5)
25 to <30	10,979 (33.7)	1990 (22.8)	6381 (41.7)	1951 (29.4)	657 (32.9)
$\geq$30	14067 (43.1)	5094 (58.4)	4706 (30.8)	3494 (52.6)	773 (38.7)
Not recorded	2658 (8.2)	886 (10.2)	1035 (6.8)	533 (8.0)	204 (10.3)
HbA1c (mmol/mol)					
<50	6927 (21.2)	2198 (25.2)	2912 (19.0)	1399 (21.1)	418 (20.9)
50 to <100	18, 368 (56.2)	4626 (53.0)	9002 (58.9)	3602 (54.3)	1138 (57.0)
$\geq$100	2561 (7.8)	627 (7.6)	970 (6.3)	795 (11.9)	169 (8.5)
Not recorded	4796 (14.7)	1276 (14.6)	2407 (15.7)	843 (12.7)	270 (13.5)
Systolic Blood Pressure (mmHg)					
<120	6396 (19.6)	1224 (14.0)	3897 (25.5)	918 (13.8)	357 (17.9)
120 to <130	7537 (23.1)	1891 (21.7)	3865 (25.3)	1315 (20.2)	430 (21.6)
130 to <140	8468 (25.9)	2404 (27.5)	3756 (24.6)	1792 (27.0)	516 (25.8)
$\geq$140	9273 (28.4)	2943 (33.7)	3348 (22.0)	2367 (35.6)	615 (30.8)
Not recorded	978 (3.0)	265 (3.0)	425 (2.8)	211 (3.2)	77 (3.9)
Total Cholesterol (mmol/L)					
<5.2	17,072 (52.3)	4537 (52.0)	8221 (53.8)	3351 (50.5)	963 (48.3)
5.2 to <6.2	8188 (25.1)	2068 (23.7)	3898 (25.5)	1719 (25.9)	503 (25.2)
$\geq$6.2	5413 (16.6)	1542 (17.7)	2359 (15.4)	1127 (16.8)	385 (19.3)
Not recorded	1979 (6.1)	580 (6.6)	813 (5.3)	442 (6.7)	144 (7.2)
eGFR (mL/min/1.73m^2)					
<60	2546 (7.8)	939 (10.8)	738 (4.8)	747 (11.3)	122 (6.2)
$\geq$60	27,832 (85.2)	7219 (82.7)	13,461 (88.0)	5438 (81.9)	1714 (85.9)
Not recorded	2274 (7.1)	569 (6.5)	1092 (7.1)	454 (6.8)	159 (7.9)

Table 1. *Cont.*

	Whole Population	White	South Asian	Black	Other
History of Cardiovascular Disease					
No	29,532 (90.4)	7339 (84.1)	14,135 (92.4)	6233 (93.8)	1825 (91.5)
Yes	3120 (9.6)	1388 (15.9)	1156 (7.4)	406 (6.2)	170 (8.5)
History of Antidiabetic					
No drug	21,779 (66.7)	6002 (66.8)	10,037 (65.6)	4409 (66.4)	1331 (66.7)
One drug	10,873 (31.3)	2725 (31.2)	5254 (34.4)	2230 (33.6)	664 (33.3)
History of Antihypertensive Medication					
No	17,620 (53.9)	3935 (45.1)	9396 (61.4)	3232 (48.7)	1057 (53.0)
Yes	15,032 (46.0)	4792 (54.9)	5895 (38.6)	3407 (51.3)	938 (47.0)
History of Statin Medication					
No	6229 (19.1)	1495 (17.1)	2553 (16.7)	1749 (26.3)	432 (21.7)
Yes	26,423 (80.9)	7232 (82.9)	12,738 (83.5)	4890 (73.7)	1563 (78.3)

Abbreviations: eGFR-Estimated glomerular filtration rate. 'Other' group: this group included Chinese, other Asian, other Black and people belong to any other ethnicity.

Table 2. Baseline characteristics of the study population—people with known type 2 diabetes.

	Whole Population	White	South Asian	Black	Other
	N = 25,564	N = 6946	N = 11,963	N = 5216	N = 1439
Age, Years					
<45	4956 (19.4)	738 (10.6)	3055 (25.5)	923 (17.7)	243 (16.9)
45 to <55	6207 (24.3)	1357 (19.5)	3297 (27.5)	1213 (23.3)	340 (23.7)
55 to <65	5860 (22.9)	1827 (26.3)	2495 (20.8)	1134 (21.7)	404 (28.0)
65 to <75	5407 (21.1)	1628 (23.4)	2218 (18.5)	1258 (24.1)	303 (21.0)
75+	3134 (12.2)	1396 (20.1)	901 (7.5)	688 (13.2)	149 (10.3)
Duration of Diabetes					
0 to <2 years (ref)	4030 (15.8)	1089 (15.7)	1875 (15.7)	806 (15.5)	260 (18.1)
2 to <5 years	6835 (26.7)	1891 (27.2)	3177 (26.6)	1368 (26.2)	399 (27.7)
5 to <10 years	8125 (31.8)	2281 (32.8)	3799 (31.7)	1642 (31.5)	403 (28.0)
≥10 years	6574 (25.7)	1685 (24.3)	3112 (26.0)	1400 (26.8)	377 (26.1)
Gender					
Male	13,289 (51.9)	3724 (53.6)	6333 (52.9)	2524 (48.4)	708 (49.2)
Female	12,275 (48.0)	3222 (46.4)	5630 (47.1)	2692 (51.6)	731 (50.8)
Townsend Score (quintiles)					
1	4812 (18.8)	1016 (14.6)	2774 (23.2)	762 (14.6)	260 (18.0)
2	4840 (18.9)	1270 (18.3)	2264 (18.9)	1021 (19.6)	285 (19.8)
3	5110 (20.0)	1622 (23.4)	2161 (18.3)	1036 (19.9)	291 (20.2)
4	5366 (20.9)	1537 (22.1)	2432 (20.3)	1151 (22.1)	246 (17.1)
5	5373 (21.0)	1488 (21.4)	2307 (19.3)	1227 (23.5)	351 (24.4)
Not recorded	63 (0.3)	13 (0.2)	25 (0.2)	19 (0.4)	6 (0.4)
Body Mass Index (kg/m^2)					
<18.5	389 (1.5)	83 (1.2)	216 (1.8)	56 (1.1)	34 (2.4)
18.5 to <25	4608 (18.0)	646 (9.3)	2999 (25.1)	675 (12.9)	288 (20.0)
25 to <30	8745 (34.2)	1778 (25.6)	4788 (40.0)	1723 (33.0)	456 (31.6)
≥30	8966 (35.1)	3466 (49.9)	2814 (23.5)	2191 (42.0)	495 (34.4)
Not recorded	2856 (11.2)	973 (14.0)	1146 (9.6)	571 (10.9)	166 (11.5)
HbA1c (mmol/mol)					
<50	5692 (22.3)	1897 (27.3)	2307 (19.3)	1177 (22.6)	312 (21.7)
50 to <100	15,362 (60.1)	3794 (54.6)	7701 (64.4)	3005 (57.6)	862 (59.9)
≥100	1644 (6.4)	373 (5.4)	702 (5.8)	474 (9.1)	95 (6.6)
Not recorded	2866 (11.2)	882 (12.7)	1254 (10.5)	560 (10.7)	170 (11.8)

Table 2. *Cont.*

Systolic Blood Pressure (mmHg)					
<120	5347 (20.9)	1129 (16.2)	3166 (26.5)	774 (14.8)	279 (19.4)
120 to <130	5733 (22.4)	1436 (20.7)	2945 (24.6)	1058 (20.3)	294 (20.5)
130 to <140	6313 (24.7)	1887 (27.2)	2699 (22.5)	1365 (26.2)	362 (25.2)
$\geq$140	7600 (29.7)	2318 (33.3)	2929 (24.5)	1881 (36.0)	472 (32.7)
Not recorded	571 (2.2)	176 (2.5)	225 (1.9)	138 (2.7)	32 (2.2)
Total Cholesterol (mmol/L)					
<5.2	18,456 (72.2)	4904 (70.6)	8949 (74.8)	3605 (69.1)	998 (69.4)
5.2 to <6.2	3237 (12.7)	883 (12.7)	1388 (11.6)	767 (14.7)	199 (13.8)
$\geq$6.2	1643 (6.4)	465 (6.7)	677 (5.7)	376 (7.2)	125 (8.7)
Not recorded	2228 (8.7)	694 (9.9)	949 (7.9)	468 (8.9)	117 (8.1)
eGFR (mL/min/1.73m^2)					
<60	2978 (11.7)	1036 (14.9)	1126 (9.4)	678 (13.0)	138 (9.6)
$\geq$60	20,160 (78.8)	5140 (74.0)	9813 (82.0)	4039 (77.4)	1168 (81.2)
Not recorded	2426 (9.5)	770 (11.1)	1024 (8.6)	499 (9.6)	133 (9.2)
History of Cardiovascular Disease					
No	21,734 (85.0)	5474 (78.8)	10,338 (86.4)	4669 (89.5)	1253 (87.1)
Yes	3830 (15.0)	1472 (21.2)	1625 (13.6)	547 (10.5)	186 (12.9)
History of Antidiabetic Drugs					
No drug	4470 (17.5)	1508 (21.7)	1750 (14.6)	954 (18.3)	258 (17.9)
One drug	7244 (28.3)	1988 (28.6)	3414 (28.5)	1403 (26.9)	439 (30.5)
Two drugs	9356 (36.6)	2242 (32.3)	4781 (40.0)	1832 (35.1)	501 (34.8)
Insulin	4494 (17.6)	1208 (17.4)	2018 (16.9)	1027 (19.7)	241 (16.8)
History of Antihypertensive Medication					
No	8880 (34.7)	2003 (28.8)	4562 (38.1)	1770 (33.9)	545 (37.9)
Yes	16,684 (65.3)	4943 (71.2)	7401 (61.9)	3446 (66.1)	894 (62.1)
History of Statin Medication—ever					
No	2528 (9.9)	741 (10.7)	763 (6.4)	831 (15.9)	193 (13.4)
Yes	23,036 (90.1)	6205 (89.3)	11,200 (93.6)	4385 (84.1)	1246 (86.6)

Abbreviations: eGFR-Estimated glomerular filtration rate. 'Other' group: this group included Chinese, other Asian, other Black and people belong to any other ethnicity.

Table 1 shows the characteristics of 32,652 people with newly diagnosed diabetes at study baseline by different ethnic group. Townsend score was recorded in 99.7%, BMI was recorded in 91% and HbA1c level was available in 85.3%, SBP was recorded in 97%, total Cholesterol level was recorded in 94% and eGFR level was recorded in 93%. A greater proportion of South Asians 6536 (43%) were in the <45 years age group compared with White (16%), Black (25%) and other (25%) indicating early onset of diabetes among South Asians. Higher proportion of South Asians were in the most affluent group 4043 (26.4%) compared with other ethnic groups with 15–21% people in this group as shown in Table 1. Almost 70% of people in all ethnic groups had a BMI value that was greater than 25 kg/m^2. Almost 50% of South Asians had lower blood pressure <130 mmHg compared with other ethnic groups with a proportion of people from 34–43% in the same blood pressure group. White ethnic groups had more than twice as high prevalence of CVD 1388 (16%) compared to other ethnic groups with prevalence of CVD ranging from 6–8%.

Table 3. Number of incident cases of DR/STDR during 10-year follow-up and incidence rates per 1000 person years in newly diagnosed and known T2DM at baseline.

	People with Newly Diagnosed T2DM at Baseline								
	DR					STDR			
Ethnicity	Total Number of People	Number of Events	Percentage with Events	Person Years	Incidence Rate per 1000 Person Years (95% CI)	Number of Events	Percentage with Events	Person Years	Incidence Rate per 1000 Person Years (95% CI)
Whole Population	32,652	8638	26.5	104,257.8	82.85 (81.12–84.61)	557	1.7	132,126.7	4.21 (3.88–4.58)
White	8727	2227	25.5	29,624.4	75.17 (72.12–78.36)	137	1.6	37,185.5	3.68 (3.11–4.35)
South Asian	15,291	4035	26.4	47,448.6	85.04 (82.45–87.70)	229	1.5	60,163.5	3.81 (3.34–4.33)
Black	6639	1833	27.6	20,953.3	87.48 (83.56–91.58)	152	2.3	26,854.9	5.66 (4.83–6.63)
Other	1995	543	27.2	6231.5	87.13 (80.11–94.78)	39	1.9	7922.8	4.92 (3.59–6.73)

	People with Known T2DM at Baseline								
	DR					STDR			
Ethnicity	Total Number of People	Number of Events	Percentage with Events	Person Years	Incidence Rate per 1000 Person Years (95% CI)	Number of Events	Percentage with Events	Person Years	Incidence Rate per 1000 Person Years (95% CI)
Whole Population	25,564	12,124	47.4	89,534.1	135.41 (133.02–137.84)	2200	8.6	137,140.3	16.04 (15.38–16. 72)
White	6946	2946	42.4	25,900.1	113.74 (109.71–117.92)	454	6.5	37,727.6	12.03 (10.97–13.19)
South Asian	11,963	6100	50.9	40,447.5	150.81 (147.07–154.64)	1145	9.6	64,136.8	17.85 (16.84–18.91)
Black	5216	2422	46.4	18,134.7	133.56 (128.37–138.98)	480	9.2	27,699.3	17.32 (15.84–18.94)
Other	1439	656	45.6	5051.9	129.85 (120.28–140.18)	121	8.4	7576.5	15.97 (13.36–19.08)

Table 2 shows the baseline characteristics of 25,564 people with prevalent T2DM at the study baseline. In this group, Townsend score was available in 99.7%, BMI was recorded in 89%, HbA1c level was recorded in 89%, SBP was available in 98%, total Cholesterol level was recorded in 91% and eGFR level was recorded in 91%. A greater proportion of South Asians 3055 (25%) were in the youngest age group whereas a greater proportion of White ethnic groups with T2DM were in the oldest age group at baseline 1396 (20%). Duration of diabetes were similar across the ethnic groups with approximately 16% of people in 0 to <2 years duration, 27% in the 2 to <5 years group, 31% in the 5 to <10 years group, 26% in the >10 years group.

3.3. Kaplan Meier Plots

Figure 1 shows the Kaplan-Meier Survival estimates for DR and STDR by different ethnic groups for newly diagnosed and people with prevalent T2DM at baseline. For DR, there was clear separation in survival probabilities over time by each ethnic group with the White group having a higher survival probability overtime compared to all other ethnic groups (log rank test; $p < 0.001$). This separation was even more visible in people with prevalent T2DM at baseline, White ethnic groups having the highest and South Asians having the lowest survival probability overtime (log rank test; $p < 0.001$).

T2DM Newly diagnosed at baseline

(a) DR **(b) STDR**

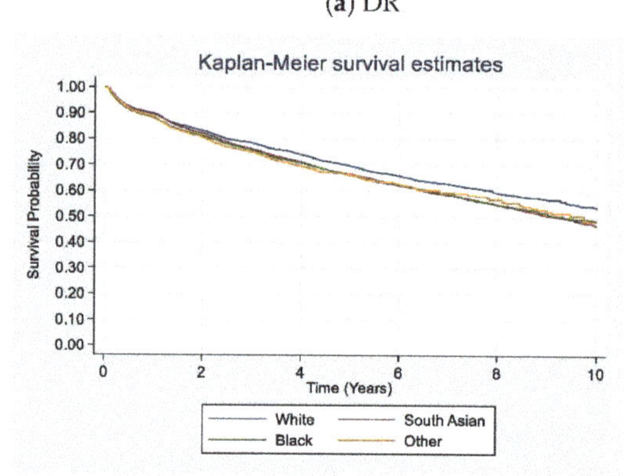

Incidence rate per 1000 person years for White–75.2 South Asian–85.0; Black–87.5; Other–87.1

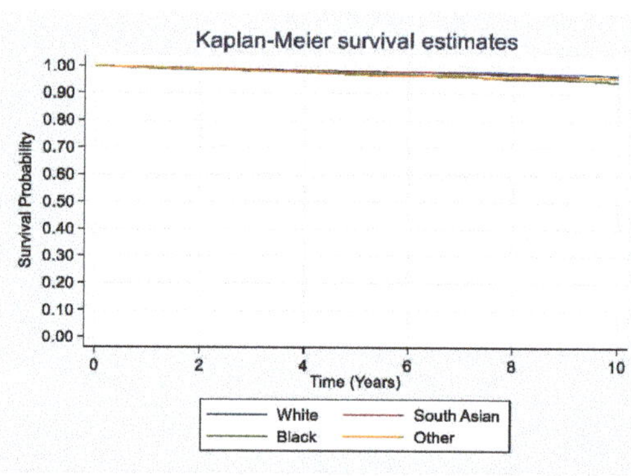

Incidence rate per 1000 person years for White–3.7 South Asian–3.8; Black – 5.7; Other–4.9

Known T2DM at baseline

(c) DR **(d) STDR**

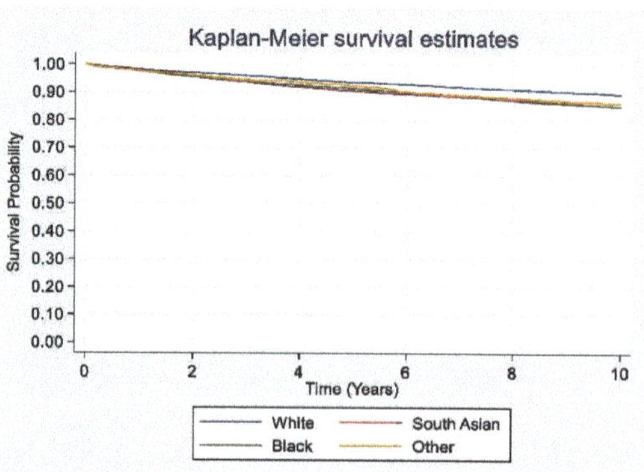

Incidence rate per 1000 person years for White–113.7 South Asian–150.8; Black–133.6; Other–129.8   Incidence rate per 1000 person years for White–12.0 South Asian–17.8; Black–17.3; Other–16.0

Figure 1. Kaplan-Meier Survival Estimates for DR and STDR.

For STDR, in the T2DM newly diagnosed group there was slight separation of survival probabilities overtime by different ethnic group (log rank test; $p < 0.001$), but there was a clear separation of survival probabilities among people with prevalent T2DM (log rank test; $p < 0.001$) with White ethnic groups having the highest survival compared with all other ethnic groups.

3.4. DR/STDR Events and Incidence Rate

Table 3 shows the number of incident cases of DR and STDR during the follow-up period and the incidence rate in each cohort. Among people with newly diagnosed T2DM at baseline, a total of 8638 people had DR events and 557 people had STDR events. For DR, White ethnic groups had the lowest incidence rate (75.2 per 1000 person years) and Black ethnic groups had the highest incidence rate (87.5 per 1000 person years). For STDR, White groups had the lowest incidence rate (3.68 per 1000 person years) and Black ethnic groups had the highest incidence rate of 5.66 per 1000 person years.

Among people with prevalent T2DM at baseline, a total of 12,124 had DR events and 2200 people had STDR events. For DR, White ethnic groups reported the lowest incidence rate (113.7 per 1000 person years) and South Asians reported the highest incidence rate of 150.8 per 1000 person years. In relation to STDR, White ethnic groups had the lowest incidence rate of 12.0 per 1000 person years and South Asian had the highest incidence rate of 17.8 per 1000 person years.

3.5. Multivariable Analysis and Ethnic Variation in Incident DR/STDR

Table 4 shows results for the multivariable Cox regression analysis for DR and STDR outcomes. Ethnicity was significantly associated with DR and STDR in both groups of people with newly diagnosed T2DM and prevalent T2DM at baseline after adjustment for all the covariates considered, which include age, gender, deprivation, BMI, Hba1c, SBP, Total Cholesterol, eGFR, history of cardiovascular disease, history of antidiabetic, antihypertensive and statin medication use.

Table 4. Multivariable Cox regression analysis of outcome measures in newly diagnosed diabetes—Newly diagnosed T2DM Cases.

	DR		STDR	
	Adjusted HR	*p* **Value**	**Adjusted HR**	*p* **Value**
Ethnic Group				
White (ref)	1		1	
South Asian				
Indian	1.10 (1.01–1.19)	0.029	0.86 (0.60–1.21)	0.377
Pakistani	1.14 (1.03–1.25)	0.011	1.36 (0.95–1.94)	0.094
Bangladeshi	1.02 (0.95–1.09)	0.733	0.80 (0.61–1.05)	0.107
Black				
Caribbean	1.07 (0.98–1.17)	0.178	1.18 (0.85–1.65)	0.315
African	1.16 (1.07–1.26)	<0.001	1.36 (1.02–1.83)	0.039
Mixed and other				
Mixed	1.02 (0.86–1.21)	0.878	1.37 (0.76–2.48)	0.296
Other	1.10 (1.03–1.19)	0.012	1.26 (0.95–1.67)	0.108
Age at Study Entry, Years				
<45	1.21 (1.09–1.35)	0.001	0.84 (0.56–1.25)	0.392
45–54	1.15 (1.04–1.27)	0.011	0.98 (0.67–1.43)	0.897
55–64	1.10 (0.99–1.22)	0.089	0.79 (0.54–1.17)	0.236
65–74	0.99 (0.89–1.10)	0.728	0.93 (0.63–1.38)	0.735
75+	1		1	
Gender				
Female	1		1	
Male	1.12 (1.07–1.17)	<0.001	1.22 (1.02–1.47)	0.001

Table 4. *Cont.*

Townsend Score (quintiles)				
1 (affluent)	1		1	
2	0.93 (0.88–1.00)	0.028	0.83 (0.63–1.09)	0.176
3	0.95 (0.88–1.01)	0.072	1.05 (0.80–1.38)	0.724
4	0.95 (0.89–1.02)	0.105	1.24 (0.96–1.62)	0.104
5 (deprived)	0.90 (0.84–0.97)	0.002	1.18 (0.90–1.54)	0.229
Not recorded	0.58 (0.38–0.90)	0.015	0.51 (0.07–3.63)	0.497
Body Mass Index (kg/m^2)				
<18.5	1		1	
18.5–25	1.00 (0.81–1.24)	0.97	0.66 (0.37–1.17)	0.157
25–30	0.90 (0.73–1.12)	0.322	0.47 (0.27–0.82)	0.008
≥30	0.87 (0.70–1.07)	0.172	0.32 (0.18–0.57)	<0.001
Not recorded	0.73 (0.58–0.91)	0.005	0.24 (0.12–0.48)	<0.001
HbA1c (mmol/mol)				
<50	1		1	
50–99	1.19 (1.12–1.26)	<0.001	1.43 (1.10–1.84)	0.007
≥100	1.70 (1.57–1.85)	<0.001	3.68 (2.73–4.95)	<0.001
Not recorded	1.07 (0.98–1.16)	0.143	1.31 (0.92–1.87)	0.13
Systolic Blood Pressure—SBP (mmHg)				
<120	1		1	
120–129	1.05 (0.98–1.12)	0.237	1.13 (0.85–1.51)	0.385
130–140	1.07 (1.01–1.15)	0.05	1.23 (0.93–1.62)	0.147
≥140	1.25 (1.17–1.33)	<0.001	1.88 (1.45–2.44)	<0.001
Not recorded	0.95 (0.80–1.14)	0.551	0.98 (0.47–2.02)	0.947
Total Cholesterol (mmol/L)				
<5.2	1		1	
5.2- 6.1	0.97 (0.92–1.02)	0.195	0.83 (0.67–1.02)	0.081
≥6.2	1.01 (0.95–1.08)	0.795	0.94 (0.75–1.19)	0.622
Not recorded	0.91 (0.80–1.03)	0.13	1.37 (0.89–2.11)	0.149
eGFR (mL/min/1.73m^2)				
≥60	1		1	
<60	1.09 (1.00–1.19)	0.057	1.32 (0.97–1.79)	0.077
Not recorded	1.00 (0.90–1.12)	0.968	1.20 (0.80–1.79)	0.375
Cardiovascular Disease History—ever				
No	1		1	
Yes	1.05 (0.97–1.13)	0.304	1.06 (0.77–1.46)	0.722
Antidiabetic Drugs History—ever Closest Record to Baseline				
No drug	1		1	
One drug	0.97 (0.93–1.02)	0.142	0.97 (0.81–1.16)	0.725
History of Antihypertensive Medication				
No	1		1	
Yes	1.02 (0.97–1.07)	0.495	0.77 (0.63–0.93)	0.008
Statin History				
No	1		1	
Yes	1.12 (1.07–1.18)	<0.001	1.20 (0.99–1.46)	0.066

Abbreviations: eGFR-Estimated glomerular filtration rate. 'Other' group: this group included Chinese, other Asian, other Black and people belong to any other ethnicity.

Among people with newly diagnosed T2DM, Indian, Pakistani, African and 'Other' ethnic groups had significantly higher risk of DR compared to White ethnic groups with African having the highest risk (HR: 1.16 95% CI 1.07–1.26) (Figure 2a). However, in relation to STDR, only Africans (HR: 1.36; 95% CI 1.02–1.83) showed an increased risk compared to White ethnic groups (Figure 2b).

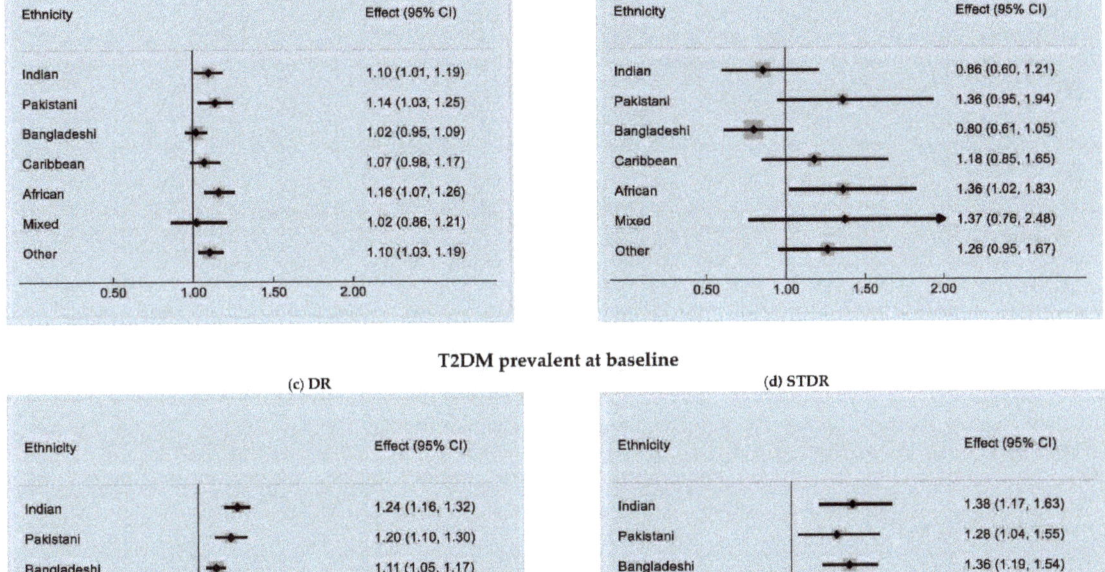

Figure 2. Risk of DR and STDR in ethnic minorities compared to white population.

Among those with prevalent T2DM at baseline, Indian, Pakistani, Bangladeshi and Caribbean ethnic groups showed an increased risk of DR compared to White ethnic groups with Indian having the highest risk (HR: 1.24 95% CI 1.16–1.32) (Figure 2c). In the same group of people, Indian, Pakistani, Bangladeshi, Caribbean and 'other' ethnic groups showed a significantly higher risk of STDR compared to White ethnic groups with Indians having the highest risk of STDR (HR: 1.38 95% CI 1.17–1.63).

3.6. Other Risk Factors Contributing to DR and STDR

There were several other risk factors that were significantly associated with DR and STDR after adjustment for all the covariates considered in this study as given Tables 4 and 5.

Table 5. Multivariable Cox regression analysis of outcome measures in people with known diabetes—KnownT2DM Cases.

	DR		STDR	
	Adjusted HR	*p* **Value**	**Adjusted HR**	*p* **Value**
Ethnic Group				
White (ref)	1		1	
South Asian				
Indian	1.24 (1.16–1.33)	<0.001	1.39 (1.18–1.63)	<0.001
Pakistani	1.20 (1.11–1.30)	<0.001	1.28 (1.05–1.55)	0.016
Bangladeshi	1.11 (1.05–1.17)	<0.001	1.36 (1.19–1.55)	<0.001
Black				
Caribbean	1.09 (1.01–1.16)	0.02	1.22 (1.04–1.43)	0.016
African	1.06 (0.98–1.14)	0.145	1.10 (0.92–1.33)	0.294
Mixed and Other				
Mixed	1.10 (0.96–1.26)	0.19	1.17 (0.85–1.61)	0.328
Other	1.03 (0.97–1.11)	0.335	1.25 (1.07–1.47)	0.006
Duration of Diabetes				
0 to <2 years (ref)	1		1	
2 to <5 years	1.17 (1.10–1.24)	<0.001	1.18 (0.99–1.40)	0.059
5 to <10 years	1.37 (1.29–1.45)	<0.001	1.40 (1.18–1.66)	<0.001
≥10 years	1.83 (1.71–1.96)	<0.001	2.46 (2.06–2.90)	<0.001
Age at Study Entry, Years				
<45	1.52 (1.40–1.66)	<0.001	1.68 (1.37–2.05)	<0.001
45–54	1.51 (1.40–1.62)	<0.001	1.91 (1.60–2.29)	<0.001
55–64	1.32 (1.23–1.42)	<0.001	1.41 (1.18–1.68)	<0.001
65–74	1.15 (1.07–1.23)	<0.001	1.35 (1.14–1.61)	0.001
75+	1		1	
Gender				
Female	1		1	
Male	1.10 (1.06–1.14)	<0.001	1.15 (1.06–1.26)	0.001
Townsend Score (quintiles)				
1	1		1	
2	0.95 (0.89–1.00)	0.067	1.06 (0.91–1.22)	0.495
3	0.97 (0.91–1.02)	0.247	1.22 (1.06–1.40)	0.005
4	0.91 (0.86–0.97)	0.003	1.22 (1.06–1.40)	0.005
5	0.95 (0.89–1.00)	0.088	1.30 (1.13–1.49)	<0.001
Not recorded	0.86 (0.58–1.28)	0.462	0.46 (0.11–1.88)	0.286
Body Mass Index (kg/m^2)				
<18.5	1		1	
18.5–25	1.03 (0.88–1.20)	0.703	0.94 (0.66–1.35)	0.756
25–30	1.00 (0.86–1.18)	0.94	0.89 (0.63–1.27)	0.551
≥30	0.91 (0.78–1.07)	0.261	0.76 (0.53–1.08)	0.132
Not recorded	0.96 (0.81–1.13)	0.63	1.06 (0.74–1.53)	0.72
HbA1c (mmol/mol)				
<50	1		1	
50–99	1.29 (1.23–1.36)	<0.001	1.62 (1.40–1.87)	<0.001
≥100	1.83 (1.69–1.98)	<0.001	3.20 (2.68–3.83)	<0.001
Not recorded	1.00 (0.92–1.09)	0.905	1.53 (1.25–1.88)	<0.001
Systolic Blood Pressure—SBP (mmHg)				
<120	1		1	
120–129	1.12 (1.05–1.18)	<0.001	1.26 (1.09–1.45)	0.001
130–140	1.16 (1.10–1.23)	<0.001	1.34 (1.17–1.55)	<0.001
≥140	1.28 (1.21–1.35)	<0.001	1.80 (1.57–2.05)	<0.001
Not recorded	0.97 (0.84–1.13)	0.709	1.17 (0.83–1.63)	0.356

Table 5. *Cont.*

Total Cholesterol (mmol/L)				
<5.2	1		1	
5.2–6.1	0.96 (0.91–1.02)	0.197	1.01 (0.88–1.15)	0.871
≥6.2	0.96 (0.89–1.03)	0.233	1.04 (0.88–1.24)	0.593
Not recorded	0.99 (0.91–1.07)	0.857	1.07 (0.89–1.27)	0.473
eGFR (mL/min/1.73m^2)				
≥60	1		1	
<60	1.10 (1.04–1.17)	0.001	1.35 (1.19–1.54)	<0.001
Not recorded	1.05 (0.98–1.14)	0.178	1.10 (0.93–1.30)	0.279
History of Cardiovascular Disease				
No	1		1	
Yes	0.98 (0.93–1.04)	0.584	0.96 (0.85–1.08)	0.514
History of Antidiabetic Drugs History-				
No drug	1		1	
One drug	1.01 (0.94–1.07)	0.854	0.82 (0.68–0.98)	0.035
Two drugs	1.26 (1.18–1.34)	<0.001	1.36 (1.15–1.60)	<0.001
Insulin	1.77 (1.65–1.89)	<0.001	2.72 (2.30–3.21)	<0.001
History of Antihypertensives				
No	1		1	
Yes	0.99 (0.95–1.04)	0.752	1.00 (0.89–1.12)	0.993
History of Statin				
No	1		1	
Yes	1.07 (1.01–1.13)	0.008	0.96 (0.85–1.08)	0.519

Abbreviations: eGFR- Estimated glomerular filtration rate. 'Other' group: this group included Chinese, other Asian, other Black and people belong to any other ethnicity.

Among people with newly diagnosed T2DM at baseline, those who were in age <55 had significantly higher risk of DR compared to those who were aged 75+ (Table 4). Men were more likely to have DR and STDR compared to women with 12% (HR: 1.12 95% CI 1.07–1.17) and 22% (HR: 1.22 95% CI 1.02–1.47) increased risk, respectively. Those with HbA1c ≥ 100mmol/mol had highest risk of DR (HR:1.70 95% CI 1.57–1.85) and STDR (HR: 3.68 95% CI 2.73–4.95) compared to those with HbA1c < 50 mmol/mol. People with SBP ≥ 140mmHg had higher risk of DR (HR: 1.25 95% CI 1.17–1.33) and STDR (HR: 1.88 95% CI 1.45–2.44) compared to those with SBP < 120 mmHg. Total cholesterol, eGFR, CVD history and history of one antidiabetic drug use were not significantly associated with the risk of DR or STDR.

Among people with prevalent T2DM at baseline, those who had duration of ≥10 years diabetes had the highest risk of any DR (HR: 1.83 95% CI 1.71–1.96) and STDR (HR: 2.46 95% CI 2.06–2.90) compared to those with 0 to 2 years of duration (Table 5). People aged 45–54 years had the highest risk of DR and STDR compared to those who are aged 75+. Men had 10% (HR: 1.10 95% CI 1.06–1.14) and 15% (HR: 1.15 95% CI 1.06–1.26) increased risk of DR and STDR compared to women, respectively. Those with HbA1c ≥ 50 mmol/mol, SBP ≥ 120 mmHg, eGFR < 60 mL/min/1.73m^2 had higher risk of DR and STDR compared to those in HbA1c < 50 mmol/mol, SBP < 120 mmHg and eGFR ≥ 60 mL/min/1.73m^2, respectively. BMI, total cholesterol, CVD history and history of antihypertensive drugs were not significantly associated with the risk of DR or STDR.

4. Discussion

We investigated the ethnic differences in developing DR and STDR among people with newly diagnosed and prevalent T2DM using a primary care dataset from East London in which more than 98% of people with T2DM had complete data on ethnicity. From this dataset, over 58,000 people were eligible to be included in this study and among people with newly diagnosed T2DM, we observed incidence rates of 82.8 and 4.2 events per 1000 person years for DR and STDR, respectively. Among people with prevalent T2DM, the incidence rates for DR and STDR were 135.4 and 16.0 events per 1000 person years, respectively. These rates varied between different ethnic groups with higher incidence rates among South Asians and Africans compared to White ethnic groups. South Asian and Black ethnic groups had increased risk of any DR and STDR compared to White ethnic groups even after adjustment for age, gender, deprivation, BMI, Hba1c, SBP, total Cholesterol, eGFR, history of cardiovascular disease, history of antidiabetic drugs, antihypertensive and statins. Among people with newly diagnosed T2DM, Africans were 36% more likely to have STDR compared to White ethnic groups (HR: 1.36 95% CI 1.02–1.83). Among people with prevalent T2DM, Indians had the highest risk of any DR with 24% increased risk (HR: 1.24 95% CI 1.16–1.32) and STDR with 38% (HR: 1.38 95% CI 1.17–1.63) increased risk compared with White ethnic groups after adjusting for all the covariates considered. Overall, people with prevalent T2DM group with a mean duration 7.4 years at baseline showed stronger association between ethnicity and DR/STDR compared to the group of people with newly diagnosed T2DM diabetes at baseline.

In existing literature, there is only a small number of studies on incidence of DR and STDR, and most of them do not consider ethnicity as a possible risk factor. These studies included a UK study conducted among 1919 people with newly diagnosed diabetes and reported that 22% of the people developed retinopathy by the end of their follow-up period of 6 years [20]. The Liverpool Diabetes Eye Study was conducted among 4770 newly diagnosed T2DM people and 3.9% people developed STDR by the end of their 5 year follow-up period [21]. In another study with 16,444 people with T2DM, 16.4% developed DR and 2.7% people developed STDR after 10 years of follow-up [22]. Two recent studies by McKay eta al [23,24] also reported DR and STDR incidence among people with T2DM using the CPRD dataset. The first study reported 8263 (13.8%) of DR cases and 832 (1.4%) STDR cases during the 3.5 year and 3.8 year follow up periods, respectively. This was equivalent to an incidence rate of 39.2 cases per 1000 person years for DR and 3.6 cases per 1000 person years for STDR [24]. The second study assessed incidence of STDR among those with mild non-proliferative DR and a total of 1037 (5.5%) people developed STDR over a mean follow-up period of 3.6 (SD 2.0) years [23]. Although it is challenging to compare these studies, the event rates of both DR and STDR were higher in our study among both groups of people with newly diagnosed as well as prevalent diabetes at baseline. This is likely to be due to the higher proportion of minority ethnic group in our study population who have a higher risk of diabetes in general [7–9] as well as higher risk of DR as shown in this study.

A study by Mathur et al. used CPRD data with a nationally representative sample of people with T2DM in the UK and reported the difference of DR and STDR incidence by ethnicity. According to this study, there was no statistically significant difference in incidence of DR among ethnic minority groups compared to White ethnic groups, however in relation to STDR, South Asians had significantly higher risk compared to White ethnic groups (HR: 1.25 95% CI 1.00–1.56) after adjustment for age, duration of diabetes, gender, deprivation and UK region [4]. Even though the CPRD study had assessed differences in DR and STDR incidence by ethnicity, one of the main limitations of this dataset was poor representation of minor ethnic groups due to lack of recording of ethnicity in CPRD data.

When we consider prevalence studies, higher prevalence of DR among South Asians and Africans have been observed compared to White ethnic groups [25]. A study that aimed to provide estimates of visual impairment (including DR) among people attending DR screening in the UK in 2012, showed that South Asians (OR 1.10, 95% CI 1.02 to 1.18) and Black ethnic groups (OR 1.79, 95CI 1.70 to 1.89) have higher prevalence of visual impairment compared to White groups [25]. Another study from the UK on DR screening has identified younger age, social deprivation, ethnicity and duration of diabetes as independent risk factor of non-attendance, and referable retinopathy confirmed the same association of increase risk of referable DR among Asians (OR 1.57, 95%CI 1.43 to 1.72) and Black (OR 1.65, 95% CI 1.49–1.83) ethnic groups compared to White groups even after adjusting for age, gender, deprivation, diabetes duration and type [26]. All these findings suggest that extra efforts should be made to identify and treat STDR in these ethnic groups and to ensure that recording of self-reported ethnic group is implemented. It is now a contractual obligation in the 2020 general practice contract and is also a requirement of hospital data [27,28]. East London general practitioners have successfully prioritised recording of self-reported ethnic group over three decades [29].

Previous studies have suggested that higher prevalence of DR and STDR among ethnic minorities is likely to be due to suboptimal control of risk factors, delayed attendance at DR screening programmes and poor treatment outcomes compared to White ethnic groups in the UK [30]. However, the role of ethnicity as an independent risk factor in developing DR and STDR is less well-researched mainly due to the lack of data from cohorts representing the three main ethnic groups in the UK. In this study, we've adjusted for all known risk factors that are found to be associated with DR and STDR and found that ethnicity is an independent risk factor for incident DR and STDR with higher risk among ethnic minorities. The initial univariable analysis showed that higher proportion of South Asians were younger, affluent and had lower blood pressure compared to other groups. Despite these characteristics of South Asians at study baseline, adjustment for all relevant covariates in the multivariable analysis revealed that they have a higher risk of DR and STDR compared to Whites and other groups.

We also ensured generalisability of our study results by utilising routinely collected data from anonymised electronic datasets of patients in the NHS in the UK. The NHS is also free to all at the point of delivery, reducing inequalities in accessing healthcare. Another strength of our study is that the primary care datasets we used also had more than 98% of completeness in relation to self-reported ethnicity recording. This routinely collected primary care data set is regularly updated and therefore can be used to provide timely information on demographic makeup of the general population in the relevant area. Moreover, the data used in this study was collected prospectively and therefore the data are less likely to be affected by recall bias, or observer bias which could be issues in survey data or retrospectively collected data.

Routinely collected primary care data only provides what is recorded by the clinician and therefore, in this dataset some of the data might be missing if they were not recorded. Data recording practices in general practice settings may also vary based on different financial incentivisation packages. In addition, incomplete data and coding errors can occur. However, we have addressed these issues by checking the data for any implausible values and excluding them and also having missing data as a separate category within the data analysis to keep any bias to a minimum.

5. Conclusions

Our study illustrates the higher risk of DR and STDR in people with T2DM from ethnic minority groups. These groups also have a higher prevalence of T2DM, further increasing their risk of DR and STDR compared to the White population. Retinopathy is another adverse health outcome therefore that is more common in people with T2DM from ethnic minority groups. Our findings illustrate the importance of improving the prevention, early diagnosis and management of T2DM in the UK to reduce the burden of ill-health from retinopathy and the other adverse outcomes of T2DM, particularly in people from ethnic minority groups.

Author Contributions: M.D.N. and S.S. had the idea for and designed the study and had full access to all of the data in the study and take responsibility for the integrity of the data and the accuracy of the data analysis. M.D.N. and S.G. did the analysis, and all authors including A.T.P., W.H. and A.M. critically revised the manuscript for important intellectual content and gave final approval for the version to be published. M.D.N., S.G., R.M. and J.R. collected the data. All authors agree to be accountable for all aspects of the work in ensuring that questions related to the accuracy or integrity of any part of the work are appropriately investigated and resolved. All authors have read and agreed to the published version of the manuscript.

Funding: This study was part of the ORNATE India project which was funded by the GCRF UKRI (MR/P027881/1). The research is supported by the NIHR Biomedical Research Centre at Moorfields Eye Hospital NHS Foundation Trust and UCL Institute of Ophthalmology. AM is supported by the NIHR Applied Research Collaboration (ARC) NW London and is an NIHR Senior Investigator. The views expressed in this publication are those of the authors and not necessarily those of the NHS, NIHR or the Department of Health and Social Care. RM is supported by a Sir Henry Wellcome Postdoctoral fellowship from the Wellcome Trust (201375/Z/16/Z).

Institutional Review Board Statement: The Moorfields Research Management Committee approved the use of this fully anonymised UK dataset for model development and validation (SIVS1047) and further ethics approval was not required. Approval was also obtained from the Caldicott guardian of this anonymised dataset in Queen Mary University London (QMUL). Individual level patient consent was not obtained, as this was an observational study using de-identified data.

Informed Consent Statement: Not applicable.

Data Availability Statement: Restrictions apply to the availability of this data. This study was conducted using de-identified data from general practice electronic health records collected in three Clinical Commissioning Groups (CCGs) in East London and data are available with permission from the Caldicott guardian of this anonymised dataset in Queen Mary University London (QMUL).

Acknowledgments: We thank the Clinical Effectiveness Group QMUL for giving access to the data which was provided from routinely collected data from general practitioner electronic health records. Grateful also to the collaboration of the general practitioners and their patients who provided access to high quality, deidentified information including self-reported ethnicity.

Conflicts of Interest: The authors declare no conflict of interest.

References

1. World Health Organization. Global Report on Diabetes 2016. 10 March 2019. Available online: https://www.who.int/publications/i/item/9789241565257 (accessed on 27 June 2021).
2. International Diabetes Federation. *IDF Diabetes Atlas*, 9th ed.; IDF: Brussels, Belgium, 2019.
3. The Royal College of Ophthalmologists. Diabetic Retinopathy Guidelines 2012. 10 March 2019. Available online: https://www.rcophth.ac.uk/wp-content/uploads/2014/12/2013-SCI-301-FINAL-DR-GUIDELINES-DEC-2012-updated-July-2013.pdf (accessed on 26 June 2021).
4. Mathur, R.; Bhaskaran, K.; Edwards, E.; Lee, H.; Chaturvedi, N.; Smeeth, L.; Douglas, I. Population trends in the 10-year incidence and prevalence of diabetic retinopathy in the UK: A cohort study in the Clinical Practice Research Datalink 2004–2014. *BMJ Open* **2017**, *7*, e014444. [CrossRef] [PubMed]
5. Thomas, R.L.; Dunstan, F.D.; Luzio, S.; Chowdhury, S.R.; North, R.; Hale, S.L.; Gibbins, R.L.; Owens, D.R. Prevalence of diabetic retinopathy within a national diabetic retinopathy screening service. *Br. J. Ophthalmol.* **2014**, *99*, 64–68. [CrossRef]
6. Office for National Statistics. Population of England and Wales 7 August 2020. Available online: https://www.ethnicity-facts-figures.service.gov.uk/uk-population-by-ethnicity/national-and-regional-populations/population-of-england-and-wales/latest (accessed on 26 June 2021).
7. Goff, L.M. Ethnicity and Type 2 diabetes in the UK. *Diabet. Med.* **2019**, *36*, 927–938. [CrossRef]
8. Pham, T.M.; Carpenter, J.R.; Morris, T.P.; Sharma, M.; Petersen, I. Ethnic Differences in the Prevalence of Type 2 Diabetes Diagnoses in the UK: Cross-Sectional Analysis of the Health Improvement Network Primary Care Database. *Clin. Epidemiol.* **2019**, *11*, 1081–1088. [CrossRef] [PubMed]
9. Whicher, C.A.; O'Neill, S.; Holt, R.I.G. Diabetes in the UK: 2019. *Diabet. Med.* **2020**, *37*, 242–247. [CrossRef] [PubMed]
10. Emanuele, N.; Sacks, J.; Klein, R.; Reda, D.; Anderson, R.; Duckworth, W.; Abraira, C. For the Veterans Affairs Diabetes Trial Group Ethnicity, Race, and Baseline Retinopathy Correlates in the Veterans Affairs Diabetes Trial. *Diabetes Care* **2005**, *28*, 1954–1958. [CrossRef]
11. Sivaprasad, S.; Gupta, B.; Crosby-Nwaobi, R.; Evans, J. Prevalence of Diabetic Retinopathy in Various Ethnic Groups: A Worldwide Perspective. *Surv. Ophthalmol.* **2012**, *57*, 347–370. [CrossRef] [PubMed]
12. Tan, G.S.; Gan, A.; Sabanayagam, C.; Tham, Y.C.; Neelam, K.; Mitchell, P.; Wang, J.J.; Lamoureux, E.L.; Cheng Ch Wong, T.Y. Ethnic Differences in the Prevalence and Risk Factors of Diabetic Retinopathy: The Singapore Epidemiology of Eye Diseases Study. *Ophthalmology* **2018**, *125*, 529–536. [CrossRef]
13. Sivaprasad, S.; Gupta, B.; Gulliford, M.; Dodhia, H.; Mohamed, M.; Nagi, D.; Evans, J.R. Ethnic Variations in the Prevalence of Diabetic Retinopathy in People with Diabetes Attending Screening in the United Kingdom (DRIVE UK). *PLoS ONE* **2012**, *7*, e32182. [CrossRef]

14. Raymond, N.T.; Varadhan, L.; Reynold, D.R.; Bush, K.; Sankaranarayanan, S.; Bellary, S.; Barnett, A.H.; Kumar, S.; O'Hare, J.P. UK Asian Diabetes Study Retinopathy Study Group. Higher prevalence of retinopathy in diabetic patients of South Asian ethnicity compared with white Euro-peans in the community: A cross-sectional study. *Diabetes Care* **2009**, *32*, 410–415. [CrossRef]

15. Nwosu, S.N. Prevalence and pattern of retinal diseases at the Guinness Eye Hospital, Onitsha, Nigeria. *Ophthalmic Epidemiol.* **2000**, *7*, 41–48. [CrossRef]

16. Leske, M.C.; Wu, S.-Y.; Nemesure, B.; Hennis, A. Causes of visual loss and their risk factors: An incidence summary from the Barbados Eye Studies. *Rev. Panam. Salud Pública* **2010**, *27*, 259–267. [CrossRef] [PubMed]

17. Haider, S.; Sadiq, S.N.; Moore, D.; Price, M.J.; Nirantharakumar, K. Prognostic prediction models for diabetic retinopathy progression: A systematic review. *Eye* **2019**, *33*, 702–713. [CrossRef] [PubMed]

18. Van der Heijden, A.A.; Nijpels, G.; Badloe, F.; Lovejoy, H.L.; Peelen, L.; Feenstra, T.; Moons, K.; Sliecker, R.; Herings, R.; Elders, P.; et al. Prediction models for development of retinopathy in people with type 2 diabetes: Systematic review and external validation in a Dutch primary care setting. *Diabetologia* **2020**, *63*, 1110–1119. [CrossRef] [PubMed]

19. Public Health England. NHS Diabetic Eye Screening Programme-Overview of Patient Pathway, Grading Pathway, Surveillance Pathways and Referral Pathways; 2017. Available online: https://assets.publishing.service.gov.uk/government/uploads/system/uploads/attachment_data/file/648658/Diabetic_Eye_Screening_pathway_overviews.pdf (accessed on 27 June 2021).

20. Stratton, I.; Kohner, E.M.; Aldington, S.; Turner, R.C.; Holman, R.R.; Manley, S.E.; Matthews, D.R. UKPDS 50: Risk factors for incidence and progression of retinopathy in Type II diabetes over 6 years from diagnosis. *Diabetologia* **2001**, *44*, 156–163. [CrossRef] [PubMed]

21. Younis, N.; Broadbent, D.M.; Vora, J.P.; Harding, S. Incidence of sight-threatening retinopathy in patients with type 2 diabetes in the Liverpool Diabetic Eye Study: A cohort study. *Lancet* **2003**, *361*, 195–200. [CrossRef]

22. Jones, C.D.; Greenwood, R.H.; Misra, A.; Bachmann, M.O. Incidence and Progression of Diabetic Retinopathy During 17 Years of a Population-Based Screening Program in England. *Diabetes Care* **2012**, *35*, 592–596. [CrossRef]

23. McKay, A.J.; Gunn, L.H.; Nugawela, M.D.; Sathish, T.; Majeed, A.; Vamos, E.; Molina, G.; Sivaprasad, S. Associations between attainment of incentivized primary care indicators and incident sight-threatening diabetic retinopathy in England: A population-based historical cohort study. *Diabetes Obes. Metab.* **2021**, *23*, 1322–1330. [CrossRef]

24. McKay, A.J.; Gunn, L.H.; Sathish, T.; Vamos, E.; Nugawela, M.; Majeed, A.; Molina, G.; Sivaprasad, S. Associations between attainment of incentivised primary care indicators and incident diabetic retinopathy in England: A population-based historical cohort study. *BMC Med.* **2021**, *19*, 1–10. [CrossRef]

25. Sivaprasad, S.; Gupta, B.; Gulliford, M.C.; Dodhia, H.; Mann, S.; Nagi, D.; Evans, J. Ethnic Variation in the Prevalence of Visual Impairment in People Attending Diabetic Retinopathy Screening in the United Kingdom (DRIVE UK). *PLoS ONE* **2012**, *7*, e39608. [CrossRef]

26. Lawrenson, J.G.; Bourmpaki, E.; Bunce, C.; Stratton, I.M.; Gardner, P.; Anderson, J. EROS Study Group Trends in diabetic retinopathy screening attendance and associations with vision impairment attributable to diabetes in a large nationwide cohort. *Diabet. Med.* **2021**, *38*, e14425. [CrossRef] [PubMed]

27. National Health Service. The National Health Service (General Medical Services Contracts and Personal Medical Services Agreements) (Amendment) (No. 3) Regulations 2020. 2020. Available online: https://www.legislation.gov.uk/uksi/2020/1415/made (accessed on 27 July 2021).

28. Nuffield Trust. Ethnicity Coding in English Health Service Datasets. 10/12/2020. 2021. Available online: https://www.nuffieldtrust.org.uk/project/ethnicity-coding-in-english-health-service-datasets#timelines (accessed on 10 December 2020).

29. Hull, S.; Mathur, R.; Boomla, K.; Chowdhury, T.A.; Dreyer, G.; Alazawi, W.; Robson, J. Research into practice: Understanding ethnic differences in healthcare usage and outcomes in general practice. *Br. J. Gen. Pr.* **2014**, *64*, 653–655. [CrossRef] [PubMed]

30. Mastropasqua, R.; Luo, Y.H.-L.; Cheah, Y.S.; Egan, C.; Lewis, J.; Da Cruz, L. Black patients sustain vision loss while White and South Asian patients gain vision following delamination or segmentation surgery for tractional complications associated with proliferative diabetic retinopathy. *Eye* **2017**, *31*, 1468–1474. [CrossRef] [PubMed]

Journal of
Personalized
Medicine

Article

Accuracy of a Machine-Learning Algorithm for Detecting and Classifying Choroidal Neovascularization on Spectral-Domain Optical Coherence Tomography

Andreas Maunz [1,*], Fethallah Benmansour [1], Yvonna Li [1], Thomas Albrecht [1], Yan-Ping Zhang [1], Filippo Arcadu [1], Yalin Zheng [2,3], Savita Madhusudhan [2,3] and Jayashree Sahni [1]

1   Pharma Research and Early Development, Roche Innovation Center, F. Hoffmann-La Roche Ltd., 4070 Basel, Switzerland; fethallah.benmansour@roche.com (F.B.); yvonna.li@roche.com (Y.L.); tom.albrecht@roche.com (T.A.); yan-ping.zhang_schaerer@roche.com (Y.-P.Z.); filippo.arcadu@roche.com (F.A.); jayashreesahni@yahoo.co.uk (J.S.)
2   Department of Eye and Vision Science, University of Liverpool, Liverpool L7 8XP, UK; Yalin.Zheng@liverpool.ac.uk (Y.Z.); Savita.Madhusudhan@liverpoolft.nhs.uk (S.M.)
3   Liverpool Ophthalmic Reading Centre (NetwORC, UK), St. Paul's Eye Unit, Royal Liverpool University Hospital, Liverpool L7 8XP, UK
*   Correspondence: andreas.maunz@roche.com

Citation: Maunz, A.; Benmansour, F.; Li, Y.; Albrecht, T.; Zhang, Y.-P.; Arcadu, F.; Zheng, Y.; Madhusudhan, S.; Sahni, J. Accuracy of a Machine-Learning Algorithm for Detecting and Classifying Choroidal Neovascularization on Spectral-Domain Optical Coherence Tomography. *J. Pers. Med.* **2021**, *11*, 524. https://doi.org/10.3390/jpm11060524

Academic Editor: Margaret M. DeAngelis

Received: 21 April 2021
Accepted: 7 June 2021
Published: 8 June 2021

Abstract: Background: To evaluate the performance of a machine-learning (ML) algorithm to detect and classify choroidal neovascularization (CNV), secondary to age-related macular degeneration (AMD) on spectral-domain optical coherence tomography (SD-OCT) images. Methods: Baseline fluorescein angiography (FA) and SD-OCT images from 1037 treatment-naive study eyes and 531 fellow eyes, without advanced AMD from the phase 3 HARBOR trial (NCT00891735), were used to develop, train, and cross-validate an ML pipeline combining deep-learning–based segmentation of SD-OCT B-scans and CNV classification, based on features derived from the segmentations, in a five-fold setting. FA classification of the CNV phenotypes from HARBOR was used for generating the ground truth for model development. SD-OCT scans from the phase 2 AVENUE trial (NCT02484690) were used to externally validate the ML model. Results: The ML algorithm discriminated CNV absence from CNV presence, with a very high accuracy (area under the receiver operating characteristic [AUROC] = 0.99), and classified occult versus predominantly classic CNV types, per FA assessment, with a high accuracy (AUROC = 0.91) on HARBOR SD-OCT images. Minimally classic CNV was discriminated with significantly lower performance. Occult and predominantly classic CNV types could be discriminated with AUROC = 0.88 on baseline SD-OCT images of 165 study eyes, with CNV from AVENUE. Conclusions: Our ML model was able to detect CNV presence and CNV subtypes on SD-OCT images with high accuracy in patients with neovascular AMD.

Keywords: age-related macular degeneration; choroidal neovascularization; classification; machine learning; optical coherence tomography

1. Introduction

Early detection of active choroidal neovascularization (CNV) is crucial for the timely treatment of neovascular age-related macular degeneration (nAMD), in order to achieve a good outcome [1]. Clinicians are increasingly switching from fluorescein angiography (FA) to optical coherence tomography (OCT) for the diagnosis and management of nAMD, due to the advantages associated with OCT, including being noninvasive, enabling quick acquisition of retinal images with minimum technician training, and providing both qualitative and quantitative information [2–4]. However, an advantage of FA is that it provides information on flow dynamics within the lesion [5], and most importantly, confirms disease activity, characterized by dye leakage. Phenotyping CNVs at baseline on FA, and sometimes additionally on indocyanine green angiography, has long been the standard of care and helps establish the management plan. For example, patients with polypoidal choroidal vasculopathy (PCV) may benefit from combination therapy [6–8]. In clinical trials, this information would help identify subgroups of patients with particularly beneficial outcomes, with novel therapies [9]. Optimal patient stratification may become increasingly important, as multiple combination therapies are poised to enter the market or are in clinical trials, especially in patients who show partial response or non-response to current first-line treatment with antiVEGF agents [10,11].

In OCT, CNV is graded based on its relationship to the retinal pigment epithelial layer [12,13], whereas in FA, en-face flow patterns within the lesion are used to phenotype the CNV [13]. By comparing the two modalities and using an automated approach to evaluate the large quantity of data from three-dimensional SD-OCT volume scans, key features mirroring the flow dynamics and substituting en-face information available in FA could be extracted. Once identified, the impact of novel and existing therapies on these key features could increase our understanding of the disease phenotype, pathophysiology, and specific response to therapy.

Machine learning has the potential to unravel high-dimensional patterns from image data (complex interactions related to a given phenotype), as opposed to features that are correlated only individually to the outcome, providing enhanced capabilities for knowledge extraction. It also provides options for automated screening and diagnosis, enhancing the speed and reproducibility of these processes.

In this study, using the phenotypic CNV definitions derived from FA as the reference standard, we developed a machine learning (ML) model capable of identifying these CNV subtypes, using OCT alone. We present the data on performance of this model for the detection and classification of CNV (as per FA) using the SD-OCT images. In addition, using a sub-symbolic approach, we identified key features on OCT that relate to particular CNV subtypes on FA. To the best of our knowledge, no previous study has leveraged this combination of ML approaches or reported findings similar to those presented in the current study.

2. Materials and Methods

2.1. Participants

This study was a retrospective analysis of prospectively collected baseline FA and SD-OCT images of the study eyes and fellow eyes of patients with nAMD, in the phase 3 HARBOR (NCT00891735) and phase 2 AVENUE (NCT02484690) trials.

The above trials adhered to the tenets of the Declaration of Helsinki, were Health Insurance Portability and Accountability Act compliant, and the protocols were approved by the relevant institutional review boards and ethics committees. Patients provided written informed consent for secondary use of data at enrolment, including future medical research, and additional analyses. In HARBOR, SD-OCT was performed using the Cirrus HD-OCT III instrument (Carl Zeiss Meditec, Dublin, CA, USA) producing $512 \times 128 \times 1024$ voxels with a size of $11.7 \times 47.2 \times 2.0$ μm^3, covering a volume of $6 \times 6 \times 2$ mm^3. In AVENUE, SD-OCT was performed using the Heidelberg Spectralis instrument (Heidelberg Engineering, Heidelberg, Germany). Study design and main outcomes of HARBOR [14,15] and AVENUE [10] have been published previously. In brief, both studies recruited patients with treatment-naive subfoveal CNV secondary to AMD, as diagnosed by a reading center (Digital Angiography Reading Center, Great Neck, NY [DARC]). In AVENUE, patients with juxtafoveal CNV on FA, with a subfoveal component on SD-OCT were also included. Eligibility for both studies was confirmed by the same central reading center (DARC), and the published standard definitions of CNV types have been used in both studies [13].

2.2. Classification of CNV

Both studies allowed recruitment of all CNV types. CNVs were classified at baseline on FA as predominantly classic, minimally classic, or occult, based on the proportion of the occult component within the CNV lesion, as previously described by the Macular Photocoagulation Study (MPS) Group [13].

2.3. Selection of Fellow Eyes

Treatment-naive fellow eyes of patients in HARBOR without advanced AMD, were also included to train the model. Fellow eyes were filtered to exclude those with prior treatment or any type of late-stage AMD.

2.4. OCT Image Processing and Analysis

2.4.1. Retinal Layer Segmentations

Twelve retinal layers (Figure 1; among them inner limiting membrane and retinal pigment epithelium [RPE]) were automatically segmented in all selected SD-OCT volumes using the Iowa reference algorithm [16]. Bruch's membrane was added as the thirteenth layer (convex hull of the RPE) and was computed using scikit-image [17]; thus, it was based on an approximation and not on a real segmentation.

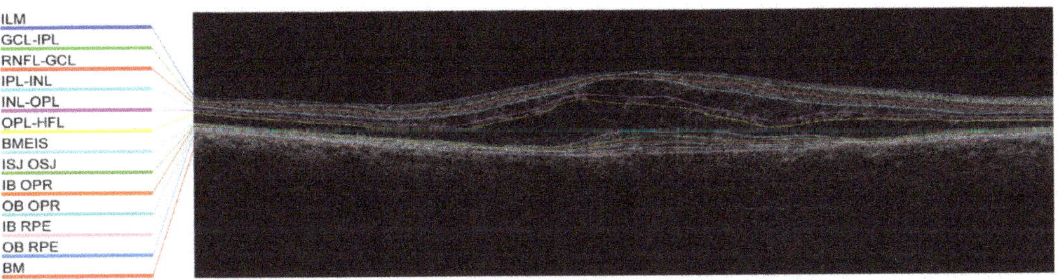

Figure 1. Example of the automated layer segmentation of 13 retinal layers. BM, Bruch's membrane; BMEIS, boundary of myoid and ellipsoid inner segments; GCL-IPL, ganglion cell layer-inner plexiform layer; IB OPR, inner boundary outer photoreceptor; IB RPE, inner boundary retinal pigment epithelium; ILM, internal limiting membrane; INL-OPL, inner nuclear layer-outer plexiform layer; IPL-INL, inner plexiform layer-inner nuclear layer; ISJ OSJ, inner segment/outer segment junction; OB OPR, outer boundary outer photoreceptor; OB RPE, outer boundary retinal pigment epithelium; OPL-HFL, outer plexiform layer-Henle's fiber layer; and RNFL-GCL, retinal nerve fiber layer-ganglion cell layer.

2.4.2. Fluid Annotations

Fluid volumes were annotated by experts from the Liverpool Ophthalmology Reading Center on the B-scan level and subjected to internal quality assurance processes (a subselection of B-scan annotations of each grader was adjudicated and reviewed by a senior clinician). Specifically, a sparse selection of 19 B-scans per volume scan across a total of 50 volume scans (950 B-scans in total), obtained from Cirrus (Carl Zeiss Meditec) OCT machines (see Supplementary Table S1 for definitions used for the annotations), were annotated by drawing contours of the intraretinal fluid (IRF; cystoid spaces), subretinal fluid (SRF), and pigment epithelial detachment (PED), as well as subretinal hyperreflective material (SHRM; a morphological feature seen on OCT as hyperreflective material located external to the neurosensory retina but internal to the RPE [18,19]).

Contours were drawn on the B-scans, stored in the raster format, and then converted to label maps of the original image dimension. The annotations were done using a Matlab software tool developed for the Liverpool reading center.

2.4.3. Fluid Segmentations

The U-Net, a convolutional neural network for biomedical image segmentation [20], was trained to recognize fluids, using the annotated volumes as a training material (pixel level semantic segmentation). All SD-OCT volumes from both HARBOR and AVENUE were segmented with the trained U-Net model, and there was no adaptation of the model to accommodate the specific characteristics of the AVENUE volumes (Heidelberg Spectralis).

2.4.4. Feature Generation

In total, 105 volume and volume-wide thickness descriptors (see Supplementary Table S2 for a detailed feature list), based on the macular subfield definitions provided by the Early Treatment Diabetic Retinopathy Study [21] grid, were automatically extracted from the automated segmentations. See Figure 2 for a sketch of the segmentation and feature extraction pipeline. Specifically, IRF, SRF, and PED, as well as SHRM, were segmented using the segmentation method described above, then reassembled into volumes, where each voxel is mapped to either one of the four segmented targets, or not mapped to any. Finally, descriptors were first individually derived for all B-scans of a volume, and then combined to form C-scan volume measurements.

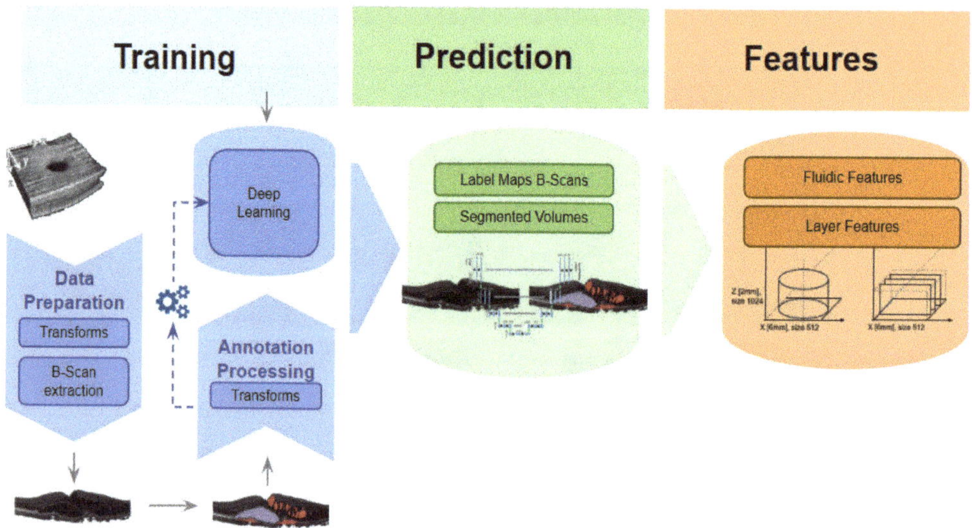

Figure 2. Segmentation pipeline. Sketch of the segmentation pipeline, involving training, prediction, and feature calculation for both fluidic and layer features. See Supplementary Table S2 for a detailed feature list.

2.4.5. Machine Learning

The SD-OCT features were profiled for their utility to predict various binary outcomes (see the Results section for outcome definitions) derived from FA. Cross-validation was used to assess the predictive performance. When no hyperparameter tuning or feature selection was used, no holdout set was put aside and every data point was predicted exactly once. The corresponding performance was reported. This approach was used for all outcomes that reached very high area under the receiver operating characteristic (AUROC) values, namely the CNV presence/absence outcome. Otherwise, a holdout set with 15% of the data was put aside first, and supervised feature elimination and hyperparameter tuning were performed in two stages, using non-nested cross-validation on 85% (Supplementary Figure S1) of the imaging data. A model was established using the best feature sets and the hyperparameter values found, and the predictive performance on the holdout set was subsequently reported.

When classifying predominantly classic versus occult CNV, the tuning process described above was applied to 100% of HARBOR data, and the established model was used to predict predominantly classic and occult CNV on SD-OCT of AVENUE study eyes

(Supplementary Figure S2). The predictions were then compared with the FA labeling in AVENUE for accuracy. See the Supplementary Methods for details on the ML methodology.

2.5. Statistical Analysis

2.5.1. Classification

AUROC was used to measure performance of the models to discriminate two classes from each other. AUROC is a measure of discriminative performance of a binary classifier predicting a numeric score (probability) for class membership. AUROC was obtained by sliding a threshold over the predicted probability to assess the tradeoff between sensitivity and specificity learned by the model. Each point thus obtained was associated with a certain sensitivity and specificity. Reported pairs of sensitivity and specificity corresponded to Youden's cutoff point. The receiver operating characteristic (ROC) curve was plotted with sensitivity and specificity on the y- and x-axes, respectively. The perfect classifier will have sensitivity and specificity = 1, and the ROC curve will pass through the top left corner of the chart, with an AUROC = 1. Confidence intervals for ROC values were obtained by bootstrapping.

2.5.2. Correlations

To validate the automatically generated features, non-parametric Spearman correlation coefficients were used, measuring the correlations between the reading center reads and an associated subset of the automated image analysis features.

2.5.3. SHapley Additive exPlanations (SHAP) Analysis

SHAP analysis [22] is a means to analyze individual predictions made by an ML model. For each feature and predicted data point, SHAP analysis explains the difference between the average model prediction of a given dataset and the individual prediction of this point. This approach explains individual predictions and contributions of each feature, as well as analyzes the overall significance (impact and bias) of the features in a trained ML model, with respect to a given population.

3. Results

3.1. Patient Characteristics

Out of a total of 1098 patients randomized in HARBOR, baseline FA and SD-OCT images were available for 1037 study eyes. In the HARBOR baseline data, CNVs were classified (per FA) as predominantly classic in 163 eyes, minimally classic in 492 eyes, and occult in 382 eyes. Out of a total of 272 patients randomized in AVENUE, baseline FA and SD-OCT images were available for 268 study eyes. CNVs in AVENUE at baseline were classified (per FA) as predominantly classic in 39 eyes, minimally classic in 103 eyes, and occult in 126 eyes. Additionally, 531 healthy fellow eyes from HARBOR without advanced AMD (neither CNV nor GA) were used as negative controls.

3.2. Feature Generation

Segmentation performance for SRF, IRF, PED, and SHRM was assessed against annotations on the HARBOR SD-OCT images and was measured via DICE scores (Sørensen–Dice similarity coefficients; Table 1). Performance for SHRM and PED was better than that for SRF and IRF. Feature evaluation was also performed on various thickness and volumetric reads from AVENUE. Automatically extracted features with the closest definitions to the reading-center-defined feature readouts were selected for comparison to the manual reads, in order to demonstrate that there was a high correlation between manual readouts and automated readouts (Table 2).

Table 1. Segmentation Performance—DICE Scores for the three fluids (IRF, SRF, PED), and SHRM.

Type	*N* (Total)	*N* (Train/Valid)	Validation DICE Mean (SD)
SRF	700	557/143	0.67 (0.05)
IRF	935	694/241	0.46 (0.12)
PED	622	508/114	0.63 (0.07)
SHRM	760	312/65	0.71 (0.06)

Table 2. Human Readouts vs. Automated Readouts for AVENUE. Many automated readouts were generated. This table demonstrates that there was a high correlation between manual readouts that most clearly correspond to the automated [MA1] readouts.

Reading Center Feature	Automated Feature	Spearman r
CENT RET THICK μm	Central subfield thickness IB RPE-to-ILM 0.5 mm min	0.84
CENT RET/LESION THICK μm	Central subfield thickness BM-to-ILM 0.5 mm min	0.79
CENT SUBFIELD THICK ILM-RPE μm	Central subfield thickness IB RPE-to-ILM 0.5 mm mean	0.93
CUBE VOL ILM-RPE mm 3.0 mm	Central subfield volume IB RPE-to-ILM 3.0 mm	0.90
LESION THICK μm	Central subfield thickness BM-to-ILM 3.0 mm max	0.83
PED THICK μm	C-scan height PED 3.0 mm	0.71
SUBRET FLUID THICK μm	C-scan height SRF 3.0 mm	0.61

Type indicates volumetric pathology type; *N* (Train/Valid) indicates number of samples in the training and validation datasets, respectively; validation DICE indicates the DICE score achieved during the validation of the model.

IRF, intraretinal fluid; PED, pigment epithelial detachment; SD, standard deviation; SHRM, subretinal hyperreflective material; and SRF, subretinal fluid.

Various thickness and volumetric measurements compared to the reading center readouts.

BM, Bruch's membrane; IB, inner boundary; ILM, inner limiting membrane; Spearman r, Spearman correlation coefficient; PED, pigment epithelial detachment; RPE, retinal pigment epithelium; and SRF, subretinal fluid.

Balance indicates the number of positive/negative cases; cutoff indicates the critical value of the predicted score corresponding to the AUROC value.

AUROC, area under the receiver operating characteristic; CNV, choroidal neovascularization; FA, fluorescein angiography; FN, number of false negatives; FP, number of false positives; and SD-OCT, spectral-domain optical coherence tomography.

3.3. HARBOR Analysis

As described in the section 'Machine Learning', we assessed the ability of ML to discern eyes with any CNV type from eyes without CNV on SD-OCT, by pooling data across the three types (predominantly classic, minimally classic, and occult) and contrasting them with feature data from the 531 healthy fellow eyes. Presence of any CNV could be almost perfectly discriminated from absence of CNV (AUROC, 0.99; 95% CI, 0.99–1.00; Table 3). Occult and predominantly classic CNV could be discriminated with high accuracy from each other with AUROC = 0.91 (95% CI, 0.89–0.94; Figure 3, Table 4). Specificity for discrimination of occult from a predominantly classic CNV was 81%, with a sensitivity of 89%, when defining occult as the positive class and predominantly classic as the negative class. There were 32 false positives and 41 false negatives out of 163 actually negative and 382 actually positive observations, respectively (Table 5). Minimally classic was discriminated from occult and predominantly classic with AUROC = 0.70 (95% CI, 0.60–0.79) and AUROC = 0.73 (95% CI, 0.61–0.85), respectively. Occult was discriminated from minimally classic and predominantly classic (when the two were pooled together) with AUROC = 0.81 (95% CI, 0.73–0.88).

Table 3. Diagnostic Accuracy of the Algorithm in Detecting FA-Defined CNV Phenotype on SD-OCT; Cross-validation results report best-tuned performance across the parameter grid. External validation reports unbiased performance against hold-out data.

Outcome	Balance	FP/FN	Cutoff	Sensitivity	Specificity	AUROC (95% CI)
Cross-Validated Performance for CNV vs. No CNV on HARBOR (No Parameter Tuning)						
Any CNV vs. none	1037/531	23/15	0.3815	0.99	0.98	1.00 (0.99–1.00)
Predominantly classic vs. none	163/531	1/7	0.9996	0.99	1.00	1.00 (1.00–1.00)
Minimally classic vs. none	492/531	10/10	0.6227	0.98	0.98	1.00 (0.99–1.00)
Minimally classic + predominantly classic vs. none	653/531	14/6	0.1831	0.99	0.98	0.99 (0.99–1.00)
Occult vs. none	382/531	17/24	0.9785	0.96	0.96	0.99 (0.99–1.00)
Holdout Performance as Measured on 15% of HARBOR						
Minimally classic + predominantly classic vs. occult	104/56	20/19	0.4036	0.67	0.81	0.81 (0.73–0.88)
Predominantly classic vs. minimally classic	22/82	22/8	0.7051	0.74	0.65	0.73 (0.61–0.85)
Occult vs. minimally classic	55/81	25/16	0.5390	0.72	0.70	0.70 (0.60–0.79)
Best-Tuned Performance on HARBOR for Predominantly Classic vs. Occult						
Predominantly classic vs. occult	163/382	32/41	0.6110	0.89	0.81	0.91 (0.89–0.94)
External Performance on AVENUE for Predominantly Classic vs. Occult						
Predominantly classic vs. occult	126/39	7/24	0.8058	0.81	0.84	0.88 (0.82–0.95)

Table 4. Resampling performance.

ROC	Sens	Spec	Resample
0.953	0.750	0.948	Fold 1
0.930	0.641	0.957	Fold 2
0.901	0.840	0.880	Fold 3
0.900	0.568	0.960	Fold 4
0.921	0.786	0.950	Fold 5

Table 5. Contingency table, counting all combinations of the predicted versus observed.

Predicted	Observed	n
POS (OCCULT)	POS (OCCULT)	341
NEG	NEG	131
POS (OCCULT)	NEG	32
NEG	POS (OCCULT)	41

ROC indicates area under the curve in percentage; and resample indicates the specific fold from the five-fold cross-validation.

Predicted indicates class predicted by the model; observed indicates class as graded on FA; and n indicates the number of samples.

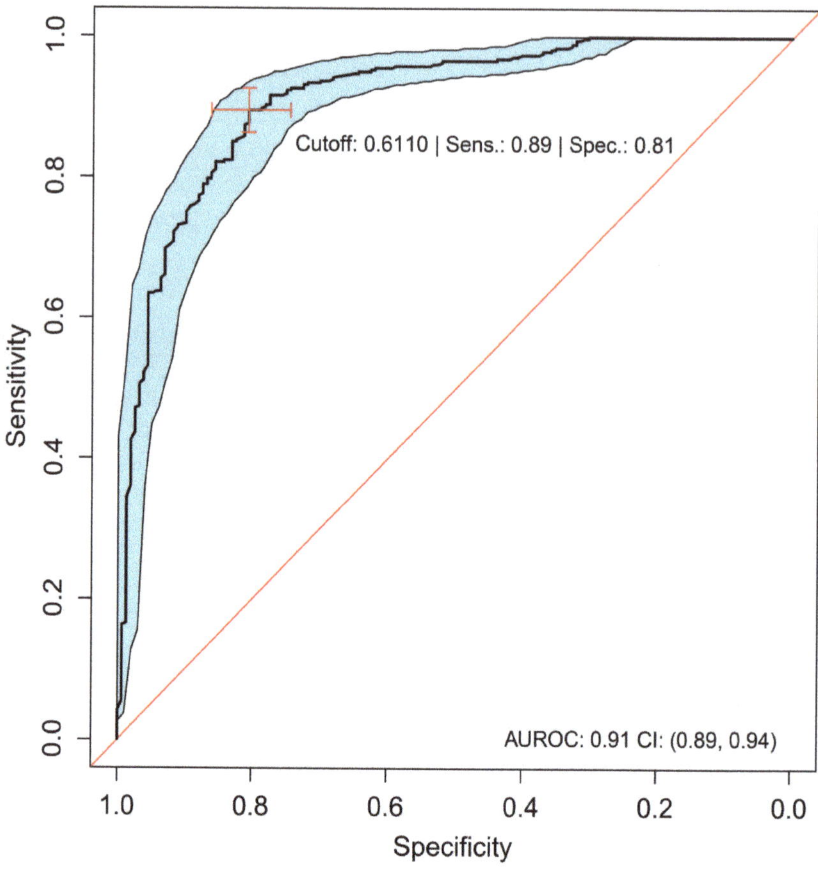

Figure 3. ROC analysis of predominantly classic versus occult (best-tuned performance). Sensitivity versus specificity for all possible ROC cutoff points with respect to the predicted occult scores in HARBOR, including 95% CIs (bootstrapped). The location of the red crosshair indicates the operating point of the model. AUROC, area under the receiver operating characteristic; FA, fluorescein angiography; NEG, negative; POS, positive; ROC, receiver operating characteristic; Sens, sensitivity; and Spec, specificity.

3.4. Recursive Feature Elimination

For discrimination of occult from predominantly classic CNV, when starting with the full 105-feature set and recursively eliminating the least important features, by repeatedly cross-validating with the reduced set, only 21 features were necessary to sustain an average AUROC = 0.91 (Figure 4A). The 20 most informative features for discrimination between occult and predominantly classic on a population level include SHRM and PED volumes (Figure 4B, Table 6). See Figure 5A,B for representative appearance of occult and predominantly classic cases on SD-OCT. SHAP analysis (Figure 6, Table 7) indicates that higher values of SHRM volumes and lower values of PED volumes are most characteristic of predominantly classic CNV, as opposed to occult CNV. Note the largely overlapping findings with feature elimination in Table 6.

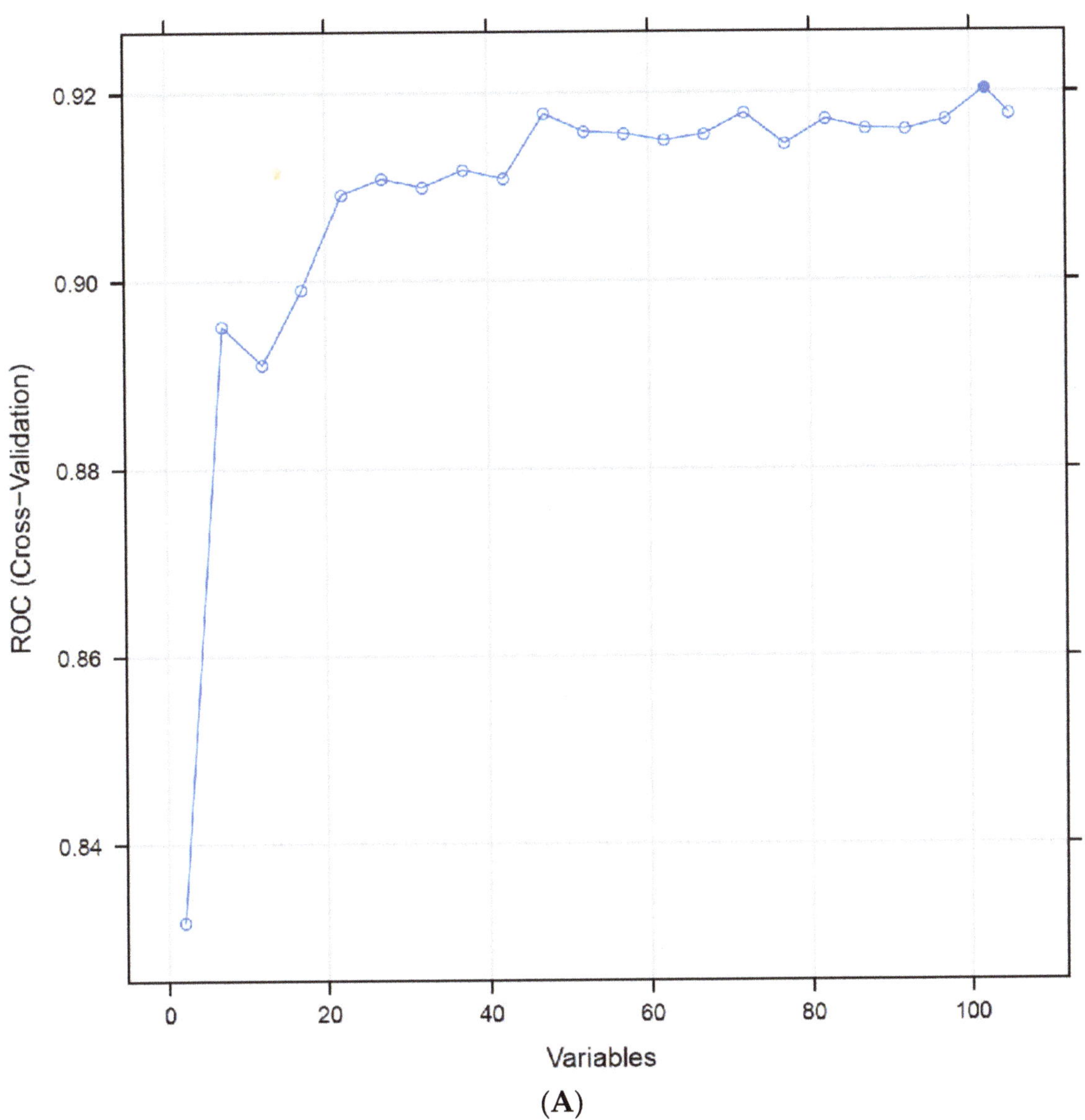

Figure 4. *Cont.*

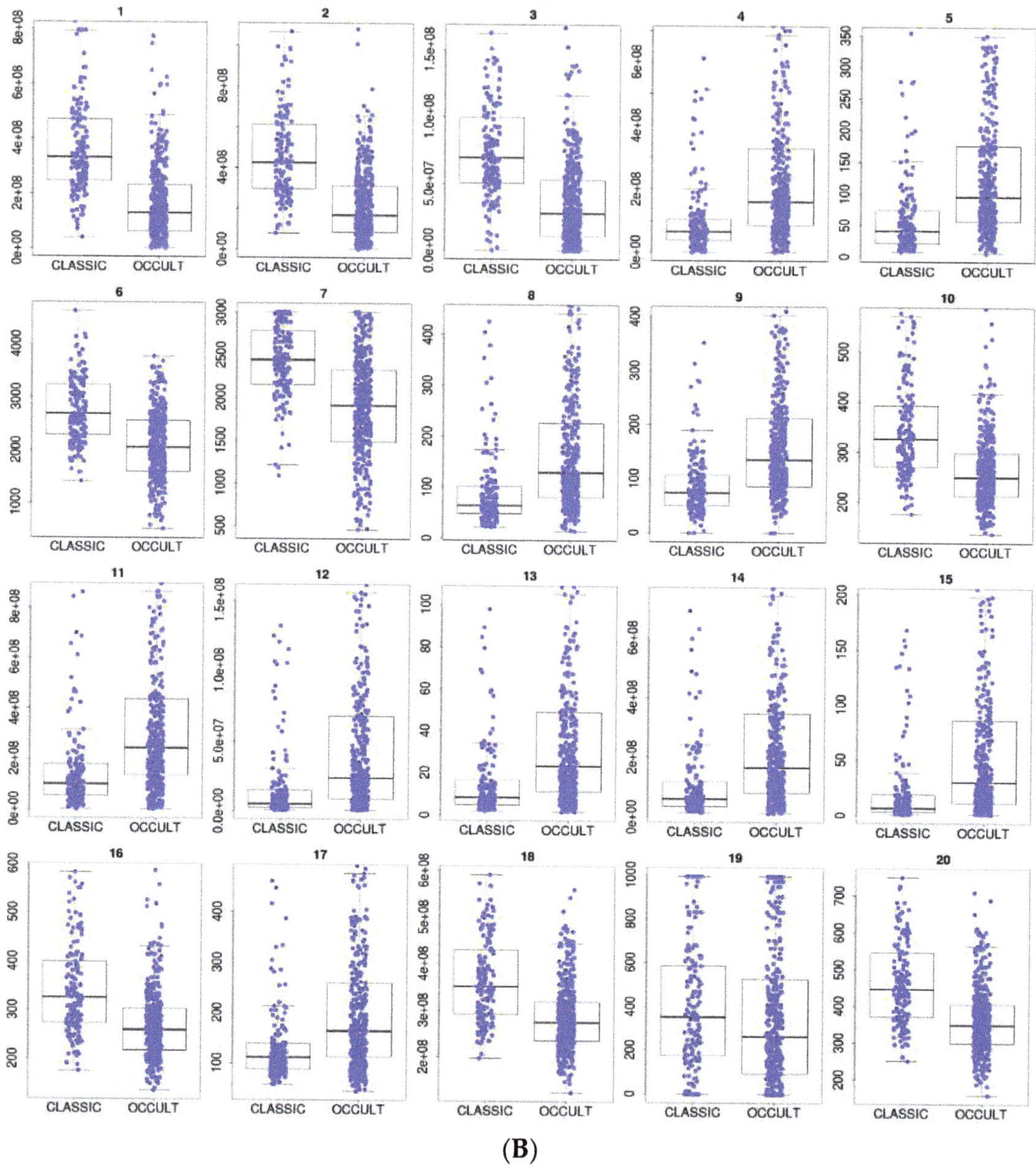

(B)

Figure 4. (A) Recursive feature elimination cross-validation. Optimal performance for predominantly classic versus occult was reached with 101 features out of 106, and only 21 features were necessary to sustain the average model performance of 91% AUROC. **(B)** Distribution of the top 20 feature values in the training data (predominantly classic vs. occult classes).

Table 6. The features from Figure 4B in descending order of importance. BM, Bruch's membrane; HFL, Henle's fiber layer; IB, inner boundary; ILM, inner limiting membrane; max, maximum; OB, outer boundary; OPL, outer plexiform layer; PED, pigment epithelial detachment; ROC, receiver operating characteristic; RPE, retinal pigment epithelium; SHRM, subretinal hyperreflective material; and SRF, subretinal fluid.

Rank	Feature Name
1	C-scan volume SHRM 1.5 mm
2	C-scan volume SHRM 3 mm
3	C-scan volume SHRM 0.5 mm
4	C-scan volume PED 1.5 mm
5	Central subfield thickness BM-to-OB_RPE 0.5 mm max
6	C-scan width SHRM 3 mm
7	C-scan width SHRM 1.5 mm
8	Central subfield thickness BM-to-OB_RPE 1.5 mm max
9	C-scan height PED 0.5 mm
10	Central subfield thickness IB_RPE-to-OPL-HFL 1.5 mm max
11	C-scan volume PED 3 mm
12	Central subfield volume BM-to-OB_RPE 0.5 mm
13	Central subfield thickness BM-to-OB_RPE 1.5 mm mean
14	Central subfield volume BM-to-OB_RPE 1.5 mm
15	Central subfield thickness BM-to-OB_RPE 0.5 mm mean
16	Central subfield thickness IB_RPE-to-OPL-HFL 3.0 mm max
17	Central subfield thickness BM-to-IB_RPE 1.5 mm max
18	Central subfield volume IB_RPE-to-ILM 0.5 mm
19	C-scan width SRF 0.5 mm
20	Central subfield thickness IB_RPE-to-ILM 0.5 mm mean

A. Representative case of predominantly classic CNV, as appearing on SD-OCT (true negative).

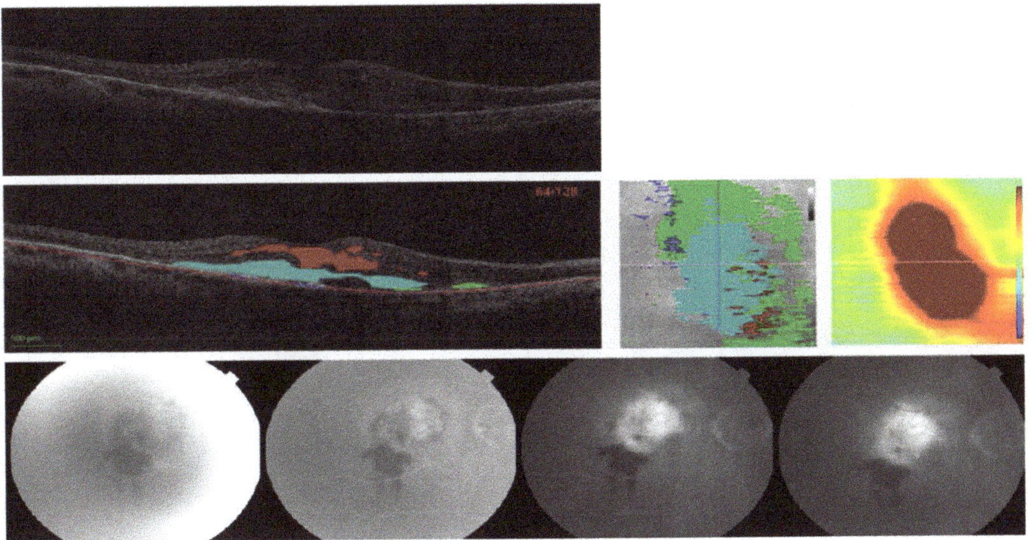

Figure 5. *Cont.*

B. Representative case of occult CNV, as appearing on SD-OCT (true positive).

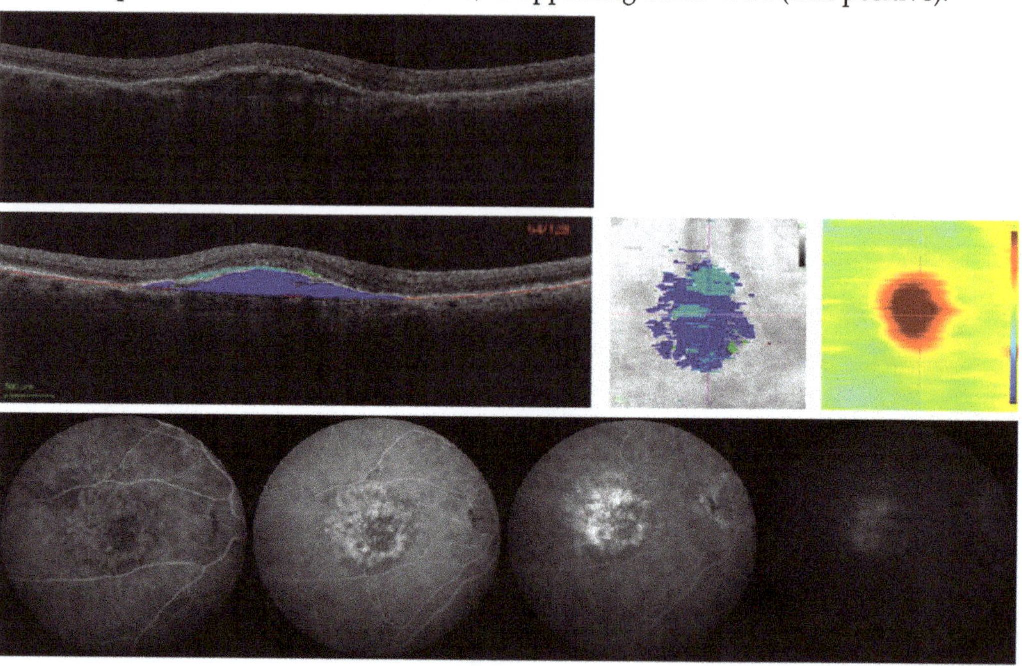

C. Classified by automated algorithm as occult CNV (false positive).

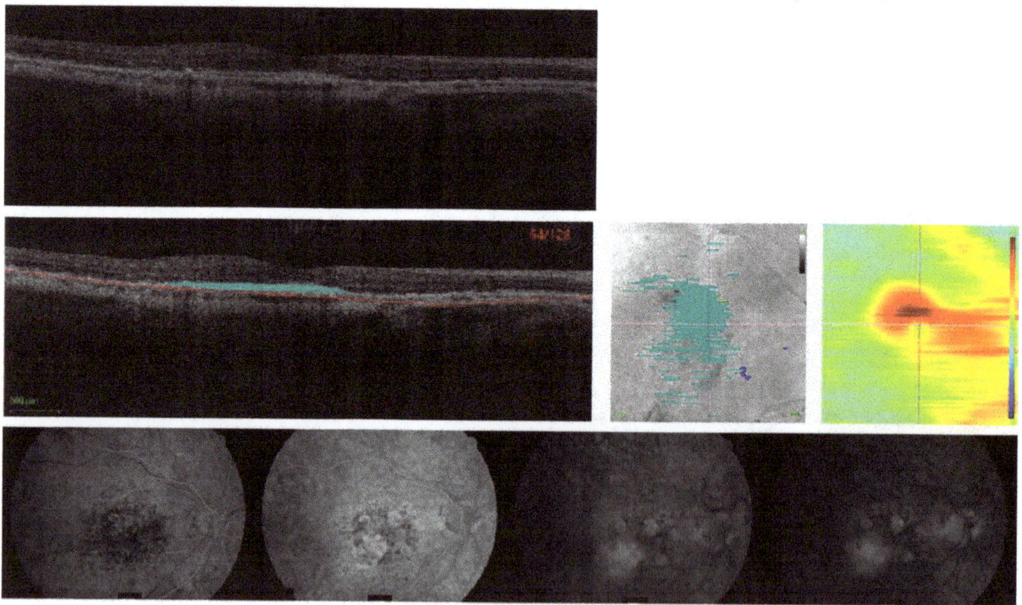

Figure 5. *Cont.*

D. Classified by automated algorithm as classic CNV (false negative).

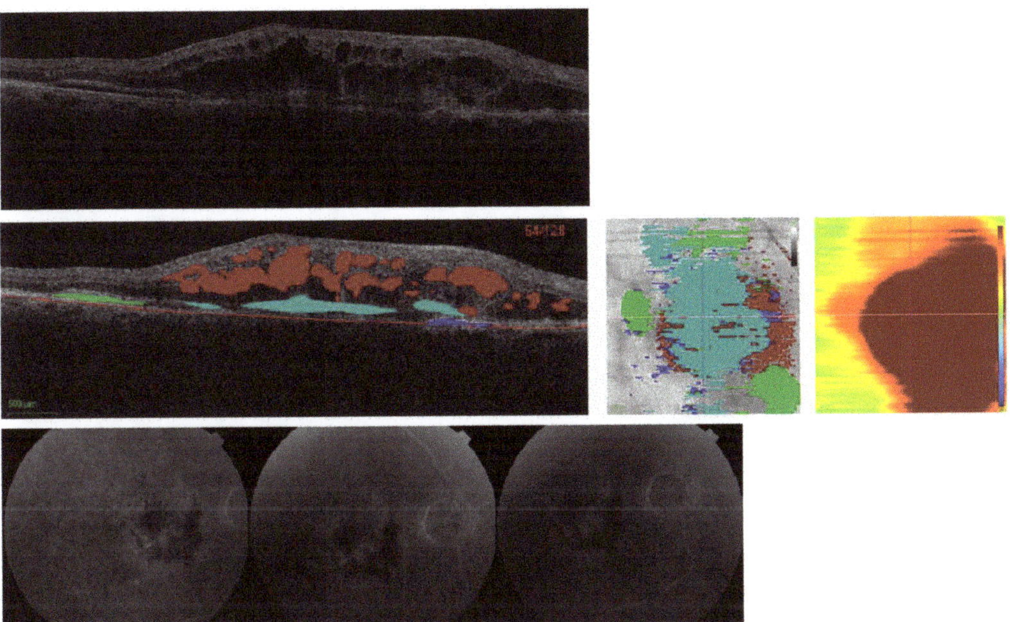

Figure 5. Representative cases showing comparison of machine algorithm with angiography. (**A–D**) Central SD-OCT B-scans (top), with segmented pixel masks of volumetric measures and Bruch's membrane (middle left), en-face projections (middle center), and thickness maps (middle right), as well as corresponding FAs (bottom). Colors on the SD-OCT images indicate volumetric measures as follows—intraretinal fluid (red), subretinal fluid (green), PED (blue), and SHRM (cyan). Bruch's membrane is shown as a red line. In (**A**), FA shows an area of hypofluorescence due to hemorrhage, and a well-demarcated area of hyperfluorescence due to a predominantly classic CNV that leaks in later frames. This was also identified as classic CNV by our ML algorithm, due to increased SHRM height and volume. In (**B**), FA demonstrates an ill-defined area of stippled hyperfluorescence, due to an occult CNV that leaks diffusely in mid and late frames, and was also identified as occult CNV by the ML algorithm, due to the presence of the PED. In (**C**), FA shows an area of well-defined hyperfluorescence in mid frames that stains and leaks in late frames due to fibrosis. The image was classified as classic CNV by the reading center, but was identified as occult CNV by the ML algorithm due to low SHRM height and volume. In (**D**), FA shows an area of hypofluorescence due to hemorrhage and a poorly demarcated area of hyperfluorescence due to the CNV. This lesion was defined as minimally classic by the reading center, but was identified as classic CNV by the ML algorithm due to the SHRM created by the hemorrhage. CNV, choroidal neovascularization; FA, fluorescein angiogram; ML, machine learning; PED, pigment epithelium detachment; SD-OCT, spectral-domain optical coherence tomography; and SHRM, subretinal hyperreflective material.

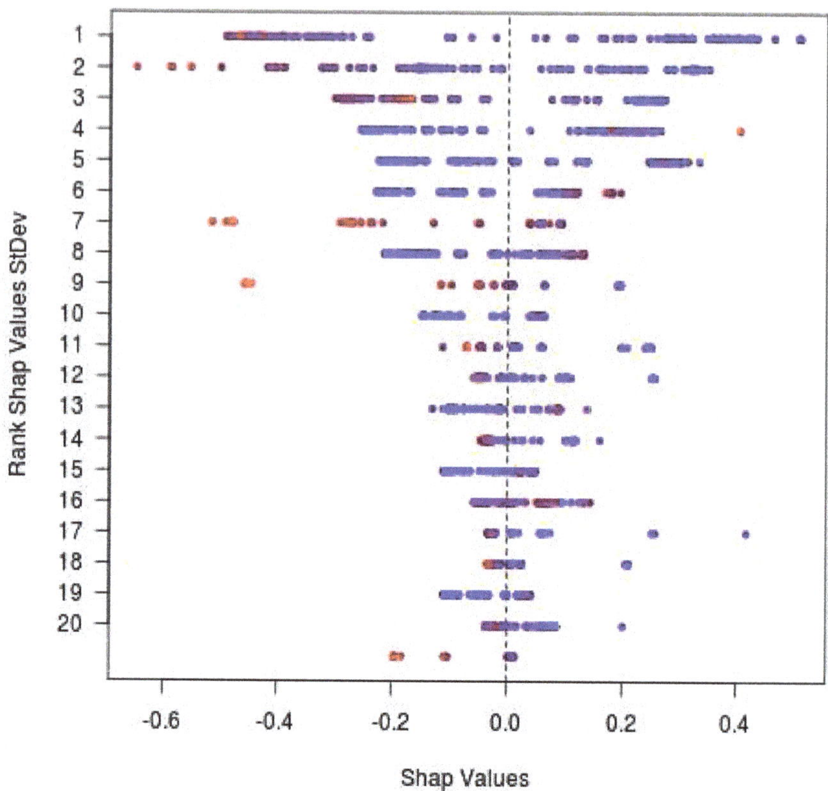

Figure 6. SHAP analysis external validation. SHAP analysis for the CNV type predictions in AVENUE. Every prediction contributes exactly one dot to each row. Blue and red colors indicate lower and higher feature values, respectively. SHAP values (x-axis) add up to the predicted probability for occult (only 20 features with highest SHAP variance shown here). BM, Bruch's membrane; CNV, choroidal neovascularization; HFL, Henle's fiber layer; IB, inner boundary; ILM, inner limiting membrane; IRF, intraretinal fluid; max, maximum; OB, outer boundary; OPL, outer plexiform layer; PED, pigment epithelial detachment; RPE, retinal pigment epithelium; SHAP, SHapley Additive exPlanations; and SHRM, subretinal hyperreflective material.

3.5. External Validation Using AVENUE Data

The performance of the model for differentiating predominantly classic from occult CNV on SD-OCT images of AVENUE was AUROC = 0.88 (95% CI, 0.79–0.94; Figure 7). Specificity was 84%, with sensitivity of 81%, when defining occult as the positive class and predominantly classic as the negative class. There were seven false positives and 24 false negatives out of 39 actually negative and 126 actually positive observations, respectively (Table 8). The most important features (according to the model internal measures) for detection of CNV were SHRM and PED (Figure 6 and Table 7).

Table 7. The top 20 features from Figure 6 in descending order of importance.

Rank	Feature Name
1	C-scan volume SHRM 1.5 mm
2	C-scan volume SHRM 3.0 mm
3	C-scan volume SHRM 0.5 mm
4	C-scan volume PED 1.5 mm
5	C-scan volume PED 3.0 mm
6	Central subfield thickness BM-to-OB_RPE 0.5 mm max
7	C-scan width SHRM 3.0 mm
8	C-scan height PED 1.5 mm
9	Central subfield thickness IB_RPE-to-ILM 1.5 mm max
10	Central subfield volume BM-to-OB_RPE 1.5 mm
11	Central subfield thickness IB_RPE-to-ILM 0.5 mm max
12	C-scan height IRF 3.0 mm
13	C-scan height PED 3.0 mm
14	Central subfield thickness BM-to-ILM 1.5 mm max
15	Central subfield thickness BM-to-OB_RPE 1.5 mm max
16	Central subfield thickness BM-to-ILM 3.0 mm min
17	Central subfield thickness OPL-HFL-to-ILM 0.5 mean
18	Central subfield thickness BM-to-ILM 0.5 mm max
19	Central subfield thickness BM-to-OB_RPE 0.5 mm mean
20	Central subfield thickness BM-to-ILM 3.0 mm max

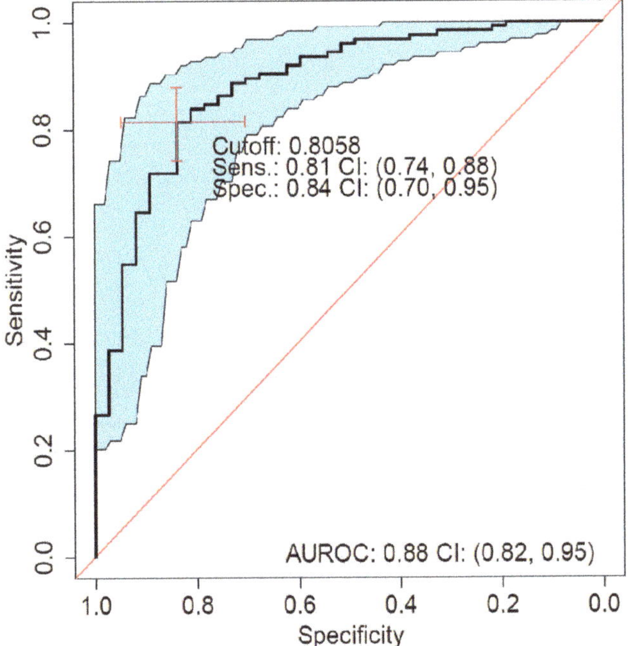

Figure 7. ROC analysis of predominantly classic versus occult external validation. Sensitivity versus specificity for all possible cutoff points with respect to predicted occult scores in AVENUE, including 95% CIs (bootstrapped). The location of the red crosshair indicates the operating point of the model. AUROC, area under the receiver operating characteristic; FA, fluorescein angiography; NEG, negative; POS, positive; ROC, receiver operating characteristic; Sens, sensitivity; and Spec, specificity.

Table 8. Contingency table, counting all combinations of the predicted versus observed. Predicted indicates class predicted by the model; observed indicates class as graded on FA; and n indicates the number of samples.

Predicted	Observed	n
POS (OCCULT)	POS (OCCULT)	102
NEG	NEG	32
POS (OCCULT)	NEG	7
NEG	POS (OCCULT)	24

4. Discussion

Until recently, FA was the reference standard to establish the diagnosis of nAMD and sometimes is also used to monitor patient response to treatment, by assessing reduction in leakage or CNV area [5,23]. However, due to the ease of image acquisition and interpretation, OCT has become the modality of choice for monitoring disease course in clinics [2–4]. As the two imaging modalities use different features and provide different data, here, we presented a bridging study that used data generated on FA to identify and subclassify CNV on SD-OCT, using ML with high accuracy. The availability of two independent sets of large and well-characterized data from the HARBOR and AVENUE clinical trials with well-defined inclusion and exclusion criteria, standardized protocols for acquisition of images, and grading of CNV, allowed the development and robust external validation of our model.

Our ML SD-OCT algorithm was trained using FA-based classification of CNV. This algorithm, developed using Zeiss Cirrus OCT images, was able to discriminate CNV absence versus CNV presence with very high accuracy (99%) and subclassify occult from predominantly classic CNV subtypes, with an accuracy of 91% AUROC. Furthermore, the performance accuracy of the ML algorithm using an external dataset was 88%, despite it being a different SD-OCT machine (Heidelberg Spectralis). Accuracy of FA-versus OCT-based approaches for detection of fluid has been explored by several researchers [24–26], but few have attempted to bridge the two technologies for the identification and classification of CNV [27–31]. Using FA as the reference standard for identification and classification of CNV, Wilde et al. [31] retrospectively evaluated 278 eyes diagnosed with CNV on SD-OCT, with their corresponding FA. They reported that while sensitivity of SD-OCT in detection of CNV was high (100%), it had a low specificity, with a 17% false-positive rate. Their findings were similar to other studies [26,28,30] that evaluated leakage on FA as a surrogate marker for CNV activity and found the sensitivity of SD-OCT to be high, but lacking specificity in comparison. Limited details of criteria for CNV identification by SD-OCT are provided in this publication, and it appeared that decision-making was mainly based on subjective criteria; features such as SHRM and PED were not included in the analysis of SD-OCT.

Our algorithm well-differentiated between occult and predominantly classic CNV types, whereas the ability to differentiate minimally classic CNV from occult or predominantly classic was lower. Our model identified the most informative features for discrimination between occult and predominantly classic, such as SHRM and PED volumes; higher SHRM volumes and lower PED volumes were most characteristic of predominantly classic CNV, as opposed to occult CNV. In contrast, absence of a well-defined SHRM in a case classified as classic CNV on FA was diagnosed as occult CNV by our model (Figure 5C), and the appearance of SHRM due to hemorrhage, resulted in our model identifying it as predominantly classic CNV, while the lesion was classified by the reading center as occult on FA (Figure 5D). Minimally classic CNV by definition has lesion components of both classic and occult CNV [13]. As the algorithm learns to find salient characteristics for either class during training, a class that combines characteristics of two other classes (instead of having its own characteristic features), posed an intrinsically harder problem to discriminate [13].

There are only a few publications that have attempted to correlate FA-defined phenotypes of CNV with features of CNV on OCT [27,29]. In a recent study by Gualino et al. [29], five retina specialists compared SD-OCT combined with color fundus photography or FA in 148 patients with treatment-naive nAMD. They classified CNV as type 1, 2, or 'other CNV', based on study-defined prespecified criteria, including features such as PED and SHRM in their decision-making. Manual readouts performed using subjective criteria developed specifically for these manuscripts/studies limit the wider application of their findings. However, it is interesting and reassuring to note that our algorithm developed on FA takes the same features into consideration as human graders in this study, to classify lesion types into type 1 and 2 on SD-OCT [12,29]. The strength of our approach is that, it is completely automated on SD-OCT and that we used the MPS standardized FA classification as the base for classification of CNV [32]. Applied to clinical practice, this automated diagnostic SD-OCT-only process may help to expand the population of patients that can benefit from CNV assessment, for example, in remote environments, without easy access to a retinal center with multiple imaging modalities.

We also presented segmentation performance for the various OCT features. For IRF and PED, the performance was poor at 0.46 ($\pm$0.12) and 0.63 ($\pm$0.07), respectively, whereas for SRF, it was better (0.67 [$\pm$0.05]); see Table 1. Interestingly, for SHRM, the model had good performance (0.71 [$\pm$0.06]). In the RETOUCH grand challenge, segmentation performance ranged between 0.57–0.85 for IRF, 0.54–0.72 for SRF, and 0.66–0.82 for PED in terms of DICE score [33]. SHRM segmentation was not part of the RETOUCH challenge. In our study, SHRM performance was assessed using a subset of our annotations. The reduced performance compared with the RETOUCH leaderboard for IRF and PED segmentation could be due to the differences in image quality and the heterogeneous conditions within the clinical trial setting (e.g., multiple study sites, imaging technicians, patient factors). In contrast, the images in RETOUCH were selected for exceptionally high quality. The challenges of manually identifying and correctly delineating these features on SD-OCT and distinguishing them from adjoining normal retina, gliosis, or other lesions in clinical trials and real-world settings cannot be underestimated. Therefore, an ML approach using a variety of images of different quality to train the model may be generalizable to a wider variety of data.

5. Limitations

ML is always impacted by variability and biases in the training data because the annotation of images is performed manually. Although two graders annotated the OCT dataset, each image was only annotated by one of them, potentially leading to a bias. Additionally, FA assessment (the previous 'reference standard' for CNV assessment), is subject to reader interpretation, however, here two graders assessed each FA image, with adjudication as needed. Additionally, the scope of the current work is limited to classification of CNV on FA as predominantly classic, minimally classic, and occult phenotypes. As information about other CNV subtypes, such as retinal angiomatous proliferation and PCV, was not available in HARBOR, these phenotypes could not be evaluated. Furthermore, novel SD-OCT classification and terminology suggested by the Consensus on Neovascular Age-Related Macular Degeneration Nomenclature (CONAN) group [12] were not available for our data; therefore, comparison with an OCT-based classification system was not possible. Moreover, as we restricted our selection of non-CNV eyes to fellow eyes without advanced AMD or other pathologies, this may not be representative of what would be encountered in the real world, and prospective validation in a broader population would be needed. Finally, features such as intraretinal hyperreflective foci have not been included in this model, as their role in diagnosis and prognosis have yet to be established [34].

6. Conclusions

Our study shows that using ML on SD-OCT images is sufficiently accurate to detect and classify nAMD. This work highlights the reduced need of FA and provides an automated alternative to manual reading of images at baseline. This in turn limits the variety of imaging data sources from which reads are drawn, reduces the need for multiple human graders, and minimizes the risk of inconsistencies in the diagnoses. Finally, automating the read will also help with a major milestone of algorithmic models, which is to streamline and standardize diagnostic processes.

Supplementary Materials: The following are available online at https://www.mdpi.com/article/10.3390/jpm11060524/s1. Figure S1: non-nested cross-validation, Figure S2, Table S1: Definitions used for the annotations, Table S2: Detailed feature list, Supplementary Methods: Details on ML methodology.

Author Contributions: Conceptualization, A.M., F.B., and J.S.; conduct of study, A.M., F.B., Y.L., Y.-P.Z., and J.S.; formal analysis, A.M., F.B., and J.S.; resources, A.M., F.B., Y.L., T.A., Y.-P.Z., Y.Z., and S.M.; writing—original draft preparation, A.M. and J.S.; writing—review and editing, A.M., F.B., Y.L., T.A., Y.-P.Z., F.A., Y.Z., S.M., and J.S. All authors have read and agreed to the published version of the manuscript.

Funding: Funding was provided by F. Hoffmann-La Roche Ltd., Basel, Switzerland, for the study and third-party writing assistance.

Institutional Review Board Statement: The study was conducted according to the guidelines of the Declaration of Helsinki, were Health Insurance Portability and Accountability Act compliant, and protocols were approved by the Institutional Review Board of ROCHE and ethics committees.

Informed Consent Statement: Informed consent was obtained from all subjects involved in the HARBOR study.

Data Availability Statement: Qualified researchers may request access to individual patient-level data through the clinical study data request platform (https://vivli.org/, accessed on 7 June 2021). Further details on Roche's criteria for eligible studies are available here (https://vivli.org/members/ourmembers/, accessed on 7 June 2021). For further details on Roche's Global Policy on the Sharing of Clinical Information and how to request access to related clinical study documents, see here (https://www.roche.com/research_and_development/who_we_are_how_we_work/clinical_trials/our_commitment_to_data_sharing.htm, accessed on 7 June 2021).

Acknowledgments: Third-party writing assistance was provided by Michelle Kelly., of Envision Pharma Group.

Conflicts of Interest: A. Maunz, F. Benmansour, Y. Li, T. Albrecht, Y.-P. Zhang, F. Arcadu, and J. Sahni, F. Hoffmann-La Roche Ltd., Basel, Switzerland (E); Y. Zheng, F. Hoffmann-La Roche Ltd., Basel, Switzerland (F); S. Madhusudhan, F. Hoffmann-La Roche Ltd., Basel, Switzerland (F).

References

1. Wong, T.Y.; Chakravarthy, U.; Klein, R.; Mitchell, P.; Zlateva, G.; Buggage, R.; Fahrbach, K.; Prost, C.; Sledge, I. The natural history and prognosis of neovascular age-related macular degeneration: A systematic review of the literature and meta-analysis. *Ophthalmology* **2008**, *115*, 116–126. [CrossRef]
2. Gess, A.J.; Fung, A.E.; Rodriguez, J.G. Imaging in neovascular age-related macular degeneration. *Semin. Ophthalmol.* **2011**, *26*, 225–233. [CrossRef]
3. Castillo, M.M.; Mowatt, G.; Elders, A.; Lois, N.; Fraser, C.; Hernandez, R.; Amoaku, W.; Burr, J.M.; Lotery, A.; Ramsay, C.R.; et al. Optical coherence tomography for the monitoring of neovascular age-related macular degeneration: A systematic review. *Ophthalmology* **2015**, *122*, 399–406. [CrossRef] [PubMed]
4. Rosenfeld, P.J. Optical coherence tomography and the development of antiangiogenic therapies in neovascular age-related macular degeneration. *Investig. Ophthalmol. Vis. Sci.* **2016**, *57*, OCT14–OCT26. [CrossRef] [PubMed]
5. Schmidt-Erfurth, U.; Chong, V.; Loewenstein, A.; Larsen, M.; Souied, E.; Schlingemann, R.; Eldem, B.; Mones, J.; Richard, G.; Bandello, F. Guidelines for the management of neovascular age-related macular degeneration by the European Society of Retina Specialists (EURETINA). *Br. J. Ophthalmol.* **2014**, *98*, 1144–1167. [CrossRef] [PubMed]
6. Koh, A.; Lee, W.K.; Chen, L.J.; Chen, S.J.; Hashad, Y.; Kim, H.; Lai, T.Y.Y.; Pilz, S.; Ruamviboonsuk, P.; Tokaji, E.; et al. EVEREST study: Efficacy and safety of verteporfin photodynamic therapy in combination with ranibizumab or alone versus ranibizumab monotherapy in patients with symptomatic macular polypoidal choroidal vasculopathy. *Retina* **2012**, *32*, 1453–1464. [CrossRef]
7. Koh, A.; Lai, T.Y.Y.; Takahashi, K.; Wong, T.Y.; Chen, L.J.; Ruamviboonsuk, P.; Tan, C.S.; Feller, C.; Margaron, P.; Lim, T.H.; et al. Efficacy and safety of ranibizumab with or without verteporfin photodynamic therapy for polypoidal choroidal vasculopathy: A randomized clinical trial. *JAMA Ophthalmol.* **2017**, *135*, 1206–1213. [CrossRef]
8. Lim, T.H.; Lai, T.Y.Y.; Takahashi, K.; Wong, T.Y.; Chen, L.J.; Ruamviboonsuk, P.; Tan, C.S.; Lee, W.K.; Cheung, C.M.G.; Ngah, N.F.; et al. Comparison of ranibizumab with or without verteporfin photodynamic therapy for polypoidal choroidal vasculopathy. The EVEREST II randomized clinical trial. *JAMA Ophthalmol.* **2020**, *138*, 935–942. [CrossRef]

9. Jaffe, G.J.; Ciulla, T.A.; Ciardella, A.P.; Devin, F.; Dugel, P.U.; Eandi, C.M.; Masonson, H.; Mones, J.; Pearlman, J.A.; Quaranta-El Maftouhi, M.; et al. Dual antagonism of PDGF and VEGF in neovascular age-related macular degeneration: A phase IIb, multicenter, randomized controlled trial. *Ophthalmology* **2017**, *124*, 224–234. [CrossRef] [PubMed]
10. Sahni, J.; Dugel, P.U.; Patel, S.S.; Chittum, M.E.; Berger, B.; Del Valle Rubido, M.; Sadikhov, S.; Szczesny, P.; Schwab, D.; Nogoceke, E.; et al. Safety and efficacy of different doses and regimens of faricimab vs ranibizumab in neovascular age-related macular degeneration: The AVENUE phase 2 randomized clinical trial. *JAMA Ophthalmol.* **2020**, *138*, 955–963. [CrossRef]
11. Khanani, A.M.; Patel, S.S.; Ferrone, P.J.; Osborne, A.; Sahni, J.; Grzeschik, S.; Basu, K.; Ehrlich, J.S.; Haskova, Z.; Dugel, P.U. Efficacy of every four monthly and quarterly dosing of faricimab vs ranibizumab in neovascular age-related macular degeneration: The STAIRWAY phase 2 randomized clinical trial. *JAMA Ophthalmol.* **2020**, *138*, 964–972. [CrossRef]
12. Spaide, R.F.; Jaffe, G.J.; Sarraf, D.; Freund, K.B.; Sadda, S.R.; Staurenghi, G.; Waheed, N.K.; Chakravarthy, U.; Rosenfeld, P.J.; Holz, F.G.; et al. Consensus nomenclature for reporting neovascular age-related macular degeneration data: Consensus on Neovascular Age-Related Macular Degeneration Nomenclature Study Group. *Ophthalmology* **2020**, *127*, 616–636. [CrossRef] [PubMed]
13. Macular Photocoagulation Study Group. Subfoveal neovascular lesions in age-related macular degeneration. Guidelines for evaluation and treatment in the Macular Photocoagulation Study. *Arch. Ophthalmol.* **1991**, *109*, 1242–1257. [CrossRef]
14. Busbee, B.G.; Ho, A.C.; Brown, D.M.; Heier, J.S.; Suner, I.J.; Li, Z.; Rubio, R.G.; Lai, P. Twelve-month efficacy and safety of 0.5 mg or 2.0 mg ranibizumab in patients with subfoveal neovascular age-related macular degeneration. *Ophthalmology* **2013**, *120*, 1046–1056. [CrossRef]
15. Ho, A.C.; Busbee, B.G.; Regillo, C.D.; Wieland, M.R.; Van Everen, S.A.; Li, Z.; Rubio, R.G.; Lai, P. Twenty-four-month efficacy and safety of 0.5 mg or 2.0 mg ranibizumab in patients with subfoveal neovascular age-related macular degeneration. *Ophthalmology* **2014**, *121*, 2181–2192. [CrossRef] [PubMed]
16. Abràmoff, M.D.; Garvin, M.K.; Sonka, M. Retinal imaging and image analysis. *IEEE Rev. Biomed. Eng* **2010**, *3*, 169–208. [CrossRef] [PubMed]
17. Van der Walt, S.; Schönberger, J.L.; Nunez-Iglesias, J.; Boulogne, F.; Warner, J.D.; Yager, N.; Gouillart, E.; Yu, T. scikit-image: Image processing in Python. *PeerJ* **2014**, *2*, e453. [CrossRef] [PubMed]
18. Ristau, T.; Keane, P.A.; Walsh, A.C.; Engin, A.; Mokwa, N.; Kirchhof, B.; Sadda, S.R.; Liakopoulos, S. Relationship between visual acuity and spectral domain optical coherence tomography retinal parameters in neovascular age-related macular degeneration. *Ophthalmologica* **2014**, *231*, 37–44. [CrossRef]
19. Willoughby, A.S.; Ying, G.S.; Toth, C.A.; Maguire, M.G.; Burns, R.E.; Grunwald, J.E.; Ebenezer, D.; Jaffe, G.J. Subretinal hyperreflective material in the Comparison of Age-Related Macular Degeneration Treatments Trials. *Ophthalmology* **2015**, *122*, 1846–1853.e5. [CrossRef] [PubMed]
20. Ronneberger, O.; Fischer, F.; Brox, T. U-Net: Convolutional networks for biomedical image segmentation. *arXiv* **2015**, arXiv:1505.04597.
21. Early Treatment Diabetic Retinopathy Study Research Group. Grading diabetic retinopathy from stereoscopic color fundus photographs—an extension of the modified Airlie House classification. ETDRS report number 10. *Ophthalmology* **1991**, *98*, 786–806. [CrossRef]
22. Lundberg, S.; Lee, S.-I. A unified approach to interpreting model predictions. *arXiv* **2017**, arXiv:1705.07874.
23. Harding, S.P. Neovascular age-related macular degeneration: Decision making and optimal management. *Eye* **2010**, *24*, 497–505. [CrossRef]
24. Khurana, R.N.; Dupas, B.; Bressler, N.M. Agreement of time-domain and spectral-domain optical coherence tomography with fluorescein leakage from choroidal neovascularization. *Ophthalmology* **2010**, *117*, 1376–1380. [CrossRef] [PubMed]
25. Salinas-Alamán, A.; García-Layana, A.; Maldonado, M.J.; Sainz-Gómez, C.; Alvárez-Vidal, A. Using optical coherence tomography to monitor photodynamic therapy in age related macular degeneration. *Am. J. Ophthalmol.* **2005**, *140*, 23–28. [CrossRef] [PubMed]
26. Van de Moere, A.; Sandhu, S.S.; Talks, S.J. Correlation of optical coherence tomography and fundus fluorescein angiography following photodynamic therapy for choroidal neovascular membranes. *Br. J. Ophthalmol.* **2006**, *90*, 304–306. [CrossRef]
27. Hee, M.R.; Baumal, C.R.; Puliafito, C.A.; Duker, J.S.; Wilkins, J.R.; Coker, J.G.; Schuman, J.S.; Swanson, E.A.; Fujimoto, J.G. Optical coherence tomography of age-related macular degeneration and choroidal neovascularization. *Ophthalmology* **1996**, *103*, 1260–1270. [CrossRef]
28. Do, D.V.; Gower, E.W.; Cassard, S.D.; Boyer, D.; Bressler, N.M.; Bressler, S.B.; Heier, J.S.; Jefferys, J.L.; Singerman, L.J.; Solomon, S.D. Detection of new-onset choroidal neovascularization using optical coherence tomography: The AMD DOC Study. *Ophthalmology* **2012**, *119*, 771–778. [CrossRef]
29. Gualino, V.; Tadayoni, R.; Cohen, S.Y.; Erginay, A.; Fajnkuchen, F.; Haouchine, B.; Krivosic, V.; Quentel, G.; Gaudric, A. Optical coherence tomography, fluorescein angiography, and diagnosis of choroidal neovascularization in age-related macular degeneration. *Retina* **2019**, *39*, 1664–1671. [CrossRef]
30. Padnick-Silver, L.; Weinberg, A.B.; Lafranco, F.P.; Macsai, M.S. Pilot study for the detection of early exudative age-related macular degeneration with optical coherence tomography. *Retina* **2012**, *32*, 1045–1056. [CrossRef]
31. Wilde, C.; Patel, M.; Lakshmanan, A.; Amankwah, R.; Dhar-Munshi, S., Amoaku, W. The diagnostic accuracy of spectral domain optical coherence tomography for neovascular age-related macular degeneration: A comparison with fundus fluorescein angiography. *Eye* **2015**, *29*, 602–609. [CrossRef] [PubMed]

32. Moisseiev, J.; Alhalel, A.; Masuri, R.; Treister, G. The impact of the Macular Photocoagulation Study results on the treatment of exudative age-related macular degeneration. *Arch. Ophthalmol* **1995**, *113*, 185–189. [CrossRef] [PubMed]
33. Bogunović, H.; Venhuizenm, F.; Klimscha, S.; Apostolopoulos, S.; Bab-Hadiashar, A.; Bagci, U.; Beg, M.F.; Bekalo, L.; Chen, Q.; Ciller, C.; et al. RETOUCH: The retinal OCT fluid detection and segmentation benchmark and challenge. *IEEE Trans. Med. Imaging* **2019**, *38*, 1858–1874. [CrossRef] [PubMed]
34. Giani, A.; Luiselli, C.; Esmaili, D.D.; Salvetti, P.; Cigada, M.; Miller, J.W.; Staurenghi, G. Spectral-domain optical coherence tomography as an indicator of fluorescein angiography leakage from choroidal neovascularization. *Investig. Ophthalmol. Vis. Sci.* **2011**, *52*, 5579–5586. [CrossRef]

Journal of
Personalized Medicine

Review

Indicators of Visual Prognosis in Diabetic Macular Oedema

Sagnik Sen [1], **Kim Ramasamy** [1] and **Sobha Sivaprasad** [2,*]

1   Department of Retina and Vitreous Services, Aravind Eye Hospital, Madurai 625020, India;
    sagnik@aravind.org (S.S.); kim@aravind.org (K.R.)
2   NIHR Moorfields Biomedical Research Centre, Moorfields Eye Hospital, 162, City Road,
    London EC1V 2PD, UK
*   Correspondence: Sobha.sivaprasad@nhs.net; Tel.: +44-2075663411

Abstract: Diabetic macular oedema (DMO) is an important cause of moderate vision loss in people with diabetes. Advances in imaging technology have shown that a significant proportion of patients with DMO respond sub-optimally to existing treatment options. Identifying associations and predictors of response before treatment is initiated may help in explaining visual prognosis to patients and aid the development of personalized treatment strategies. Imaging features, such as central subfoveal thickness, photoreceptor integrity, disorganization of retinal inner layers, choroidal changes, and macular perfusion, have been reported to be prognostic factors of visual acuity (VA) in DMO. In this review we evaluated each risk factor to understand their relative importance in visual prognostication of DMO eyes post-treatment. Although individually, some of these factors may not be significant predictors, in combination they may form phenotypes that can inform visual prognosis. Stratification based on these phenotypes needs to be developed to progress to personalized medicine for DMO.

Keywords: diabetic retinopathy; diabetic macular oedema; visual prognosis; indicators; personalized medicine

Citation: Sen, S.; Ramasamy, K.;
Sivaprasad, S. Indicators of Visual
Prognosis in Diabetic Macular
Oedema. *J. Pers. Med.* **2021**, *11*, 449.
https://doi.org/10.3390/jpm11060449

Academic Editors: Peter D.
Westenskow and Andreas Ebneter

Received: 25 April 2021
Accepted: 20 May 2021
Published: 22 May 2021

Publisher's Note: MDPI stays neutral
with regard to jurisdictional claims in
published maps and institutional affiliations.

1. Introduction

Diabetic macular oedema (DMO) is the most frequent cause of moderate vision loss in people with diabetes. In 2019, there were approximately 28 million people with DMO globally [1]. For at least 40 years, patients and ophthalmologists were satisfied with the outcomes of the Early Treatment Diabetic Retinopathy Study (ETDRS) study that showed that the risk of moderate visual loss can be reduced by about 50% in laser treated individuals [2]. Over the last decade, intravitreal anti-vascular endothelial growth factor agents (anti-VEGF) have replaced macular laser and intravitreal steroids as the main treatment option for visual impairment due to centre-involving DMO (CI-DMO) [3]. Expectations of patients and providers have increased with the availability of anti-VEGF agents as approximately 50% of patients with visual impairment treated with anti-VEGF improve by two lines of visual acuity (VA) on ETDRS visual acuity charts by two years, if treated optimally [3,4]. However, DMO is not always associated with visual impairment, and some may resolve spontaneously or with treatment, while others progress to irreversible visual impairment. In this review, we evaluate the potential risk factors that can be considered in future prognostic models on visual impairment due to DMO, so that management of DMO can be personalised based on risk of visual loss. We broadly classified the prognostic factors into subjective and objective and then further grouped the imaging features into definite and possible, based on current evidence. However, the predicted visual outcome based on baseline prognostic factors may also be influenced by the drug given and treatment frequency.

2. Subjective Factors

Presenting VA

Presenting VA is a definite prognostic marker. However, there are some caveats. Firstly, ceiling effect in eyes with good VA means that change in VA is not a good indicator of visual outcome in this group and the outcome in these eyes needs to focus on prevention of visual loss. Secondly, eyes with new-onset CI-DMO do not always present with visual impairment. However, if left untreated, visual deterioration may occur in some of these eyes. In DRCR Protocol V trial, 18.6% of eyes with good vision at presentation had a reduction of VA by 10 ETDRS letters or more after an observation period of 2 years and needed injections [5]. The exact timepoint or reason for incident visual impairment in these cases and not others remain unclear. Control arms of the RISE and RIDE studies also showed that on average, patients with visual impairment due to untreated DMO can remain stable for a further two years [6]. However, this group did not attain similar VA gains when initiated on ranibizumab after two years compared to those who were initiated early, suggesting that irreversible structural changes do occur if CI-DMO is left untreated and that these are not visible on OCT or probably, these characteristic changes have been unidentifiable to date [7]. When we synthesize the results of RISE and RIDE and Protocol V together, we can conclude that eyes with CI-DMO that experience visual deterioration over the two years of observation may not gain full visual potential compared to those who were started on anti-VEGF early. Although significant emphasis is made on presenting VA, these observations reinforce that structural changes precede decreases in visual function. However, further research is required to identify the imaging phenotype of this high-risk group that might explain irreversible visual loss while being observed for deterioration of CI-DMO.

On the contrary, in DMO eyes with visual impairment of 6/12 or worse (Snellen), baseline VA is a predictor of response to anti-VEGF treatment. The RESTORE study showed that patients with poorer starting VA had greater gains in VA with ranibizumab than those with better presenting VA [8]. The post-hoc analysis of Protocol T also confirmed that the treatment effect varied according to the baseline VA, after adjusting for both VA and CST and the interactions with treatment [9]. However, in real-life studies with ranibizumab, eyes with baseline VA less than 37 letters fail to achieve good VA that met driving standards. This outcome was also noted in a study on early response to aflibercept after 3 loading injections in real-life suggesting that there are more eyes with irreversible structural changes in eyes with VA less than 37 letters [10]. Based on available evidence, early treatment with anti-VEGF agents is recommended. Waiting for visual deterioration in eyes with CI-DMO with good presenting VA in routine clinical practice is itself a poor prognostic indicator of final visual outcome.

3. Objective Factors

3.1. Optical Coherence Tomography Imaging

3.1.1. Central Subfoveal Thickness (CST)

A decrease in CST on OCT is the most reliable objective measure of anatomical treatment response of DMO. It is also the most common secondary anatomical outcome measure used to substantiate primary visual outcomes in clinical trials on interventions for DMO. However, reduction of CST does not mirror final VA outcome and is not a definite predictor of VA [11–16]. The RESTORE study shows that CST $\geq 400\ \mu m$ resulted in more profound reduction of DMO with ranibizumab compared to laser, while the effects of macular laser and ranibizumab were similar in DMO eyes with CST $< 400\ \mu m$ [4]. However, the baseline predictors of the VIVID and VISTA study did not reveal such a cut-off. In fact, early treatment with aflibercept in eyes with CI-DME with CST $< 400\ \mu m$ resulted in significantly more gains in vision than those treated with laser [17]. Post-hoc analyses of RISE/RIDE and DRCR.net data suggest that continued treatment with frequent intravitreal anti-VEGF injections may improve functional outcome in non-responders who initially show suboptimal reduction of CST. These suggest that CST reduction has to be qualified with other morphological OCT features to predict visual outcome. For example, CST of similar thickness may have varying neuronal atrophy and Müller cell dysfunction that may relate better to visual outcomes [18]. Although CST is a prognostic indicator for fluid resolution in DMO, it is not a definite visual prognostic indicator.

3.1.2. Morphological Phenotypes

There are several OCT morphological phenotypes of DMO. These include diffuse retinal thickening (DRT), cystoid macular oedema (CMO) and neurosensory detachment (NSD) [19]. These may be present in isolation or in combination. These features may affect different retinal layers. For example, sponge-like DRT may be seen with cystoid spaces present in the inner nuclear layer (INL) and outer plexiform layer (OPL) [20]. The cysts in CMO may be small to large in size and display variable fluorescein angiography findings ranging from honey-combing to petaloid pooling [20,21]. The size, location and angiographic appearance of cysts do not influence visual outcome in DMO.

However, association of cysts with other OCT features determine visual outcome. Cysts that are associated with photoreceptor damage negatively affect visual outcomes [22–24]. In addition, cysts that cause Müller cell dysfunction or loss are also detrimental to vision [25–27]. Early histological studies of DMO eyes have revealed that reversible edema develops from fluid accumulation within Müller cells [28]. The number of bridges between or within large cysts also determines VA. These bridges are presumed to be composed of Müller cells and bipolar cells [29]. Chronic fluid accumulation can lead to death of Müller cells and manifest as absence of bridging tissue between inner and outer retina. Therefore, although individually large cysts do not offer any prognostic significance, cysts in combination with ellipsoid zone loss or large cysts with lack of intervening bridges carry poor prognosis [30–32]. Large central cysts may also be a sign of macular ischaemia [33,34] but it is not a definite prognostic factor. Therefore, when we consider a prognostic model on cystoid spaces, the presence of large cysts with less bridging between inner and outer retina, foveal cysts, cysts with hyper-reflective material or hyper-reflective wall are all possible poor visual prognostic indicators [35–37].

Subretinal fluid (SRF) accumulation has been reported to be present in 15–30% of eyes with DMO [38]. The presence of NSD correlates with higher choroidal thickness, more hyperreflective foci, external limiting membrane (ELM) disruption but there are insufficient data to suggest that NSD is a visual prognostic marker.

3.1.3. Photoreceptor Integrity

Photoreceptors may be affected by the overlying capillary non-perfusion or continual fluid accumulation can secondarily affect the outer layers [39,40]. Enhanced resolution of OCT has enabled distinct imaging of the photoreceptor inner segment/outer segment (IS/OS) junction, now known as the ellipsoid zone (EZ) [41]. The EZ marks the increased mitochondrial content in the photoreceptor inner segment ellipsoid, critical to photoreceptor function [41,42]. It is now understood that EZ discontinuity (loss of outer segments) precedes disruption of ELM (loss of photoreceptor cell bodies), EZ loss being representative of extensive photoreceptor cell body damage, with poorer visual prognosis [41,42]. The relative importance of ELM and EZ in preserving VA is unclear. An important consideration with the growing literature regarding the EZ is how to objectively and consistently evaluate disruption. Few studies have analyzed the percentage disruption of each layer, few have classified the EZ disruption into less than or more than 200 microns, while others have merely comment on the integrity [42–45]. Considering all the evidence on the effect of outer retinal changes on OCT on visual outcome, it can be concluded that a visible sub-foveal loss of ellipsoid layer or ELM and the percentage disruption are definite poor visual prognostic indicators [46–50].

Microperimetry data have revealed that loss of EZ may lead to 3.28 dB reduction in retinal sensitivity [51]. Some authors have found correlation between EZ disruption and significant reduction of macular sensitivity in DMO eyes. Wang et al. have reported the macular integrity index (MI) as a useful benchmark of reflecting functional status in DMO [52]. MI denotes the percentage of threshold drop to measure local functional deterioration, and has been found to be significantly and independently correlated with EZ disruption. However, one should bear in mind that visualisation of EZ and ELM in eyes with gross oedema may be challenging. Hence, the presence of subfoveal intact EZ and ELM are definite good prognostic indicators but their absence in the presence of gross oedema may not be a poor prognostic indicator and it is prudent to wait for fluid resolution before commenting on the integrity of EZ and ELM in these eyes.

3.1.4. Intraretinal Hyperreflective Foci/Dots

Dot-like hyperreflective lesions within retinal layers have been described in OCT of DMO eyes, known as hyperreflective foci or dots (HRF/ HRD). These HRF have been proposed as lipoprotein exudates which pass into the interstitium after a breach in the inner blood-retinal barrier, and may possibly be precursors of hard exudates [53,54]. Hyperreflective foci may also indicate damaged photoreceptors and retinal pigment epithelium hyperplasia or metaplasia in other retinal diseases [54,55]. HRF are seen to be present initially in the inner retinal layers, from where they migrate into the outer layers. Subretinal HRF may end up as subfoveal hard exudates after resolution of NSD [56]. HRF present inside retinal cysts or lining cyst borders are believed to correspond to an advanced morphologic pattern of DMO which may be refractory to treatments [57,58]. Overall, HRF are believed to represent markers of inflammation in the retina [59].

A reduction of HRF has been seen with anti-VEGF injections [37,60]. The association of HRF and VA is controversial [53,60–64]. Whilst some reports suggest that eyes with HRF should be treated with intravitreal steroids, others have found similar response to either anti-VEGF or steroid. HRF responsive to anti-VEGF agents may be seen predominantly in the inner retinal layers; it has been proposed that the outer retinal layer HRFs may actually be a different entity. Anti-VEGF responsive inner retinal HRFs have been proposed to be of microglial origin [21,65]. On the contrary, the outer retinal HRFs, which are probable precursors of hard exudates, may not regress with treatment [56,66–68].

In summary, HRFs seem to be a bridge between the subclinical breakdown of the inner blood–retina barrier and the clinical swelling of retinal layers, and may indicate microvascular damage. However, more research is required to understand its relevance as a visual prognostic indicator [69]. Moreover, different reports have defined HRF based on their reflectivity either similar to the retinal pigment epithelium or to the surrounding tissue, and there is a need for standardization of the same [61,70].

3.1.5. Disorganization of Retinal Inner Layers

Disorganization of inner retinal layers (DRIL) is believed to be a secondary morphological change of the retina when the synaptic connections in the bipolar, horizontal and amacrine cells are affected, leading to loss of transmission between the inner retina and outer retina [71]. In DMO, both ischaemia and inflammation may lead to neuronal and glial degeneration that result in DRIL [72,73]. Visual outcome depends on the intactness and organization of retinal pathways, and the presence of DRIL may be an indirect measure of intactness of the local neural connections in the retina. Unlike other OCT based morphological parameters of the inner retina that do not correlate well with functional outcomes, DRIL is a better predictor of visual outcome [20].

DRIL has been proposed as a robust and surrogate biomarker of visual function in existing or resolved DMO [71,74]. DRIL may not be specific for DR or DMO, and may be a common response to retinal stress in presence of ischaemia. However, it may not always be associated with poor prognosis, highlighting that there may be further ultrastructural changes within DRIL that may more accurately define visual outcome. The presence of DRIL may also be associated with other OCT changes like EZ and ELM disruption, enlarged foveal avascular zone (FAZ) on OCTA. DRIL is associated with other functional changes including subnormal multifocal electroretinogram, impaired contrast sensitivity and visual field tests [75–81].

DRIL at the parafoveal area may be associated with a worse baseline VA in DMO eyes. Presence of DRIL at initiation of treatment, new-onset DRIL or increase in DRIL have all been reported as poor prognostic indicator of visual outcome in DMO eyes [71,82,83]. Recently, authors have noted that for each 100-μm increase in DRIL, there is a reduction in VA by approximately six letters, that is more than one line on the ETDRS chart [84]. The baseline volume of the intraretinal fluid in DMO may also show positive correlation with DRIL, along with poorer final VA after treatment with bevacizumab [85]. More importantly, DRIL may reverse with anti-VEGF treatment and the amount of change over initial few months may affect the final VA, independent of CST [71,79,82,85].

It is important that DRIL reversal be re-examined. However, these findings show that there may be a minimum area of DRIL required before visual function is affected. Reversal of DRIL is suggestive of decompression of the retina or realignment of neuronal and glial cells after transient disorganization in DMO. The definition or grading of DRIL in eyes with DMO may be challenging. Previous studies have assessed DRIL using SDOCT imaging via a Heidelberg Spectralis system (Heidelberg Engineering, Heidelberg, Germany), using a standard imaging protocol of 49 B-scans spanning a 20 × 20 frame in high-resolution mode [79]. Investigators have observed good correlations among DRIL and other OCT variables on different OCT platforms [86]. Understanding the variability in estimation of DRIL extent and ways to measure the reversal of DRIL may help us set meaningful thresholds in regard to its correlation with functional outcomes in DMO eyes. In summary, the presence of parafoveal DRIL is a definite poor visual prognostic indicator.

3.1.6. Choroid

There are contradicting reports on the changes in choroidal thickness and DMO. Firstly, investigators have shown increased choroidal thickness in DMO in keeping with the fact that choroidal vasculature is dependent on VEGF [87]. A higher choroidal thickness at baseline and its response to anti-VEGF may be a surrogate for good response to treatment [88]. Conversely, a thick choroid post anti-VEGF may also indicate an eye that is a poor responder to anti-VEGF and an indicator of chronicity of DMO. However, thick choroid is not an indicator for DMO recurrence or higher number of injections [89]. On the contrary, other investigators have suggested that choroidal blood volume, flow and velocity decreases in diabetic eyes with DMO, which may lead to hypoxia of the outer retina, thereby leading to increase in VEGF levels [90–92]. The choroid in DMO is also not affected by systemic factors such as increased HbA1c, blood pressure, cholesterol or abnormal renal function [93]. At the present time, there is insufficient evidence to suggest that changes in central choroidal thickness, subfoveal choroidal thickness or choroidal vascularity index influence visual outcome in DMO [94–104].

3.2. OCT-Angiography

Optical coherence tomography angiography (OCTA) is a noninvasive technique by which signals detected from blood flow in retinal and choroidal vasculature can be used to yield blood flow maps and quantify retinal perfusion in all vascular layers of the retina and the choroid [105]. Currently, the delineation of the capillary plexuses is done automatically by each OCTA software into the superior capillary plexus (SCP) which is embedded in the ganglion cell layer and nerve fiber layer, the deep capillary plexus (DCP) in the inner nuclear layer, and the choriocapillaris (CC).

Macular oedema may affect the reproducibility and effectiveness of OCTA in the measurement of FAZ area, perimeter and circularity, flow in CC and vascular density (VD) indices, especially in DCP [106,107]. Non-perfusion areas in DCP with VD reduction and flow-voids in CC layer correspond to photoreceptor/ellipsoid zone disruption in macular ischemia cases [108–110]. If the resolution of DMO persists for over 12 months, photoreceptors may show long-term recovery along with improved visual outcome, especially in eyes which have better baseline DCP integrity [111]. Thus, baseline VD of DCP is a predictor of photoreceptor recovery (ellipsoid zone integrity) and subsequent visual outcome. Moreover, areas of ischemia or capillary dropouts on OCTA may correspond to areas of lower macular sensitivity on microperimetry [112]. DMO eyes may have more microaneurysms (MA) in both the capillary layers, with poor responders to anti-VEGF showing more MAs in DCP [113,114]. The significance of the location of MAs in SCP versus DCP is unclear. However, as DCP also contributes to the blood supply of the photoreceptor layer, MAs in the DCP may be associated with loss of outer retinal integrity compared to MAs in SCP, which is associated with poorer visual outcomes. However, further research is required in this area [115–117].

RISE/RIDE study data suggest that anti-VEGF is efficacious in DMO irrespective of macular perfusion status at baseline [118]. There is no difference in VD of SCP and DCP in both treatment responders and non-responders in DMO, and there is no correlation between treatment response and baseline or final VD, suggesting that macular perfusion may not be a predictor of treatment response [119]. A higher resolution of DCP is required to evaluate if changes in DCP may be used as a prognostic marker for anti-VEGF responsiveness in DMO eyes [111].

The role of macular perfusion in final visual prognosis is certain but the best parameter to measure macular perfusion on OCT-A remains to be understood.

3.3. Colour Fundus Photography

The association of baseline diabetic retinopathy severity score (DRSS) score with visual outcome of patients with CI-DME is also of interest. Patients who show a limited early response to treatment may have better baseline VA, thinner CMT and lesser severity of diabetic retinopathy [120]. The overall impact of DRSS score on visual outcome of CI-DME patients treated in the VIVID and VISTA study showed that the VA outcomes did not differ based on baseline DRSS score stratified into ETDRS severity levels $\leq 43, 47$, and ≥ 53 [120,121].

4. Other Factors Affecting Visual Outcome
Role of Systemic Risk Factors

There is significant evidence that systemic risk factors such as hyperglycaemia, hypertension and hyperlipidemia need to be optimally controlled before and after the onset of DMO and this needs to be emphasized to each patient. However, people with optimally controlled systemic risk factors do develop visual impairment due to DMO and some with poorly controlled risk factors do not develop DMO throughout their whole diabetes history. There is insufficient evidence that suboptimal control of any of these systemic risk factors influences visual prognosis due to DMO. These observations suggest that local ocular features are relatively more important in determining the visual status of eyes with DMO.

5. Conclusions

This review on factors determining visual impairment in DMO or visual prognostic factors in eyes post-treatment suggest the need for a systematic evaluation of both subjective and objective definite predictive factors in a well characterized cohort of sufficient sample size. The relative importance of these factors is unclear. Baseline VA is an important predictor of final visual gain, and patients need to be treated early, to prevent irreversible VA loss. However, even if patients present with visual loss at a later stage, initiating prompt treatment may still help preservation of remaining vision. OCT has significant importance in deciding initiation and continuation of treatment in DMO eyes. Although a change in central subfoveal thickness is a secondary outcome in clinical trials and is used for monitoring anatomical treatment response, it is not a visual prognostic indicator. Ultrastructural features, such as photoreceptor integrity and the presence of DRIL, may serve as better prognostic indicators. Although poor macular perfusion is likely to have a negative impact, further studies on it are required to ascertain the best parameter to measure macular perfusion. Systemic factors do not seem to have a significant role as a visual prognostic indicator. Although individual factors may not be of significance, the combination of these may form response phenotypes that are of prognostic significance. Prognostic models on visual impairment should be developed before personalised medicine can be provided for patients with DMO.

Author Contributions: Conception and design: S.S. (Sobha Sivaprasad) and S.S. (Sagnik Sen). Analysis and interpretation: S.S. (Sagnik Sen), K.R., and S.S. (Sobha Sivaprasad). Data collection: S.S. (Sagnik Sen). Obtained funding: K.R. and S.S. (Sobha Sivaprasad). Overall responsibility: S.S. (Sagnik Sen), K.R., and S.S. (Sobha Sivaprasad). All authors have read and agreed to the published version of the manuscript.

Funding: The ORNATE India project is funded by the GCRF UKRI (MR/P207881/1). The research is supported by the NIHR Biomedical Research Centre at Moorfields Eye Hospital NHS Foundation Trust and UCL Institute of Ophthalmology.

Institutional Review Board Statement: Not applicable.

Informed Consent Statement: Not applicable.

Data Availability Statement: Not applicable.

Conflicts of Interest: Sagnik Sen and Kim Ramasamy declare no conflict of interest; Sobha Sivaprasad reported receiving research grants from Novartis, Bayer, Allergan, Roche, Boehringer Ingelheim, and Optos Plc, travel grants from Novartis and Bayer, speaker fees from Novartis, Bayer, and Optos Plc, and attending advisory board meetings for Novartis, Bayer, Allergan, Roche, Boehringer Ingelheim, Optos Plc, Oxurion, Opthes, Apellis, Oculis and Heidelberg Engineering.

References

1. International Diabetes Federation. IDF Diabetes Atlas, 9th ed. Brussels, Belgium: International Diabetes Federation, 2019. Available online: https://www.diabetesatlas.org/en/resources/ (accessed on 19 April 2021).
2. Early Photocoagulation for Diabetic Retinopathy. ETDRS Report Number 9. Early Treatment Diabetic Retinopathy Study Research Group. *Ophthalmology* **1991**, *98* (Suppl. 5), 766–785. [CrossRef]
3. Virgili, G.; Parravano, M.; Evans, J.R.; Gordon, I.; Lucenteforte, E. Anti-Vascular Endothelial Growth Factor for Diabetic Macular Oedema: A Network Meta-Analysis. *Cochrane Database Syst. Rev.* **2018**, *10*, CD007419. [CrossRef] [PubMed]
4. Mitchell, P.; Bandello, F.; Schmidt-Erfurth, U.; Lang, G.E.; Massin, P.; Schlingemann, R.O.; Sutter, F.; Simader, C.; Burian, G.; Gerstner, O.; et al. The RESTORE Study: Ranibizumab Monotherapy or Combined with Laser versus Laser Monotherapy for Diabetic Macular Edema. *Ophthalmology* **2011**, *118*, 615–625. [CrossRef] [PubMed]
5. Baker, C.W.; Glassman, A.R.; Beaulieu, W.T.; Antoszyk, A.N.; Browning, D.J.; Chalam, K.V.; Grover, S.; Jampol, L.M.; Jhaveri, C.D.; Melia, M.; et al. Effect of Initial Management with Aflibercept vs Laser Photocoagulation vs Observation on Vision Loss Among Patients with Diabetic Macular Edema Involving the Center of the Macula and Good Visual Acuity: A Randomized Clinical Trial. *JAMA* **2019**, *321*, 1880–1894. [CrossRef] [PubMed]
6. Nguyen, Q.D.; Brown, D.M.; Marcus, D.M.; Boyer, D.S.; Patel, S.; Feiner, L.; Gibson, A.; Sy, J.; Rundle, A.C.; Hopkins, J.J.; et al. Ranibizumab for Diabetic Macular Edema: Results from 2 Phase III Randomized Trials: RISE and RIDE. *Ophthalmology* **2012**, *119*, 789–801. [CrossRef] [PubMed]
7. Brown, D.M.; Nguyen, Q.D.; Marcus, D.M.; Boyer, D.S.; Patel, S.; Feiner, L.; Schlottmann, P.G.; Rundle, A.C.; Zhang, J.; Rubio, R.G.; et al. Long-Term Outcomes of Ranibizumab Therapy for Diabetic Macular Edema: The 36-Month Results from Two Phase III Trials: RISE and RIDE. *Ophthalmology* **2013**, *120*, 2013–2022. [CrossRef] [PubMed]
8. Mitchell, P.; Chong, V.; Group, R.S. Baseline Predictors of 3-Year Responses to Ranibizumab and Laser Photocoagulation Therapy in Patients with Visual Impairment Due to Diabetic Macular Edema (DME): The RESTORE Study. *Invest. Ophthalmol. Vis. Sci.* **2013**, *54*, 2373.
9. Wells, J.A.; Glassman, A.R.; Jampol, L.M.; Aiello, L.P.; Antoszyk, A.N.; Baker, C.W.; Bressler, N.M.; Browning, D.J.; Connor, C.G.; Elman, M.J.; et al. Association of Baseline Visual Acuity and Retinal Thickness With 1-Year Efficacy of Aflibercept, Bevacizumab, and Ranibizumab for Diabetic Macular Edema. *JAMA Ophthalmol.* **2016**, *134*, 127–134. [CrossRef]
10. Halim, S.; Gurudas, S.; Chandra, S.; Greenwood, J.; Sivaprasad, S. Evaluation of Real-World Early Response of DMO to Aflibercept Therapy to Inform Future Clinical Trial Design of Novel Investigational Agents. *Sci. Rep.* **2020**, *10*, 16499. [CrossRef]
11. Diabetic Retinopathy Clinical Research Network; Browning, D.J.; Glassman, A.R.; Aiello, L.P.; Beck, R.W.; Brown, D.M.; Fong, D.S.; Bressler, N.M.; Danis, R.P.; Kinyoun, J.L.; et al. Relationship between Optical Coherence Tomography-Measured Central Retinal Thickness and Visual Acuity in Diabetic Macular Edema. *Ophthalmology* **2007**, *114*, 525–536.

12. Scott, I.U.; VanVeldhuisen, P.C.; Oden, N.L.; Ip, M.S.; Blodi, B.A.; Jumper, J.M.; Figueroa, M.; SCORE Study Investigator Group. SCORE Study Report 1: Baseline Associations between Central Retinal Thickness and Visual Acuity in Patients with Retinal Vein Occlusion. *Ophthalmology* **2009**, *116*, 504–512. [CrossRef] [PubMed]
13. Keane, P.A.; Liakopoulos, S.; Chang, K.T.; Wang, M.; Dustin, L.; Walsh, A.C.; Sadda, S.R. Relationship between Optical Coherence Tomography Retinal Parameters and Visual Acuity in Neovascular Age-Related Macular Degeneration. *Ophthalmology* **2008**, *115*, 2206–2214. [CrossRef] [PubMed]
14. Ou, W.C.; Brown, D.M.; Payne, J.F.; Wykoff, C.C. Relationship Between Visual Acuity and Retinal Thickness During Anti-Vascular Endothelial Growth Factor Therapy for Retinal Diseases. *Am. J. Ophthalmol.* **2017**, *180*, 8–17. [CrossRef] [PubMed]
15. Maggio, E.; Sartore, M.; Attanasio, M.; Maraone, G.; Guerriero, M.; Polito, A.; Pertile, G. Anti-Vascular Endothelial Growth Factor Treatment for Diabetic Macular Edema in a Real-World Clinical Setting. *Am. J. Ophthalmol.* **2018**, *195*, 209–222. [CrossRef]
16. Midena, E.; Gillies, M.; Katz, T.A.; Metzig, C.; Lu, C.; Ogura, Y. Impact of Baseline Central Retinal Thickness on Outcomes in the VIVID-DME and VISTA-DME Studies. *J. Ophthalmol.* **2018**, *2018*, 3640135. [CrossRef]
17. Ziemssen, F.; Schlottman, P.G.; Lim, J.I.; Agostini, H.; Lang, G.E.; Bandello, F. Initiation of Intravitreal Aflibercept Injection Treatment in Patients with Diabetic Macular Edema: A Review of VIVID-DME and VISTA-DME Data. *Int. J. Retina Vitreous* **2016**, *2*, 16. [CrossRef]
18. Pelosini, L.; Hull, C.C.; Boyce, J.F.; McHugh, D.; Stanford, M.R.; Marshall, J. Optical Coherence Tomography May Be Used to Predict Visual Acuity in Patients with Macular Edema. *Investig. Ophthalmol. Vis. Sci.* **2011**, *52*, 2741–2748. [CrossRef]
19. Otani, T.; Kishi, S.; Maruyama, Y. Patterns of Diabetic Macular Edema with Optical Coherence Tomography. *Am. J. Ophthalmol.* **1999**, *127*, 688–693. [CrossRef]
20. Otani, T.; Kishi, S. Correlation between Optical Coherence Tomography and Fluorescein Angiography Findings in Diabetic Macular Edema. *Ophthalmology* **2007**, *114*, 104–107. [CrossRef]
21. Byeon, S.H.; Chu, Y.K.; Hong, Y.T.; Kim, M.; Kang, H.M.; Kwon, O.W. New Insights into the Pathoanatomy of Diabetic Macular Edema: Angiographic Patterns and Optical Coherence Tomography. *Retina* **2012**, *32*, 1087–1099. [CrossRef]
22. Murakami, T.; Nishijima, K.; Akagi, T.; Uji, A.; Horii, T.; Ueda-Arakawa, N.; Muraoka, Y.; Yoshimura, N. Optical Coherence Tomographic Reflectivity of Photoreceptors beneath Cystoid Spaces in Diabetic Macular Edema. *Investig. Ophthalmol. Vis. Sci.* **2012**, *53*, 1506–1511. [CrossRef] [PubMed]
23. Deák, G.G.; Bolz, M.; Ritter, M.; Prager, S.; Benesch, T.; Schmidt-Erfurth, U.; Diabetic Retinopathy Research Group Vienna. A Systematic Correlation between Morphology and Functional Alterations in Diabetic Macular Edema. *Investig. Ophthalmol. Vis. Sci.* **2010**, *51*, 6710–6714. [CrossRef] [PubMed]
24. Reznicek, L.; Cserhati, S.; Seidensticker, F.; Liegl, R.; Kampik, A.; Ulbig, M.; Neubauer, A.S.; Kernt, M. Functional and Morphological Changes in Diabetic Macular Edema over the Course of Anti-Vascular Endothelial Growth Factor Treatment. *Acta Ophthalmol.* **2013**, *91*, e529–e536. [CrossRef] [PubMed]
25. Bressler, S.B.; Ayala, A.R.; Bressler, N.M.; Melia, M.; Qin, H.; Ferris, F.L.; Flaxel, C.J.; Friedman, S.M.; Glassman, A.R.; Jampol, L.M.; et al. Persistent Macular Thickening After Ranibizumab Treatment for Diabetic Macular Edema With Vision Impairment. *JAMA Ophthalmol.* **2016**, *134*, 278–285. [CrossRef]
26. Bressler, S.B.; Glassman, A.R.; Almukhtar, T.; Bressler, N.M.; Ferris, F.L.; Googe, J.M.; Gupta, S.K.; Jampol, L.M.; Melia, M.; Wells, J.A.; et al. Five-Year Outcomes of Ranibizumab With Prompt or Deferred Laser Versus Laser or Triamcinolone Plus Deferred Ranibizumab for Diabetic Macular Edema. *Am. J. Ophthalmol.* **2016**, *164*, 57–68. [CrossRef]
27. Boyer, D.S. Treatment of Moderately Severe to Severe Nonproliferative Diabetic Retinopathy with Intravitreal Aflibercept Injection: 52-Week Results from the Phase 3 PANORAMA Study. *Investig. Ophthalmol. Vis. Sci.* **2019**, *60*, 1731.
28. Yeung, L.; Lima, V.C.; Garcia, P.; Landa, G.; Rosen, R.B. Correlation between Spectral Domain Optical Coherence Tomography Findings and Fluorescein Angiography Patterns in Diabetic Macular Edema. *Ophthalmology* **2009**, *116*, 1158–1167. [CrossRef]
29. Yanoff, M.; Fine, B.S.; Brucker, A.J.; Eagle, R.C. Pathology of Human Cystoid Macular Edema. *Surv. Ophthalmol.* **1984**, *28*, 505–511. [CrossRef]
30. Yamamoto, S.; Yamamoto, T.; Hayashi, M.; Takeuchi, S. Morphological and Functional Analyses of Diabetic Macular Edema by Optical Coherence Tomography and Multifocal Electroretinograms. *Graefes. Arch. Clin. Exp. Ophthalmol.* **2001**, *239*, 96–101. [CrossRef]
31. Al Faran, A.; Mousa, A.; Al Shamsi, H.; Al Gaeed, A.; Ghazi, N.G. Spectral Domain Optical Coherence Tomography Predictors of Visual Outcome in Diabetic Cystoid Macular Edema after Bevacizumab Injection. *Retina* **2014**, *34*, 1208–1215. [CrossRef]
32. Ehlers, J.P.; Uchida, A.; Hu, M.; Figueiredo, N.; Kaiser, P.K.; Heier, J.S.; Brown, D.M.; Boyer, D.S.; Do, D.V.; Gibson, A.; et al. Higher-Order Assessment of OCT in Diabetic Macular Edema from the VISTA Study: Ellipsoid Zone Dynamics and the Retinal Fluid Index. *Ophthalmol. Retina* **2019**, *3*, 1056–1066. [CrossRef] [PubMed]
33. Yalçın, G.; Özdek, Ş.; Baran Aksakal, F.N. Defining Cystoid Macular Degeneration in Diabetic Macular Edema: An OCT-Based Single-Center Study. *Turk. J. Ophthalmol.* **2019**, *49*, 315–322. [CrossRef] [PubMed]
34. Yalçın, N.G.; Özdek, Ş. The Relationship Between Macular Cyst Formation and Ischemia in Diabetic Macular Edema. *Turk. J. Ophthalmol.* **2019**, *49*, 194–200. [CrossRef]
35. Murakami, T.; Suzuma, K.; Uji, A.; Yoshitake, S.; Dodo, Y.; Fujimoto, M.; Yoshitake, T.; Miwa, Y.; Yoshimura, N. Association between Characteristics of Foveal Cystoid Spaces and Short-Term Responsiveness to Ranibizumab for Diabetic Macular Edema. *Jpn J. Ophthalmol.* **2018**, *62*, 292–301. [CrossRef] [PubMed]

36. Terada, N.; Murakami, T.; Uji, A.; Dodo, Y.; Mori, Y.; Tsujikawa, A. Hyperreflective Walls in Foveal Cystoid Spaces as a Biomarker of Diabetic Macular Edema Refractory to Anti-VEGF Treatment. *Sci. Rep.* **2020**, *10*, 7299. [CrossRef]
37. Zur, D.; Iglicki, M.; Busch, C.; Invernizzi, A.; Mariussi, M.; Loewenstein, A.; International Retina Group. OCT Biomarkers as Functional Outcome Predictors in Diabetic Macular Edema Treated with Dexamethasone Implant. *Ophthalmology* **2018**, *125*, 267–275. [CrossRef]
38. Gaucher, D.; Sebah, C.; Erginay, A.; Haouchine, B.; Tadayoni, R.; Gaudric, A.; Massin, P. Optical Coherence Tomography Features during the Evolution of Serous Retinal Detachment in Patients with Diabetic Macular Edema. *Am. J. Ophthalmol.* **2008**, *145*, 289–296. [CrossRef]
39. Scarinci, F.; Jampol, L.M.; Linsenmeier, R.A.; Fawzi, A.A. Association of Diabetic Macular Nonperfusion with Outer Retinal Disruption on Optical Coherence Tomography. *JAMA Ophthalmol.* **2015**, *133*, 1036–1044. [CrossRef]
40. Freyler, H.; Prskavec, F.; Stelzer, N. Diabetic choroidopathy–a retrospective fluorescein angiography study. Preliminary report. *Klin. Monbl. Augenheilkd.* **1986**, *189*, 144–147. [CrossRef] [PubMed]
41. Mohamed, E.H.E.S.; Nabil, K.M.; Gomaa, A.R.; Haddad, O.H.E.H. External limiting membrane and ellipsoid zone integrity and presenting visual acuity in treatment naïve center involved diabetic macular edema. *EC Ophthalmology.* **2018**, *9*, 408–421.
42. Shah, V.A.; Brown, J.S.; Mahmoud, T.H. Correlation of Outer Retinal Microstucture and Foveal Thickness with Visual Acuity after Pars Plana Vitrectomy for Complications of Proliferative Diabetic Retinopathy. *Retina* **2012**, *32*, 1775–1780. [CrossRef] [PubMed]
43. Muftuoglu, I.K.; Mendoza, N.; Gaber, R.; Alam, M.; You, Q.; Freeman, W.R. Integrity of outer retinal layers after resolution of central involved diabetic macular edema. *Retina* **2017**, *37*, 2015–2024. [CrossRef]
44. Maheshwary, A.S.; Oster, S.F.; Yuson, R.M.S.; Cheng, L.; Mojana, F.; Freeman, W.R. The Association between Percent Disruption of the Photoreceptor Inner Segment-Outer Segment Junction and Visual Acuity in Diabetic Macular Edema. *Am. J. Ophthalmol.* **2010**, *150*, 63–67.e1. [CrossRef]
45. Hannouche, R.Z.; de Avila, M.P.; Isaac, D.L.C.; Silva, R.S.; Rassi, A.R. Correlation between Central Subfield Thickness, Visual Acuity and Structural Changes in Diabetic Macular Edema. *Arq. Bras. Oftalmol.* **2012**, *75*, 183–187. [CrossRef] [PubMed]
46. Shin, H.J.; Lee, S.H.; Chung, H.; Kim, H.C. Association between Photoreceptor Integrity and Visual Outcome in Diabetic Macular Edema. *Graefes. Arch. Clin. Exp. Ophthalmol.* **2012**, *250*, 61–70. [CrossRef] [PubMed]
47. Chhablani, J.K.; Kim, J.S.; Cheng, L.; Kozak, I.; Freeman, W. External Limiting Membrane as a Predictor of Visual Improvement in Diabetic Macular Edema after Pars Plana Vitrectomy. *Graefes. Arch. Clin. Exp. Ophthalmol.* **2012**, *250*, 1415–1420. [CrossRef]
48. Shen, Y.; Liu, K.; Xu, X. Correlation Between Visual Function and Photoreceptor Integrity in Diabetic Macular Edema: Spectral-Domain Optical Coherence Tomography. *Curr. Eye Res.* **2016**, *41*, 391–399. [CrossRef] [PubMed]
49. Kaya, M.; Karahan, E.; Ozturk, T.; Kocak, N.; Kaynak, S. Effectiveness of Intravitreal Ranibizumab for Diabetic Macular Edema with Serous Retinal Detachment. *Korean J. Ophthalmol.* **2018**, *32*, 296–302. [CrossRef]
50. Mori, Y.; Murakami, T.; Suzuma, K.; Ishihara, K.; Yoshitake, S.; Fujimoto, M.; Dodo, Y.; Yoshitake, T.; Miwa, Y.; Tsujikawa, A. Relation between Macular Morphology and Treatment Frequency during Twelve Months with Ranibizumab for Diabetic Macular Edema. *PLoS ONE* **2017**, *12*, e0175809. [CrossRef]
51. Yohannan, J.; Bittencourt, M.; Sepah, Y.J.; Hatef, E.; Sophie, R.; Moradi, A.; Liu, H.; Ibrahim, M.; Do, D.V.; Coulantuoni, E.; et al. Association of Retinal Sensitivity to Integrity of Photoreceptor Inner/Outer Segment Junction in Patients with Diabetic Macular Edema. *Ophthalmology* **2013**, *120*, 1254–1261. [CrossRef]
52. Wang, J.-W.; Jie, C.-H.; Tao, Y.-J.; Meng, N.; Hu, Y.-C.; Wu, Z.-Z.; Cai, W.-J.; Gong, X.-M. Macular Integrity Assessment to Determine the Association between Macular Microstructure and Functional Parameters in Diabetic Macular Edema. *Int. J. Ophthalmol.* **2018**, *11*, 1185–1191. [PubMed]
53. Uji, A.; Murakami, T.; Nishijima, K.; Akagi, T.; Horii, T.; Arakawa, N.; Muraoka, Y.; Ellabban, A.A.; Yoshimura, N. Association between Hyperreflective Foci in the Outer Retina, Status of Photoreceptor Layer, and Visual Acuity in Diabetic Macular Edema. *Am. J. Ophthalmol.* **2012**, *153*, 710–717, 717.e1. [CrossRef] [PubMed]
54. Bolz, M.; Schmidt-Erfurth, U.; Deak, G.; Mylonas, G.; Kriechbaum, K.; Scholda, C.; Diabetic Retinopathy Research Group Vienna. Optical Coherence Tomographic Hyperreflective Foci: A Morphologic Sign of Lipid Extravasation in Diabetic Macular Edema. *Ophthalmology* **2009**, *116*, 914–920. [CrossRef] [PubMed]
55. Schuman, S.G.; Koreishi, A.F.; Farsiu, S.; Jung, S.; Izatt, J.A.; Toth, C.A. Photoreceptor Layer Thinning over Drusen in Eyes with Age-Related Macular Degeneration Imaged in Vivo with Spectral-Domain Optical Coherence Tomography. *Ophthalmology* **2009**, *116*, 488–496.e2. [CrossRef] [PubMed]
56. Coscas, G.; De Benedetto, U.; Coscas, F.; Li Calzi, C.I.; Vismara, S.; Roudot-Thoraval, F.; Bandello, F.; Souied, E. Hyperreflective Dots: A New Spectral-Domain Optical Coherence Tomography Entity for Follow-up and Prognosis in Exudative Age-Related Macular Degeneration. *Ophthalmologica* **2013**, *229*, 32–37. [CrossRef] [PubMed]
57. Ota, M.; Nishijima, K.; Sakamoto, A.; Murakami, T.; Takayama, K.; Horii, T.; Yoshimura, N. Optical Coherence Tomographic Evaluation of Foveal Hard Exudates in Patients with Diabetic Maculopathy Accompanying Macular Detachment. *Ophthalmology* **2010**, *117*, 1996–2002. [CrossRef]
58. Jonas, J.B.; Jonas, R.A.; Neumaier, M.; Findeisen, P. Cytokine Concentration in Aqueous Humor of Eyes with Diabetic Macular Edema. *Retina* **2012**, *32*, 2150–2157. [CrossRef]
59. Midena, E.; Pilotto, E.; Bini, S. Hyperreflective Intraretinal Foci as an OCT Biomarker of Retinal Inflammation in Diabetic Macular Edema. *Investig. Ophthalmol. Vis. Sci.* **2018**, *59*, 5366. [CrossRef]

60. Kang, J.-W.; Chung, H.; Chan Kim, H. Correlation of optical coherence tomographic hyperreflective foci with visual outcomes in different patterns of diabetic macular edema. *Retina* **2016**, *36*, 1630–1639. [CrossRef]
61. Hwang, T.S.; Jia, Y.; Gao, S.S.; Bailey, S.T.; Lauer, A.K.; Flaxel, C.J.; Wilson, D.J.; Huang, D. Optical coherence tomography angiography features of diabetic retinopathy. *Retina* **2015**, *35*, 2371–2376. [CrossRef]
62. Li, B.; Zhang, B.; Chen, Y.; Li, D. Optical Coherence Tomography Parameters Related to Vision Impairment in Patients with Diabetic Macular Edema: A Quantitative Correlation Analysis. *J. Ophthalmol.* **2020**, *2020*, 5639284. [CrossRef] [PubMed]
63. Vujosevic, S.; Torresin, T.; Bini, S.; Convento, E.; Pilotto, E.; Parrozzani, R.; Midena, E. Imaging Retinal Inflammatory Biomarkers after Intravitreal Steroid and Anti-VEGF Treatment in Diabetic Macular Oedema. *Acta Ophthalmol.* **2017**, *95*. [CrossRef] [PubMed]
64. Vujosevic, S.; Berton, M.; Bini, S.; Casciano, M.; Cavarzeran, F.; Midena, E. Hyperreflective retinal spots and visual function after anti-vascular endothelial growth factor treatment in center-involving diabetic macular edema. *Retina* **2016**, *36*, 1298–1308. [CrossRef] [PubMed]
65. Zeng, H.; Green, W.R.; Tso, M.O.M. Microglial Activation in Human Diabetic Retinopathy. *Arch. Ophthalmol.* **2008**, *126*, 227–232. [CrossRef]
66. Domalpally, A.; Ip, M.S.; Ehrlich, J.S. Effects of Intravitreal Ranibizumab on Retinal Hard Exudate in Diabetic Macular Edema: Findings from the RIDE and RISE Phase III Clinical Trials. *Ophthalmology* **2015**, *122*, 779–786. [CrossRef] [PubMed]
67. Davoudi, S.; Papavasileiou, E.; Roohipoor, R.; Cho, H.; Kudrimoti, S.; Hancock, H.; Hoadley, S.; Andreoli, C.; Husain, D.; James, M.; et al. Optical coherence tomography characteristics of macular edema and hard exudates and their association with lipid serum levels in type 2 diabetes. *Retina* **2016**, *36*, 1622–1629. [CrossRef] [PubMed]
68. Framme, C.; Wolf, S.; Wolf-Schnurrbusch, U. Small Dense Particles in the Retina Observable by Spectral-Domain Optical Coherence Tomography in Age-Related Macular Degeneration. *Investig. Ophthalmol. Vis. Sci.* **2010**, *51*, 5965–5969. [CrossRef] [PubMed]
69. Framme, C.; Schweizer, P.; Imesch, M.; Wolf, S.; Wolf-Schnurrbusch, U. Behavior of SD-OCT-Detected Hyperreflective Foci in the Retina of Anti-VEGF-Treated Patients with Diabetic Macular Edema. *Investig. Ophthalmol. Vis. Sci.* **2012**, *53*, 5814–5818. [CrossRef]
70. Schreur, V.; Altay, L.; van Asten, F.; Groenewoud, J.M.M.; Fauser, S.; Klevering, B.J.; Hoyng, C.B.; de Jong, E.K. Hyperreflective Foci on Optical Coherence Tomography Associate with Treatment Outcome for Anti-VEGF in Patients with Diabetic Macular Edema. *PLoS ONE* **2018**, *13*, e0206482. [CrossRef]
71. Sun, J.K.; Lin, M.M.; Lammer, J.; Prager, S.; Sarangi, R.; Silva, P.S.; Aiello, L.P. Disorganization of the Retinal Inner Layers as a Predictor of Visual Acuity in Eyes with Center-Involved Diabetic Macular Edema. *JAMA Ophthalmol.* **2014**, *132*, 1309–1316. [CrossRef]
72. Bek, T. Transretinal Histopathological Changes in Capillary-Free Areas of Diabetic Retinopathy. *Acta Ophthalmol. (Copenh)* **1994**, *72*, 409–415. [CrossRef] [PubMed]
73. Barber, A.J.; Lieth, E.; Khin, S.A.; Antonetti, D.A.; Buchanan, A.G.; Gardner, T.W. Neural Apoptosis in the Retina during Experimental and Human Diabetes. Early Onset and Effect of Insulin. *J. Clin. Investig.* **1998**, *102*, 783–791. [CrossRef] [PubMed]
74. Grewal, D.S.; O'Sullivan, M.L.; Kron, M.; Jaffe, G.J. Association of Disorganization of Retinal Inner Layers With Visual Acuity In Eyes With Uveitic Cystoid Macular Edema. *Am. J. Ophthalmol.* **2017**, *177*, 116–125. [CrossRef] [PubMed]
75. Nadri, G.; Saxena, S.; Stefanickova, J.; Ziak, P.; Benacka, J.; Gilhotra, J.S.; Kruzliak, P. Disorganization of Retinal Inner Layers Correlates with Ellipsoid Zone Disruption and Retinal Nerve Fiber Layer Thinning in Diabetic Retinopathy. *J. Diabetes Complicat.* **2019**, *33*, 550–553. [CrossRef]
76. Onishi, A.C.; ashraf, M.; soetikno, B.T.; fawzi, A.A. Multilevel ischemia in disorganization of the retinal inner layers on projection-resolved optical coherence tomography angiography. *Retina* **2019**, *39*, 1588–1594. [CrossRef]
77. Joltikov, K.A.; Sesi, C.A.; de Castro, V.M.; Davila, J.R.; Anand, R.; Khan, S.M.; Farbman, N.; Jackson, G.R.; Johnson, C.A.; Gardner, T.W. Disorganization of Retinal Inner Layers (DRIL) and Neuroretinal Dysfunction in Early Diabetic Retinopathy. *Investig. Ophthalmol. Vis. Sci.* **2018**, *59*, 5481–5486. [CrossRef]
78. Moein, H.-R.; Novais, E.A.; Rebhun, C.B.; Cole, E.D.; Louzada, R.N.; Witkin, A.J.; Baumal, C.R.; Duker, J.S.; Waheed, N.K. Optical coherence tomography angiography to detect macular capillary ischemia in patients with inner retinal changes after resolved diabetic macular edema. *Retina* **2018**, *38*, 2277–2284. [CrossRef]
79. Radwan, S.H.; Soliman, A.Z.; Tokarev, J.; Zhang, L.; van Kuijk, F.J.; Koozekanani, D.D. Association of Disorganization of Retinal Inner Layers with Vision After Resolution of Center-Involved Diabetic Macular Edema. *JAMA Ophthalmol.* **2015**, *133*, 820–825. [CrossRef]
80. Eraslan, S.; Yıldırım, Ö.; Dursun, Ö.; Dinç, E.; Orekici Temel, G. Relationship Between Final Visual Acuity and Optical Coherence Tomography Findings in Patients with Diabetic Macular Edema Undergoing Anti-VEGF Therapy. *Turk. J. Ophthalmol.* **2020**, *50*, 163–168. [CrossRef]
81. Khojasteh, H.; Riazi-Esfahani, H.; Khalili Pour, E.; Faghihi, H.; Ghassemi, F.; Bazvand, F.; Mahmoudzadeh, R.; Salabati, M.; Mirghorbani, M.; Riazi Esfahani, M. Multifocal Electroretinogram in Diabetic Macular Edema and Its Correlation with Different Optical Coherence Tomography Features. *Int. Ophthalmol.* **2020**, *40*, 571–581. [CrossRef]
82. Sun, J.K.; Radwan, S.H.; Soliman, A.Z.; Lammer, J.; Lin, M.M.; Prager, S.G.; Silva, P.S.; Aiello, L.B.; Aiello, L.P. Neural Retinal Disorganization as a Robust Marker of Visual Acuity in Current and Resolved Diabetic Macular Edema. *Diabetes* **2015**, *64*, 2560–2570. [CrossRef] [PubMed]

83. Zur, D.; Iglicki, M.; Sala-Puigdollers, A.; Chhablani, J.; Lupidi, M.; Fraser-Bell, S.; Mendes, T.S.; Chaikitmongkol, V.; Cebeci, Z.; Dollberg, D.; et al. Disorganization of Retinal Inner Layers as a Biomarker in Patients with Diabetic Macular Oedema Treated with Dexamethasone Implant. *Acta Ophthalmol.* **2020**, *98*, e217–e223. [CrossRef] [PubMed]

84. Das, R.; Spence, G.; Hogg, R.E.; Stevenson, M.; Chakravarthy, U. Disorganization of Inner Retina and Outer Retinal Morphology in Diabetic Macular Edema. *JAMA Ophthalmol.* **2018**, *136*, 202–208. [CrossRef]

85. De, S.; Saxena, S.; Kaur, A.; Mahdi, A.A.; Misra, A.; Singh, M.; Meyer, C.H.; Akduman, L. Sequential Restoration of External Limiting Membrane and Ellipsoid Zone after Intravitreal Anti-VEGF Therapy in Diabetic Macular Oedema. *Eye (Lond)* **2020**. [CrossRef]

86. Sampani, K.; Abdulaal, M.; Peiris, T.; Lin, M.M.; Pitoc, C.; Ledesma, M.; Lammer, J.; Silva, P.S.; Aiello, L.P.; Sun, J.K. Comparison of SDOCT Scan Types for Grading Disorganization of Retinal Inner Layers and Other Morphologic Features of Diabetic Macular Edema. *Transl. Vis. Sci. Technol.* **2020**, *9*, 45. [CrossRef] [PubMed]

87. Funatsu, H.; Noma, H.; Mimura, T.; Eguchi, S.; Hori, S. Association of Vitreous Inflammatory Factors with Diabetic Macular Edema. *Ophthalmology* **2009**, *116*, 73–79. [CrossRef] [PubMed]

88. Kim, M.; Cho, Y.J.; Lee, C.H.; Lee, S.C. Effect of Intravitreal Dexamethasone Implant on Retinal and Choroidal Thickness in Refractory Diabetic Macular Oedema after Multiple Anti-VEGF Injections. *Eye (Lond)* **2016**, *30*, 718–725. [CrossRef] [PubMed]

89. Mathis, T.; Mendes, M.; Dot, C.; Bouteleux, V.; Machkour-Bentaleb, Z.; El Chehab, H.; Agard, E.; Denis, P.; Kodjikian, L. Increased Choroidal Thickness: A New Indicator for Monitoring Diabetic Macular Oedema Recurrence. *Acta Ophthalmol.* **2020**, *98*, e968–e974. [CrossRef] [PubMed]

90. Schocket, L.S.; Brucker, A.J.; Niknam, R.M.; Grunwald, J.E.; DuPont, J.; Brucker, A.J. Foveolar Choroidal Hemodynamics in Proliferative Diabetic Retinopathy. *Int. Ophthalmol.* **2004**, *25*, 89–94. [CrossRef]

91. Nagaoka, T.; Kitaya, N.; Sugawara, R.; Yokota, H.; Mori, F.; Hikichi, T.; Fujio, N.; Yoshida, A. Alteration of Choroidal Circulation in the Foveal Region in Patients with Type 2 Diabetes. *Br. J. Ophthalmol.* **2004**, *88*, 1060–1063. [CrossRef]

92. Cao, J.; McLeod, S.; Merges, C.A.; Lutty, G.A. Choriocapillaris Degeneration and Related Pathologic Changes in Human Diabetic Eyes. *Arch. Ophthalmol.* **1998**, *116*, 589–597. [CrossRef] [PubMed]

93. Kase, S.; Endo, H.; Takahashi, M.; Ito, Y.; Saito, M.; Yokoi, M.; Katsuta, S.; Sonoda, S.; Sakamoto, T.; Ishida, S.; et al. Alteration of Choroidal Vascular Structure in Diabetic Macular Edema. *Graefes. Arch. Clin. Exp. Ophthalmol.* **2020**, *258*, 971–977. [CrossRef] [PubMed]

94. Gupta, P.; Thakku, S.G.; Sabanayagam, C.; Tan, G.; Agrawal, R.; Cheung, C.M.G.; Lamoureux, E.L.; Wong, T.-Y.; Cheng, C.-Y. Characterisation of Choroidal Morphological and Vascular Features in Diabetes and Diabetic Retinopathy. *Br. J. Ophthalmol.* **2017**, *101*, 1038–1044. [CrossRef]

95. Endo, H.; Kase, S.; Takahashi, M.; Saito, M.; Yokoi, M.; Sugawara, C.; Katsuta, S.; Ishida, S.; Kase, M. Relationship between Diabetic Macular Edema and Choroidal Layer Thickness. *PLoS ONE* **2020**, *15*, e0226630. [CrossRef] [PubMed]

96. Eliwa, T.F.; Hegazy, O.S.; Mahmoud, S.S.; Almaamon, T. Choroidal Thickness Change in Patients with Diabetic Macular Edema. *Ophthalmic. Surg. Lasers Imaging Retina* **2017**, *48*, 970–977. [CrossRef]

97. Lee, K.F.; Lim, J.W.; Shin, M.C. Comparison of Choroidal Thickness in Patients with Diabetes by Spectral-domain Optical Coherence Tomography. *Korean J. Ophthalmol.* **2013**, *27*, 433–439. [CrossRef] [PubMed]

98. Kim, M.; Ha, M.J.; Choi, S.Y.; Park, Y.-H. Choroidal Vascularity Index in Type-2 Diabetes Analyzed by Swept-Source Optical Coherence Tomography. *Sci. Rep.* **2018**, *8*, 70. [CrossRef] [PubMed]

99. Rewbury, R.; Want, A.; Varughese, R.; Chong, V. Subfoveal Choroidal Thickness in Patients with Diabetic Retinopathy and Diabetic Macular Oedema. *Eye (Lond)* **2016**, *30*, 1568–1572. [CrossRef]

100. Sudhalkar, A.; Chhablani, J.K.; Venkata, A.; Raman, R.; Rao, P.S.; Jonnadula, G.B. Choroidal Thickness in Diabetic Patients of Indian Ethnicity. *Indian J. Ophthalmol.* **2015**, *63*, 912–916. [CrossRef] [PubMed]

101. Laíns, I.; Talcott, K.E.; Santos, A.R.; Marques, J.H.; Gil, P.; Gil, J.; Figueira, J.; Husain, D.; Kim, I.K.; Miller, J.W.; et al. Choroidal thickness in diabetic retinopathy assessed with swept-source optical coherence tomography. *Retina* **2018**, *38*, 173–182. [CrossRef]

102. Galgauskas, S.; Laurinavičiūtė, G.; Norvydaitė, D.; Stech, S.; Ašoklis, R. Changes in Choroidal Thickness and Corneal Parameters in Diabetic Eyes. *Eur. J. Ophthalmol.* **2016**, *26*, 163–167. [CrossRef] [PubMed]

103. Xu, J.; Xu, L.; Du, K.F.; Shao, L.; Chen, C.X.; Zhou, J.Q.; Wang, Y.X.; You, Q.S.; Jonas, J.B.; Wei, W.B. Subfoveal Choroidal Thickness in Diabetes and Diabetic Retinopathy. *Ophthalmology* **2013**, *120*, 2023–2028. [CrossRef] [PubMed]

104. Agrawal, R.; Gupta, P.; Tan, K.-A.; Cheung, C.M.G.; Wong, T.-Y.; Cheng, C.-Y. Choroidal Vascularity Index as a Measure of Vascular Status of the Choroid: Measurements in Healthy Eyes from a Population-Based Study. *Sci. Rep.* **2016**, *6*, 21090. [CrossRef] [PubMed]

105. Or, C.; Sabrosa, A.S.; Sorour, O.; Arya, M.; Waheed, N. Use of OCTA, FA, and Ultra-Widefield Imaging in Quantifying Retinal Ischemia: A Review. *Asia Pac. J. Ophthalmol. (Phila)* **2018**, *7*, 46–51. [CrossRef]

106. Gill, A.; Cole, E.D.; Novais, E.A.; Louzada, R.N.; de Carlo, T.; Duker, J.S.; Waheed, N.K.; Baumal, C.R.; Witkin, A.J. Visualization of Changes in the Foveal Avascular Zone in Both Observed and Treated Diabetic Macular Edema Using Optical Coherence Tomography Angiography. *Int. J. Retina Vitreous* **2017**, *3*, 19. [CrossRef]

107. Tarassoly, K.; Miraftabi, A.; Soltan Sanjari, M.; Parvaresh, M.M. The relationship between foveal avascular zone area, vessel density, and cystoid changes in diabetic retinopathy: An optical coherence tomography angiography study. *Retina* **2018**, *38*, 1613–1619. [CrossRef] [PubMed]

108. Coscas, G.; Lupidi, M.; Coscas, F. Optical Coherence Tomography Angiography in Diabetic Maculopathy. *Dev. Ophthalmol.* **2017**, *60*, 38–49. [CrossRef]
109. Cennamo, G.; Romano, M.R.; Nicoletti, G.; Velotti, N.; de Crecchio, G. Optical Coherence Tomography Angiography versus Fluorescein Angiography in the Diagnosis of Ischaemic Diabetic Maculopathy. *Acta Ophthalmol.* **2017**, *95*, e36–e42. [CrossRef]
110. Dodo, Y.; Suzuma, K.; Ishihara, K.; Yoshitake, S.; Fujimoto, M.; Yoshitake, T.; Miwa, Y.; Murakami, T. Clinical Relevance of Reduced Decorrelation Signals in the Diabetic Inner Choroid on Optical Coherence Tomography Angiography. *Sci. Rep.* **2017**, *7*, 5227. [CrossRef]
111. Moon, B.G.; Um, T.; Lee, J.; Yoon, Y.H. Correlation between Deep Capillary Plexus Perfusion and Long-Term Photoreceptor Recovery after Diabetic Macular Edema Treatment. *Ophthalmol. Retina* **2018**, *2*, 235–243. [CrossRef]
112. Pereira, F.; Godoy, B.R.; Maia, M.; Regatieri, C.V. Microperimetry and OCT Angiography Evaluation of Patients with Ischemic Diabetic Macular Edema Treated with Monthly Intravitreal Bevacizumab: A Pilot Study. *Int. J. Retina Vitreous* **2019**, *5*, 24. [CrossRef] [PubMed]
113. Lee, J.; Moon, B.G.; Cho, A.R.; Yoon, Y.H. Optical Coherence Tomography Angiography of DME and Its Association with Anti-VEGF Treatment Response. *Ophthalmology* **2016**, *123*, 2368–2375. [CrossRef] [PubMed]
114. Hasegawa, N.; Nozaki, M.; Takase, N.; Yoshida, M.; Ogura, Y. New Insights Into Microaneurysms in the Deep Capillary Plexus Detected by Optical Coherence Tomography Angiography in Diabetic Macular Edema. *Investig. Ophthalmol. Vis. Sci.* **2016**, *57*, OCT348–OCT355. [CrossRef] [PubMed]
115. Ishibazawa, A.; Nagaoka, T.; Takahashi, A.; Omae, T.; Tani, T.; Sogawa, K.; Yokota, H.; Yoshida, A. Optical Coherence Tomography Angiography in Diabetic Retinopathy: A Prospective Pilot Study. *Am. J. Ophthalmol.* **2015**, *160*, 35–44.e1. [CrossRef]
116. Yu, S.; Lu, J.; Cao, D.; Liu, R.; Liu, B.; Li, T.; Luo, Y.; Lu, L. The Role of Optical Coherence Tomography Angiography in Fundus Vascular Abnormalities. *BMC Ophthalmol.* **2016**, *16*, 107. [CrossRef]
117. Peres, M.B.; Kato, R.T.; Kniggendorf, V.F.; Cole, E.D.; Onal, S.; Torres, E.; Louzada, R.; Belfort, R.; Duker, J.S.; Novais, E.A.; et al. Comparison of Optical Coherence Tomography Angiography and Fluorescein Angiography for the Identification of Retinal Vascular Changes in Eyes With Diabetic Macular Edema. *Ophthalmic Surg. Lasers Imaging Retina* **2016**, *47*, 1013–1019. [CrossRef] [PubMed]
118. Reddy, R.K.; Pieramici, D.J.; Gune, S.; Ghanekar, A.; Lu, N.; Quezada-Ruiz, C.; Baumal, C.R. Efficacy of Ranibizumab in Eyes with Diabetic Macular Edema and Macular Nonperfusion in RIDE and RISE. *Ophthalmology* **2018**, *125*, 1568–1574. [CrossRef]
119. Sorour, O.A.; Sabrosa, A.S.; Yasin Alibhai, A.; Arya, M.; Ishibazawa, A.; Witkin, A.J.; Baumal, C.R.; Duker, J.S.; Waheed, N.K. Optical Coherence Tomography Angiography Analysis of Macular Vessel Density before and after Anti-VEGF Therapy in Eyes with Diabetic Retinopathy. *Int. Ophthalmol.* **2019**, *39*, 2361–2371. [CrossRef]
120. Pieramici, D.; Singh, R.P.; Gibson, A.; Saroj, N.; Vitti, R.; Berliner, A.J.; Zeitz, O.; Metzig, C.; Soo, Y.; Zhu, X.; et al. Outcomes of Diabetic Macular Edema Eyes with Limited Early Response in the VISTA and VIVID Studies. *Ophthalmol. Retina* **2018**, *2*, 558–566. [CrossRef]
121. Staurenghi, G.; Feltgen, N.; Arnold, J.J.; Katz, T.A.; Metzig, C.; Lu, C.; Holz, F.G.; VIVID-DME and VISTA-DME study investigators. Impact of Baseline Diabetic Retinopathy Severity Scale Scores on Visual Outcomes in the VIVID-DME and VISTA-DME Studies. *Br. J. Ophthalmol.* **2018**, *102*, 954–958. [CrossRef]

Journal of
Personalized
Medicine

Article

Diabetic Macular Edema Treated with 577-nm Subthreshold Micropulse Laser: A Real-Life, Long-Term Study

Luisa Frizziero [1], Andrea Calciati [1], Tommaso Torresin [1], Giulia Midena [2], Raffaele Parrozzani [1], Elisabetta Pilotto [1] and Edoardo Midena [1,2,*]

[1] Department of Neuroscience—Ophthalmology, University of Padova, 35128 Padova, Italy; lfrizziero@gmail.com (L.F.); andrear.cal@gmail.com (A.C.); tommasotorresin@gmail.com (T.T.); raffaele.parrozani@unipd.it (R.P.); elisabetta.pilotto@unipd.it (E.P.)

[2] IRCCS—Fondazione Bietti, 00120 Rome, Italy; giulia.midena@gmail.com

* Correspondence: edoardo.midena@unipd.it; Tel.: +39-049-821-2110

Abstract: The aim of this study was to evaluate the long-term efficacy and safety of 577-nm subthreshold micropulse laser (SMPL) treatment in a large population of patients affected by mild diabetic macular edema (DME) in a real-life setting. We retrospectively evaluated 134 eyes affected by previously untreated center-involving mild DME, and treated with 577-nm SMPL, using fixed parameters. Retreatment was performed at 3 months, in case of persistent retinal thickening. Optical coherence tomography (OCT), along with short and near-infrared fundus autofluorescence, were used to confirm long-term safety. At the end of at least one year follow-up, a significant improvement in visual acuity was documented, compared to baseline (77.3 ± 4.5 and 79.4 ± 4.4 ETDRS score at baseline and at final follow-up, respectively), as well as a reduction in the mean retinal thickness of the thickest ETDRS macular sector at baseline. A reduction in the central retinal thickness and the mean thickness of the nine ETDRS sectors was also found, without reaching statistical significance. No patients required intravitreal injections. No adverse effects were detected. This study suggests that 577-nm SMPL is a safe and repeatable treatment for mild DME that may be applied to real-life clinical settings using fixed parameters and protocols.

Keywords: subthreshold micropulse laser; 577-nm laser; laser fixed parameters; diabetic retinopathy; diabetic macular edema; optical coherence tomography; autofluorescence; real-life

Citation: Frizziero, L.; Calciati, A.; Torresin, T.; Midena, G.; Parrozzani, R.; Pilotto, E.; Midena, E. Diabetic Macular Edema Treated with 577-nm Subthreshold Micropulse Laser: A Real-Life, Long-Term Study. *J. Pers. Med.* **2021**, *11*, 405. https://doi.org/10.3390/jpm11050405

Academic Editors: Peter D. Westenskow and Andreas Ebneter

Received: 19 April 2021
Accepted: 11 May 2021
Published: 13 May 2021

Publisher's Note: MDPI stays neutral with regard to jurisdictional claims in published maps and institutional affiliations.

1. Introduction

The World Health Organization has estimated that about 460 million people worldwide are affected by diabetes mellitus, and both the number of cases and the prevalence of diabetes have been steadily increasing over the past few decades [1]. Despite considerable progress made in recent decades in understanding its pathogenesis and implementing effective therapeutic strategies, diabetic retinal involvement, affecting approximately one-third of diabetic subjects, remains a major public health problem with important socioeconomic implications [1]. Diabetic macular edema (DME) is the earliest and most common cause of visual loss in patients with DR [1,2]. Although VEGF has been found to be a key molecule to the development of macular edema, DME is multifactorial and there are numerous potential biochemical pathways involved in its pathogenesis, such as inflammation [3]. Intravitreal anti-VEGF (Aflibercept, Ranibizumab and Bevacizumab) and steroids (Dexamethasone and Fluocinolone implant) injections are the current standard of care in DME treatment; however, they cause a significant burden for both the patients and the health care system because of the necessity of retreatments and follow-ups [4]. Moreover, some patients respond incompletely, or are complete non-responders to anti-VEGF injections. Lastly, the safety of intravitreal corticosteroids has limitations in many countries [5].

The Early Treatment Diabetic Retinopathy Study (ETDRS) demonstrated the effectiveness of macular laser photocoagulation in preventing progressive visual loss in eyes affected by DME [6]. However, side effects related to external retina necrosis in treated areas moved this treatment modality to limited cases, mainly with extra-foveal thickening [7]. Subthreshold micropulse laser (SMPL) combines the value of subthreshold treatment with micropulse technology to produce subthreshold retinal changes (invisible to any current imaging and functional diagnostic technology) without any evidence of retinal damage [8]. Yellow 577-nm SMPL has recently gained increasing diffusion in clinical practice. However, at present, standard and widely accepted indication criteria, setting parameters and treatment protocols to apply in clinical practice are still lacking [4,9,10].

Therefore, the aim of this study was to evaluate the long-term efficacy and safety of 577-nm SMPL treatment in a large population of patients affected by mild DME in a real-life clinical setting using a fixed setting of laser parameters, follow-up and retreatment modalities.

2. Materials and Methods

2.1. Study Population

This was a retrospective longitudinal study. Patients were consecutively enrolled among those who underwent SMPL for DME involving the foveal area between January 2017 and December 2019. A written consent form was obtained from all patients, as well as the approval from our institutional ethics committee (protocol number 3194/AO/14).

The research was carried out in accordance with the Declaration of Helsinki. The clinical records of all identified subjects were retrospectively reviewed. Inclusion criteria were: type 1 or 2 diabetes mellitus (DM) with good metabolic control (HbA1C < 8%), previously untreated center-involving mild macular edema with central retinal thickness (CRT) $\leq$ 400 μm and a follow-up period, after the first SMPL treatment, of at least 12 months [11]. Exclusion criteria were: proliferative DR, vision-limiting ocular conditions other than DR (such as amblyopia, age-related macular degeneration, myopic degeneration, retinal dystrophies, optic neuropathies, retinal vascular diseases, advanced glaucoma or corneal opacity of any cause), CRT > 400 μm and/or a history of previous macular laser treatment, previous intravitreal injection therapy, history of previous ocular trauma or surgery, except for uncomplicated cataract extraction.

Patients underwent a full ophthalmological evaluation at baseline and every three months after the first SMPL treatment, including spectral domain-optical coherence tomography OCT (SD-OCT) and fundus autofluorescence (FAF) using Spectralis (Spectralis HRA+OCT, Heidelberg Engineering, Heidelberg, Germany). Short-wavelength and near-infrared FAF were performed on a 30° field of view centered onto the fovea, as previously described [12]. Briefly, images were taken with a confocal scanning laser ophthalmoscope. The images' resolution was 768 × 768 pixels. Argon laser light (488 nm) was used for short-wavelength FAF. A band-pass filter with a cut-off at 500 nm, included in the system, was inserted in front of the detector. A diode infrared laser light (790 nm) was used for near-infrared FAF. A band-pass filter with a cut-off at 830 nm, included in the system, was inserted in front of the detector [12]. An OCT macular map scan pattern was used, with a 20° × 20° scan area centered onto the fovea. Forty-nine horizontal scans 120 μm apart from each other were obtained in an automated pattern. At each follow-up examination, the follow-up modality was enabled, allowing for the examination to be repeated according to the previous baseline examination.

Retinal thickness was automatically calculated by the device in nine ETDRS sectors: a 1 mm diameter circular zone centered on the fovea (CRT) with two concentric outer rings of 3 and 6 mm diameters, each divided into 4 sectors. Mean total retinal thickness was recorded in each of the nine ETDRS sectors.

2.2. Laser Treatment Delivery

Prior to treatment, topical anesthesia with proparacaine 0.5% drops was administered. Yellow micropulse laser was delivered through an Ocular Mainster Focal/Grid lens (Ocular Instruments, Washington, DC, USA) using the Iridex IQ 577 (Iridex Corporation,

Mountain view, CA, USA) instrument, with the following standard settings: spot size of 100 μm, power of 250 mW, duration of each spot 200 ms and 5% duty cycle, as previously described [9,13]. Confluent, non-overlapping spots, were applied over the whole macular area. Retreatment was performed every 3 months, according to the persistence of intraretinal fluid.

2.3. Statistical Analysis

The study parameters were analyzed using the usual descriptive statistic indicators: mean and standard deviation for quantitative variables, and absolute and relative (percentage) frequency for qualitative ones.

The following result parameters were considered in the statistical analysis of retinal thickness values gathered from the OCT exam: thickness of the central sector, mean thickness (arithmetic mean of 9 ETDRS sectors) and thickness of the sector that had the highest value at baseline.

Variations from the baseline to the end of follow-up were evaluated for each parameter by means of an ANOVA model adjusted for total follow-up length and the replication of measurements for patients who contributed with both eyes.

Then, the subgroups of eyes observed at different follow-up intervals were used to analyze the relationship between retinal thickness (the three parameters separately) and the number of laser treatments in relation to the length of follow-up. This analysis was performed by means of a multivariate linear regression model using retinal thickness as a dependent variable (one model for each parameter previously mentioned), and the number of laser treatments and the interval time as independent variables. Such models were adjusted for replication of measurements in patients who contributed with both eyes. The relationship was judged as noteworthy when the regression coefficient of the number of treatments variable resulted in statistical significance. Because of replication of testing, Bonferroni correction criterion was applied.

All the analyses were performed using SAS© 9.4 software (SAS Institute, Cary, NC, USA).

3. Results

A total of 134 eyes of 94 patients were enrolled. Mean diabetes duration was 21.2 ± 13.7 years and mean HbA1c was 7.5 ± 1.1 at baseline. Diabetic retinopathy was mild to moderate in 120 eyes and severe (for the presence of intraretinal microvascular abnormalities) in 14 eyes. Mean age at the beginning of follow-up was 66.7 ± 9.6 years. Patients' clinical and demographic characteristics are summarized in Table 1.

Table 1. Patients' clinical and demographic characteristics at baseline.

N. eyes/patients	134/94
Age, years, mean $\pm$ SD	66.7 ± 9.6
Sex, F:M	33:61
Type of diabetes, 1:2	21:73
BCVA, ETDRS score $\pm$ SD	77.3 ± 4.5
Diabetes duration, years $\pm$ SD	21.2 ± 13.7
HbA1c, mean $\pm$ SD	7.5 ± 1.1
Follow-up, months, mean $\pm$ SD	16.6 ± 6.5

N: number; F: female; M: male; SD: standard deviation; BCVA: best corrected visual acuity; ETDRS: early treatment diabetic retinopathy study.

Mean follow-up duration was 16.6 ± 6.5 months with an average of 2.3 ± 1.3 laser treatments per patient. At the end of follow-up, no significant change in HbA1c was detected. BCVA was 79.4 ± 4.4 ($p < 0.001$ compared to baseline) and a reduction of CRT and mean thickness of the nine ETDRS sectors was found, compared to baseline, without reaching statistical significance (-4.4 ± 49.5 μm, $p = 0.3009$ and -2.4 ± 22.9 μm, $p = 0.235$ for CRT and the nine sectors' mean retinal thickness, respectively). A significant reduction of the mean retinal thickness of the thickest ETDRS sector at baseline was documented

$(-12.7 \pm 36.2 \ \mu m, \ p < 0.0001)$ (Table 2, Figure 1). This reduction was significant at any follow-up visit.

Table 2. Morphologic results (whole follow-up).

	Baseline Thickness (μm, Mean ± SD)	End of Follow-Up Thickness (μm, Mean ± SD)	Thickness Variation (μm, Mean ± SD)	*p*-Value
CRT	309.2 ± 45.8	304.7 ± 54.6	−4.4 ± 49.5	0.3009
RetMean	327.2 ± 21.3	324.9 ± 27.2	−2.4 ± 22.9	0.2358
RetMax	373.3 ± 34.2	360 ± 41.8	−12.7 ± 36.2	<0.0001

SD: standard deviation; CRT: Central retinal thickness; RetMean: mean retinal thickness in 9 ETDRS sectors; RetMax: retinal thickness in the thickest ETDRS sector at baseline.

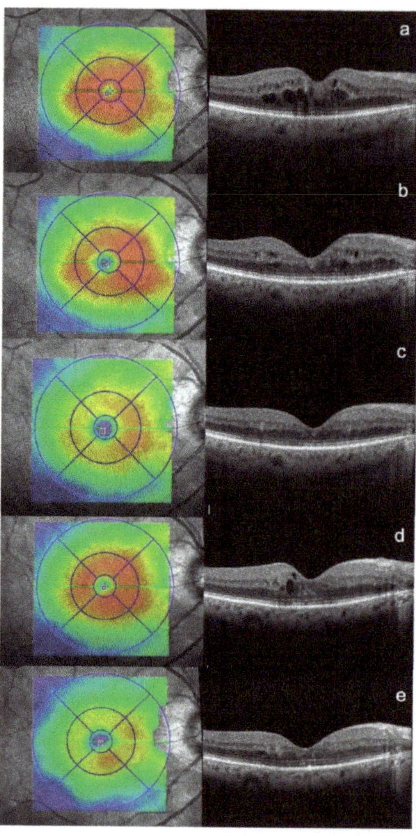

Figure 1. Optical coherence tomography (OCT) scans of a patient affected by diabetic macular edema: at (**a**) baseline, (**b**) 3 months after subthreshold micropulse macular laser (SMPL) 1st treatment, (**c**) 3 months after SMPL 2nd treatment, (**d**) additional 3 months after SMPL 2nd treatment, (**e**) 3 months after SMPL 3rd treatment.

A significant relationship was also found between a higher number of treatments and final lower thickness value at longer follow-up, for both mean retinal thickness in the nine ETDRS sectors and retinal thickness of the thickest sector at baseline (for longer follow-ups). Moreover, a greater number of laser treatments was significantly correlated with greater reduction in thickness in the thickest sector at baseline. After Bonferroni correction, significance was maintained for the relationship with the retinal thickness of the thickest sector at baseline. The analysis of laser effect on retinal thickness is summarized in Tables 3 and 4.

Table 3. Morphologic results at follow-up intervals: mean retinal thickness in 9 ETDRS sectors.

Follow-Up Visit	Laser Treatments (Mean ± SD)	Retinal thickness Variation from Baseline (Micrometers, Mean ± SD (p-Value))	Effect of Number of Laser Treatments on Retinal thickness [1]	Effect of Number of Laser Treatments on Retinal Thickness Variation [2]
1	1	−2.7 ± 8.3 (0.0071)	−0.6925 (0.7204)	1.5149 (0.2945)
2	1.4 ± 0.7	−3.0 ± 13.9 (0.0246)	−3.5751 (0.0663)	−1.3270 (0.5054)
3	1.7 ± 0.8	−2.1 ± 19.7 (0.2283)	−2.0692 (0.1096)	−2.1302 (0.3151)
4	1.9 ± 0.9	−1.3 ± 23.1 (0.5009)	−1.2346 (0.2912)	−2.7467 (0.2051)
5	2.0 ± 1.0	−1.3 ± 23.2 (0.5273)	−1.5412 (0.1260)	−2.9853 (0.1286)
6	2.2 ± 1.2	−2.0 ± 22.7 (0.3015)	−1.5063 (0.0645)	−2.5123 (0.1316)
7	2.2 ± 1.2	−2.2 ± 22.7 (0.2706)	−1.4373 (0.0635)	−2.2309 (0.1653)
8	2.2 ± 1.3	−2.3 ± 22.8 (0.2504)	−1.5357 (0.0370)	−2.2271 (0.1500)
9	2.3 ± 1.3	−2.4 ± 22.9 (0.2358)	−1.5711 (0.0309)	−2.2756 (0.1397)

[1] Regression coefficients (p-value) of multivariate linear regression model adjusted for repeated measures on the same eye. [2] Regression coefficients (p-value) of multivariate linear regression model adjusted for baseline value of retinal thickness. Statistically significant p-values according to Bonferroni adjustment due to multiplicity testing are reported in bold characters.

Table 4. Morphologic results at follow-up intervals: thickest sector at baseline.

Follow-Up Visit	Laser Treatments (Mean ± SD)	Retinal Thickness Variation from Baseline (Micrometers, Mean ± SD (p-Value))	Effect of Number of Laser Treatments on Retinal Thickness [1]	Effect of Number of Laser Treatments on Retinal Thickness Variation [2]
1	1	−9.8 ± 22.4 (**0.0005**)	3.7394 (0.4313)	9.7915 (0.0099)
2	1.4 ± 0.7	−10.2 ± 28.9 (**0.0003**)	−1.2129 (0.7553)	2.5570 (0.5176)
3	1.7 ± 0.8	−10.3 ± 34.5 (**0.0008**)	−3.8937 (0.0929)	−1.4021 (0.6968)
4	1.9 ± 0.9	−10.4 ± 35.9 (**0.0010**)	−3.3133 (0.0816)	−3.3168 (0.3137)
5	2.0 ± 1.0	−9.8 ± 36.1 (**0.0020**)	−3.3935 (0.0430)	−4.1176 (0.1761)
6	2.2 ± 1.2	−12.1 ± 35.8 (**0.0001**)	−4.6878 (**0.0005**)	−5.6862 (0.0288)
7	2.2 ± 1.2	−12.7 ± 36.2 (**<0.0001**)	−4.9631 (**<0.0001**)	−5.7742 (0.0219)
8	2.2 ± 1.3	−13.1 ± 36.5 (**<0.0001**)	−5.2394 (**<0.0001**)	−5.8335 (0.0163)
9	2.3 ± 1.3	−13.3 ± 36.5 (**<0.0001**)	−5.1858 (**<0.0001**)	−5.6725 (0.0188)

[1] Regression coefficients (p-value) of multivariate linear regression model adjusted for repeated measures on the same eye. [2] Regression coefficients (p-value) of multivariate linear regression model adjusted for baseline value of retinal thickness. Statistically significant p-values according to Bonferroni adjustment due to multiplicity testing are reported in bold characters.

OCT imaging, analyzed layer by layer, showed no laser side effects and no signs of outer retinal structure damage. Short-wave and near-infrared FAF acquired on a 30° field centered onto the fovea did not show any laser-related lesions in our study population, nor did the ophthalmoscopic fundus examination.

4. Discussion

The current devices for SMPL treatment are the infrared (810-nm) diode laser and the yellow laser, which emits at 577-nm [9,14]. Both wavelengths have been safely employed in treating patients affected by center-involving DME and comparable efficacy and safety have been reported [9,14]. The cornerstones of subthreshold retinal laser therapy were established and defined using an 810-nm near-infrared micropulse laser [15,16]. The 577-nm yellow laser was later introduced and, although at present it is the most commercially available device, its efficacy has been evaluated by fewer studies involving smaller sample sizes and shorter follow-up periods [9,13,15–22]. At present, a wide array of different duty cycle durations, spot sizes and power settings are employed by different clinicians, limiting, in our opinion, the diffuse clinical application of this technique [4,9,10,19]. This is a long-term follow-up retrospective report of patients treated with 577-nm SMPL for mild DME, with the aim of assessing the long-term morphologic efficacy and safety of repeated SMPL treatments using a fixed setting of laser parameters and follow-up and retreatment modalities.

A recent analysis compared fixed parameters treatment and titrated treatment for 577-nm yellow laser: the first one was recommended based on its greater speed of setup and the implicit avoidance of any possibility of titration errors [13]. Titration has also been discouraged due to the absence of any reliable, safe and scientifically-validated titration algorithms [23]. Fixed, previously validated parameters have been employed for all eyes included in this study, namely: 5% duty cycle, 200 µs duration, 250 mW power and 100 µm spot size [9,13]. Moreover, the whole macular area was treated in order to obtain a widespread "mass effect" involving the different cellular elements strictly correlated to each other in the macula, as suggested by Lavinsky et al. [14,24,25]. In fact, a correctly performed subthreshold micropulse laser treatment produces targeted metabolic retinal effects, without any necrotic (useless) effect [16]. Therefore, an extensive and confluent treatment is required to obtain a "mass effect" of cellular deactivation/activation [14,24]. No side effects have been reported both at OCT and FAF for all treated eyes.

As regards treatment indications, the efficacy of SMPL in severe edema has proven to be limited, possibly owing to the scattered distribution of laser energy throughout the target tissues [18,26]. Therefore, all treated eyes in this study were affected by mild DME with CRT inferior or equal to 400 µm [26]. Moreover, previous studies showed that the first morphological and functional results after SMPL appear at about three months [20]. Luttrull et al. found that macular thickness did not change significantly in the first 2 months after treatment. At 3 months, however, reduction of macular thickness was observed [27]. Lavinsky et al. reported a significant increase in BCVA from the third month after high-density SMPL [25]. A significant increase in central retinal sensitivity was also reported from the third month of follow-up after SMPL treatment [9]. Furthermore, in these studies, patients continued to improve over a long-term follow-up period (1 year) [21,25]. Therefore, at present, SMPL might be recommended in mild DME, with retreatment at least 3 months from the previous SMPL session.

Analyzing the functional and morphological results of our large number of patients treated with these standardized protocols and parameters, we found, at the end of the follow-up period (13.6 ± 6.5 months), improvement in visual acuity compared to baseline, which is associated with a reduction of retinal thickness that became significant only when considering the thickest ETDRS sector at baseline. This may be explained by the higher sensitivity of this last analysis: in case of mild edema, significant thickness variations may be difficult to detect, since a DME-related increase in thickness does not always affect the whole macular area and subsequent modifications can be "drowned out" by the relatively stable (by virtue of their already small baseline thickness) surrounding sectors, which will not shrink beyond a certain point. Considering the retinal thickness of the thickest ETDRS sector at baseline may allow for easier detection of thickness changes in the area that is most involved by the pathologic process at the beginning of the observation period. Moreover, these results suggest that the treatment of the whole macular area allows for the normalization of the most affected sectors and the stabilization of the other sectors, which are maintained over a long-term follow-up period.

Continuous improvement over a long-term follow-up period has already been suggested after SMPL treatment; moreover, we found that the number of treatments is also correlated to a greater reduction in macular thickness, and the relationship between retinal thickness and number of laser treatments increased in magnitude and statistical significance with the length of the observation period [21,25]. This result may be attributed to the slow and complex effect of SMPL that involves a multitude of retinal elements: repeating the treatment probably reinforces the normalizing process in the targeted retinal tissue and may be done without any significant safety concerns [21,28]. In the early stages of its application, it was speculated that the main target for SMPL action was RPE, and that the principal advantage of micropulse over continuous-wave lasers was the sparing of retinal and choroidal tissue adjacent to RPE [8,29]. More recently, aqueous humor (AH) analysis in eyes treated with 577-nm yellow SMPL for DME showed a significant reduction in AH concentration of chemotactic and pro-inflammatory cytokines after SMPL, as well as of specific Müller-cells related proteins (namely GFAP and Kir 4.1) [21,28]. Müller cells play a crucial role in the maintenance of the physiological retinal structure and function; they undergo activation and proliferation in response to different kinds of stressors and retinal injuries, including oxidative damage and hyperglycemia, and release pro-inflammatory and vaso-active substances that contribute to local inflammation and vascular permeability increase [30,31]. A reduction in retinal thickness accompanied by fluid reabsorption can also be related to a de-activating effect on Müller cells [20]. Thus, SMPL probably initiates a wide curtailing of local inflammatory processes that is accompanied by cellular de-activation and return to morphological integrity, with the restoration of the retinal microvascular network, as evidenced by a reduction of the foveal avascular zone area in the deep capillary plexus [22]. In this study, we did not analyze macular and peripheral perfusion status. Previous studies have reported the efficacy of both intravitreal steroids and anti-VEGF in reducing DME (and peripheral non-perfusion), regardless of the peripheral retinal perfusion [32,33]. However, further studies are needed to investigate the influence of retinal perfusion in mild DME and SMPL treatment. In conclusion, SMPL seems to cause a slow and generalized restoration of retinal homeostasis that includes blood-retinal barrier repair, reduction of local inflammation and widespread cellular normalization over the long term, as confirmed by our morphological results. All patients had good glycemic control and HbA1c remained unchanged during the study, showing that changes in mild DME were unrelated to the metabolic control.

Another noteworthy result is that, over the course of the follow-up period, no patients required intravitreal therapy. This confirms the previously reported potential for SMPL to control intraretinal fluid and alleviate the burden of intravitreal treatments [4,34,35]; SMPL was effective in keeping retinal thickness under control, stabilizing macular retinal homeostasis. A recent study has shown that visual acuity (in term of loss of five or more ETDRS letters) is not significantly different among eyes initially managed with intravitreal aflibercept, laser photocoagulation or observation at 2 years' follow-up, in patients with good visual acuity (>79 ETDRS score) at baseline [36]. However, the proportion of eyes with visual acuity of 20/20 or better was significantly greater with aflibercept than observation, but not with laser photocoagulation, and eyes in the laser photocoagulation group had a lower likelihood of receiving aflibercept injections compared to eyes in the observation group [36]. This difference seems to suggest a possible benefit of laser photocoagulation in reducing the need for anti-VEGF. As already mentioned, SMPL proved to be as effective as laser photocoagulation, without its adverse effects (macular scarring and visual scotomas) [37]. Therefore, SMPL may have a role as an effective long-term retinal maintenance therapy, including in patients with good visual acuity, minimizing the risks and side effects associated with laser photocoagulation and intravitreal injections.

With regards to safety, OCT analysis did not detect any sign of disruption of the outer retinal layers' integrity following 577-nm SMPL treatment; short-wave and near-infrared FAF images failed to disclose any secondary effects of the laser treatment on the macular area, and no visible result of the applied laser spots was apparent on ophthalmoscopy. No harmful effect of the laser treatment was detected in any eye over the course of follow-up, irrespective of the number of repeated laser treatments or total observation time.

The main limitation of this study is its retrospective nature. However, this was a real-life report on a large population of patients with mild center-involving DME, all treated by 577-nm SMPL with fixed parameters and retreatment modalities. It allowed us to prove the efficacy of treatment repetition over a long-term follow-up period, using a standardized treatment on the whole macular area. This may suggest the necessity of planning long-term treatment protocols with repeated SMPL sessions at pre-planned intervals of at least 3 months. This approach would also limit the need for control examinations, reducing the burden for patients and the health care system, and thus improving adherence to the treatment schedule.

Author Contributions: Conceptualization, E.M., T.T. and L.F.; methodology, E.M., L.F and A.C.; validation, E.P. and E.M.; formal analysis, E.M., L.F. and T.T.; investigation, E.M., L.F., G.M., T.T., R.P. and A.C.; resources E.M. and E.P.; data curation, R.P., T.T., G.M., E.M. and L.F.; writing—original draft preparation, LF, A.C. and G.M.; writing—review and editing, T.T., E.M. and R.P.; supervision, E.M, E.P. and R.P.; All authors have read and agreed to the published version of the manuscript.

Funding: This research received no external funding.

Institutional Review Board Statement: The study was conducted according to the guidelines of the Declaration of Helsinki, and approved by the Institutional Review Board of the Azienda Sanitaria di Padova.

Informed Consent Statement: Informed consent was obtained from all subjects involved in the study.

Data Availability Statement: The data presented in this study are available in the article. Eventual additional data are available on request from the corresponding author.

Acknowledgments: The research contribution by the G.B. Bietti Foundation was supported by Fondazione Roma and Ministry of Health. Luisa Frizziero is supported by Fondazione Umberto Veronesi.

Conflicts of Interest: The authors declare no conflict of interest.

References

1. Teo, Z.L.; Tham, Y.-C.; Yu, M.C.Y.; Chee, M.L.; Rim, T.H.; Cheung, N.; Bikbov, M.M.; Wang, Y.X.; Tang, Y.; Lu, Y.; et al. Global Prevalence of Diabetic Retinopathy and Projection of Burden through 2045: Systematic Review and Meta-analysis. *Ophthalmology* **2021**, *S0161–6420*, 00321–00323. [CrossRef]
2. Lee, R.; Wong, T.Y.; Sabanayagam, C. Epidemiology of diabetic retinopathy, diabetic macular edema and related vision loss. *Eye Vis.* **2015**, *2*, 17. [CrossRef]
3. Santos, A.R.; Ribeiro, L.; Bandello, F.; Lattanzio, R.; Egan, C.; Frydkjaer-Olsen, U.; García-Arumí, J.; Gibson, J.; Grauslund, J.; Harding, S.P.; et al. Functional and Structural Findings of Neurodegeneration in Early Stages of Diabetic Retinopathy: Cross-sectional Analyses of Baseline Data of the EUROCONDOR Project. *Diabetes* **2017**, *66*, 2503–2510. [CrossRef] [PubMed]
4. Moisseiev, E.; Abbassi, S.; Thinda, S.; Yoon, J.; Yiu, G.; Morse, L.S. Subthreshold micropulse laser reduces anti-VEGF injection burden in patients with diabetic macular edema. *Eur. J. Ophthalmol.* **2018**, *28*, 68–73. [CrossRef]
5. Blinder, K.J.; Dugel, P.U.; Chen, S.; Jumper, J.M.; Walt, J.G.; Hollander, A.D.; Scott, L.C. Anti-VEGF treatment of diabetic macular edema in clinical practice: Effectiveness and patterns of use (ECHO Study Report 1). *Clin. Ophthalmol.* **2017**, *11*, 393–401. [CrossRef] [PubMed]
6. Early Treatment Diabetic Retinopathy Study Research Group. Treatment Techniques and Clinical Guidelines for Photocoagulation of Diabetic Macular Edema. *Ophthalmology* **1987**, *94*, 761–774. [CrossRef]
7. Writing Committee for the Diabetic Retinopathy Clinical Research Network; Fong, D.S.; Strauber, S.F.; Aiello, L.P.; Beck, R.W.; Callanan, D.G.; Danis, R.P.; Davis, M.D.; Feman, S.S.; Ferris, F.; et al. Writing Committee for the Diabetic Retinopathy Clinical Research Network Comparison of the Modified Early Treatment Diabetic Retinopathy Study and Mild Macular Grid Laser Photocoagulation Strategies for Diabetic Macular Edema. *Arch. Ophthalmol.* **2007**, *125*, 469–480. [CrossRef] [PubMed]
8. Dorin, G. Subthreshold and micropulse diode laser photocoagulation. *Semin. Ophthalmol.* **2003**, *18*, 147–153. [CrossRef] [PubMed]
9. Vujosevic, S.; Martini, F.; Longhin, E.; Convento, E.; Cavarzeran, F.; Midena, E. Subthreshold Micropulse Yellow Laser Versus Subthreshold Micropulse Infrared Laser in Center-Involving Diabetic Macular Edema. *Retina* **2015**, *35*, 1594–1603. [CrossRef]
10. Chhablani, J.; Alshareef, R.; Kim, D.T.; Narayanan, R.; Goud, A.; Mathai, A. Comparison of different settings for yellow subthreshold laser treatment in diabetic macular edema. *BMC Ophthalmol.* **2018**, *18*, 168. [CrossRef] [PubMed]
11. Qaseem, A.; Wilt, T.J.; Kansagara, D.; Horwitch, C.; Barry, M.J.; Forciea, M.A. For the Clinical Guidelines Committee of the American College of Physicians Hemoglobin A1c Targets for Glycemic Control With Pharmacologic Therapy for Nonpregnant Adults With Type 2 Diabetes Mellitus: A Guidance Statement Update From the American College of Physicians. *Ann. Intern. Med.* **2018**, *168*, 569–576. [CrossRef] [PubMed]
12. Pilotto, E.; Vujosevic, S.; Melis, R.; Convento, E.; Sportiello, P.; Alemany-Rubio, E.; Segalina, S.; Midena, E. Short wavelength fundus autofluorescence versus near-infrared fundus autofluorescence, with microperimetric correspondence, in patients with geographic atrophy due to age-related macular degeneration. *Br. J. Ophthalmol.* **2010**, *95*, 1140–1144. [CrossRef]
13. Donati, M.C.; Murro, V.; Mucciolo, D.P.; Giorgio, D.; Cinotti, G.; Virgili, G.; Rizzo, S. Subthreshold yellow micropulse laser for treatment of diabetic macular edema: Comparison between fixed and variable treatment regimen. *Eur. J. Ophthalmol.* **2020**, 1120672120915169. [CrossRef]
14. Sivaprasad, S.; Dorin, G. Subthreshold diode laser micropulse photocoagulation for the treatment of diabetic macular edema. *Expert Rev. Med. Devices* **2012**, *9*, 189–197. [CrossRef]
15. Luttrull, J.K.; Musch, D.C.; Mainster, A.M. Subthreshold diode micropulse photocoagulation for the treatment of clinically significant diabetic macular oedema. *Br. J. Ophthalmol.* **2005**, *89*, 74–80. [CrossRef]
16. Luttrull, J.K. Subthreshold Diode Micropulse Laser Photocoagulation (SDM) as Invisible Retinal Phototherapy for Diabetic Macular Edema: A Review. *Curr. Diabetes Rev.* **2012**, *8*, 274–284. [CrossRef]
17. Latalska, M.; Prokopiuk, A.; Wróbel-Dudzinska, D.; Mackiewicz, J. Subthreshold micropulse yellow 577 nm laser therapy of diabetic macular oedema in rural and urban patients of south-eastern Poland. *Ann. Agric. Environ. Med.* **2017**, *24*, 96–99. [CrossRef] [PubMed]

18. Citirik, M. The impact of central foveal thickness on the efficacy of subthreshold micropulse yellow laser photocoagulation in diabetic macular edema. *Lasers Med. Sci.* **2018**, *34*, 907–912. [CrossRef]
19. Kwon, Y.H.; Lee, D.K.; Kwon, O.W. The Short-term Efficacy of Subthreshold Micropulse Yellow (577-nm) Laser Photocoagulation for Diabetic Macular Edema. *Korean J. Ophthalmol.* **2014**, *28*, 379–385. [CrossRef] [PubMed]
20. Vujosevic, S.; Frizziero, L.; Martini, F.; Bini, S.; Convento, E.; Cavarzeran, F.; Midena, E. Single Retinal Layer Changes After Subthreshold Micropulse Yellow Laser in Diabetic Macular Edema. *Ophthalmic Surg. Lasers Imaging Retin.* **2018**, *49*, e218–e225. [CrossRef]
21. Midena, E.; Bini, S.; Martini, F.; Enrica, C.; Pilotto, E.; Micera, A.; Esposito, G.; Vujosevic, S. Changes of Aqueous Humor Müller Cells' Biomarkers in Human Patients Affected by Diabetic Macular Edema After Subthreshold Micropulse Laser Treatment. *Retinal* **2020**, *40*, 126–134. [CrossRef] [PubMed]
22. Vujosevic, S.; Toma, C.; Villani, E.; Brambilla, M.; Torti, E.; Leporati, F.; Muraca, A.; Nucci, P.; De Cilla, S. Subthreshold Micropulse Laser in Diabetic Macular Edema: 1-Year Improvement in OCT/OCT-Angiography Biomarkers. *Transl. Vis. Sci. Technol.* **2020**, *9*, 31. [CrossRef] [PubMed]
23. Chang, D.B.; Luttrull, J.K. Comparison of Subthreshold 577 and 810 nm Micropulse Laser Effects on Heat-Shock Protein Activation Kinetics: Implications for Treatment Efficacy and Safety. *Transl. Vis. Sci. Technol.* **2020**, *9*, 23. [CrossRef] [PubMed]
24. Luttrull, J.K.; Kent, D. Modern retinal laser for neuroprotection in open-angle glaucoma. In *New Concepts in Glaucoma Surgery*; Kugler Publications: Amsterdam, The Netherlands, 2019; pp. 255–274.
25. Lavinsky, D.; Cardillo, J.A., Jr.; Melo, L.A.S.; Dare, A.; Farah, M.E.; Belfprt, R., Jr. Randomized Clinical Trial Evaluating mETDRS versus Normal or High-Density Micropulse Photocoagulation for Diabetic Macular Edema. *Investig. Opthalmol. Vis. Sci.* **2011**, *52*, 4314–4323. [CrossRef]
26. Mansouri, A.; Sampat, K.M.; Malik, K.J.; Steiner, J.N.; Glaser, B.M. Efficacy of subthreshold micropulse laser in the treatment of diabetic macular edema is influenced by pre-treatment central foveal thickness. *Eye* **2014**, *28*, 1418–1424. [CrossRef] [PubMed]
27. Luttrull, J.K.; Spink, C.J. Serial Optical Coherence Tomography of Subthreshold Diode Laser Micropulse Photocoagulation for Diabetic Macular Edema. *Ophthalmic Surg. Lasers Imaging Retin.* **2006**, *37*, 370–377. [CrossRef]
28. Midena, E.; Micera, A.; Frizziero, L.; Pilotto, E.; Esposito, G.; Bini, S. Sub-threshold micropulse laser treatment reduces inflammatory biomarkers in aqueous humour of diabetic patients with macular edema. *Sci. Rep.* **2019**, *9*, 1–9. [CrossRef]
29. Scholz, P.; Altay, L.; Fauser, S. A Review of Subthreshold Micropulse Laser for Treatment of Macular Disorders. *Adv. Ther.* **2017**, *34*, 1528–1555. [CrossRef]
30. Tackenberg, M.A.; Tucker, B.A.; Swift, J.S.; Jiang, C.; Redenti, S.; Greenberg, K.P.; Flannery, J.G.; Reichenbach, A.; Young, M.J. Müller cell activation, proliferation and migration following laser injury. *Mol. Vis.* **2009**, *15*, 1886–1896.
31. Vujosevic, S.; Micera, A.; Bini, S.; Berton, M.; Esposito, G.; Midena, E. Aqueous Humor Biomarkers of Müller Cell Activation in Diabetic Eyes. *Investig. Opthalmol. Vis. Sci.* **2015**, *56*, 3913–3918. [CrossRef] [PubMed]
32. Sędziak-Marcinek, B.; Teper, S.; Chełmecka, E.; Wylęgała, A.; Marcinek, M.; Bas, M.; Wylęgała, E. Diabetic Macular Edema Treatment with Bevacizumab Does Not Depend on the Retinal Nonperfusion Presence. *J. Diabetes Res.* **2021**, *2021*, 1–15. [CrossRef]
33. Borrelli, E.; Parravano, M.; Querques, L.; Sacconi, R.; Giorno, P.; De Geronimo, D.; Bandello, F.; Querques, G. One-year follow-up of ischemic index changes after intravitreal dexamethasone implant for diabetic macular edema: An ultra-widefield fluorescein angiography study. *ACTA Diabetol.* **2020**, *57*, 543–548. [CrossRef] [PubMed]
34. Akhlaghi, M.; Dehghani, A.; Pourmohammadi, R.; Asadpour, L.; Pourazizi, M. Effects of subthreshold diode micropulse laser photocoagulation on treating patients with refractory diabetic macular edema. *J. Curr. Ophthalmol.* **2019**, *31*, 157–160. [CrossRef] [PubMed]
35. Kanar, H.S.; Arsan, A.; Altun, A.; Akı, S.F.; Hacısalihoglu, A. Can subthreshold micropulse yellow laser treatment change the anti-vascular endothelial growth factor algorithm in diabetic macular edema? A randomized clinical trial. *Indian J. Ophthalmol.* **2020**, *68*, 145–151. [CrossRef] [PubMed]
36. Baker, C.W.; Glassman, A.R.; Beaulieu, W.T.; Antoszyk, A.N.; Browning, D.J.; Chalam, K.V.; Grover, S.; Jampol, L.M.; Jhaveri, C.D.; Melia, M.; et al. Effect of Initial Management with Aflibercept vs Laser Photocoagulation vs Observation on Vision Loss Among Patients With Diabetic Macular Edema Involving the Center of the Macula and Good Visual Acuity. *JAMA* **2019**, *321*, 1880–1894. [CrossRef] [PubMed]
37. Vujosevic, S.; Bottega, E.; Casciano, M.; Pilotto, E.; Convento, E.; Midena, E. Microperimetry and Fundus Autofluorescence in Diabetic Macular Edema. *Retina* **2010**, *30*, 908–916. [CrossRef]

MDPI
St. Alban-Anlage 66
4052 Basel
Switzerland
Tel. +41 61 683 77 34
Fax +41 61 302 89 18
www.mdpi.com

Journal of Personalized Medicine Editorial Office
E-mail: jpm@mdpi.com
www.mdpi.com/journal/jpm